COMETS

ASTROPHYSICS AND SPACE SCIENCE LIBRARY

VOLUME 328

COMETS

Nature, Dynamics, Origin, and their Cosmogonical Relevance

by

JULIO ANGEL FERNÁNDEZ
Astronomy Department, Faculty of Sciences,
Montevideo, Uruguay

Springer

A C.I.P. Catalogue record for this book is available from the Library of Congress.

ISBN-10 1-4020-3490-3 (HB)
ISBN-13 978-1-4020-3490-9 (HB)
ISBN-10 1-4020-3495-4 (e-book)
ISBN-13 978-1-4020-3495-4 (e-book)

Published by Springer,
P.O. Box 17, 3300 AA Dordrecht, The Netherlands.

www.springeronline.com

Image "Wild2a.jpg" : Artist concept depicting a view of comet Wild 2 as seen from NASA's Stardust spacecraft during its flyby of the comet on January 2, 2004 (Courtesy NASA/JPL-Caltech)

Image "Wild2b.jpg" : Composite image of comet Wild 2 taken by the navigation camera of Stardust on January 2, 2004 flyby. A short exposure image showing the nucleus with great surface detail is overlain on a long exposure image, taken just 10 sec later, that could pick details of the faint coma and jets surrounding the nucleus (Courtesy NASA/JPL-Caltech)

Printed on acid-free paper

Printed in the Netherlands.

Table of contents

1 **Preface** **ix**

1 **Early ideas about comets** **1**
- 1.1 Early records of comet apparitions 2
- 1.2 Heavenly bodies or atmospheric phenomena? 5
- 1.3 The confirmation of their celestial nature 8
- 1.4 The determination of their trajectories 11
- 1.5 Interstellar visitors or members of the solar system? 13

2 **Statistics of comet orbits and magnitudes** **19**
- 2.1 The orbital elements and Kepler's laws 20
- 2.2 Periodic and nonperiodic orbits 22
- 2.3 Naming procedure 24
- 2.4 Total and nuclear magnitudes 25
- 2.5 Selection effects 31
- 2.6 The distribution of perihelion distances 33
- 2.7 The discovery rate and the flux of new and LP comets 36

3 **The understanding of their physical nature** **39**
- 3.1 Comet sizes and masses 39
- 3.2 'Sand bank' versus 'dirty snowball' model 41
- 3.3 The solid nucleus 42
- 3.4 Sublimation of volatiles 44
- 3.5 Spectroscopic observations 49
- 3.6 Inventory of parent molecules 53
- 3.7 Distribution of parent molecules and radicals in the coma 56
- 3.8 The hydrogen corona 59
- 3.9 The nature of dust particles 63
- 3.10 The dust to gas mass ratio 67
- 3.11 Formation of the tails 70

4 **Dynamics of LP comets entering the inner planetary region** **77**
- 4.1 The original orbit 77
- 4.2 Change in the orbital energy 78
- 4.3 Nongravitational forces 81

4.4 Resonant and non-resonant motion 85
4.5 Random-walk in the energy space 87
4.6 The diffusion equation 90
4.7 Dynamical time scale for capture into a periodic orbit 91
4.8 Dynamical losses 92
4.9 The capture problem 95
4.10 From "new" comets to "old" comets. The problem of physical decay 96
4.11 Changes in the inclination-distribution of LP comets with the dynamical age 99

5 The Oort cloud 103
5.1 The distribution of original energies 104
5.2 Stellar perturbations 106
5.3 Galactic tidal forces 110
5.4 Penetrating encounters with giant molecular clouds 119
5.5 Randomization of orbital inclinations 121
5.6 Distribution of eccentricities of a thermalized population 126
5.7 The outer boundary of the Oort cloud 131

6 The flux of new comets: quiescent and excited stages 133
6.1 The drift of the perihelion distance under external perturbers 133
6.2 The Jupiter-Saturn barrier 136
6.3 The "inner" and "outer" Oort cloud. The possible existence of an inner core 138
6.4 The fluxes of Oort cloud comets in the inner and outer planetary region 141
6.5 Comet showers 144
6.6 Do comet showers reflect in the impact cratering record? 150
6.7 The current passage rate of Oort cloud comets: does it represent a quiescent stage or an excited one? 153

7 The Jupiter family 155
7.1 Statistics of comet discoveries and orbital properties 155
7.2 Determination of comet sizes: methods 159
7.3 The size distribution 167
7.4 Nongravitational forces and masses 172
7.5 Lightcurves, shapes and rotation periods 175
7.6 Color, albedo and phase function 180

7.7 Dynamics of close encounters 183
7.8 The population size 187
7.9 The source region 189

8 The trans-neptunian belt 193
8.1 The puzzle of the Jupiter family comet population 195
8.2 The discovery 196
8.3 The naming debate 198
8.4 Resonance dynamics 199
8.5 Dynamical structure and transfer mechanisms 204
8.6 Transient objects 209
8.7 Population size and size distribution 211
8.8 Binary TNOs 216
8.9 The outer edge of the belt 218
8.10 Dynamical and physical erosion of the belt 220
8.11 The trans-neptunian belt as a replenishment source of the Oort cloud 225
8.12 Physical properties of trans-neptunian objects and Centaurs 228

9 Physical end states of comets 237
9.1 The sublimation of the volatile material 237
9.2 Formation of a dust mantle 240
9.3 Can defunct or dormant comets be disguised as asteroids? 246
9.4 Splitting events 249
9.5 Possible scenarios for the physical evolution and decay of a comet 256
9.6 Small comets 259
9.7 Collision with the Sun 262
9.8 Collisions with the planets 265
9.9 Collisions with asteroids and interplanetary boulders 267

10 Leftovers of the solar system formation 273
10.1 Early phases of star formation 273
10.2 Circumstellar disks 276
10.3 The protoplanetary disk of the early Sun 278
10.4 Grain assemblage and formation of planetesimals 281
10.5 From planetesimals to planets 286

10.6 Outward transport of planetesimals by resonant coupling with the outward-drifting proto-Neptune: The buildup of the trans-neptunian belt 292
10.7 The scattering of residual planetesimals and the buildup of the Oort cloud 297
10.8 The early galactic environment of the solar system 299
10.9 The physical structure of icy planetesimals 303
10.10 Chemistry of icy planetesimals 306
10.11 The collisional history of comets 310
10.12 How well preserved is the comet material? 311

11 Comets and life 315
11.1 The volatile content of the Earth: endogenous or exogenous? 315
11.2 The post-accretion heavy bombardment 317
11.3 Comet contribution to the atmospheres of the terrestrial planets 320
11.4 Comet contribution to the impact rate for the last 3800 Myr 323
11.5 The deuterium problem 328
11.6 May comets harbor lifeforms? 331
11.7 Space missions to comets 334

Appendix 1. Table A1: Astronomical and physical constants 341

Appendix 2. Table A2: Absolute nuclear magnitudes and radii of JF comets 343

References 347

Index 375

Preface

Comets are small bodies, but of great cosmic relevance. Given its pristine nature, they may preserve valuable and unique information on the chemical and physical processes that took place in the early solar system, and that may be occurring in the formation of other planetary systems. They might have even played a very important role in the origin of life on Earth. Beyond that, since ancient times comets have inspired awe, superstition, and also curiosity and debate. Their sudden apparitions challenged the long-held view of the immutability of the heavens, which triggered a long debate on whether comets had a heavenly or terrestrial nature. Therefore, comets have a prominent role in the history of scientific thought, that goes back to the most ancient civilizations.

The last apparition of comet Halley in 1986 was a landmark since it arouse a great expectation in the scientific community and in the public at large. For the first time, a flotilla of spacecrafts visited a comet. A great number of popular and technical books were written on Halley, and comets in general, around the mid-eighties. The interest in comets never subsided after Halley's passage which is reflected in the large volume of printed material on these bodies. I have taken the challenge to write a new book on comets that summarizes most of the recent advances on the subject, including my own work developed during the last 25 years. I tried to cover dynamical as well as physical aspects of comets, highlighting their importance as relics of the accretion processes in the early solar system and, perhaps, as carriers of water and organics that permitted the development of life on our planet.

The book has perhaps a major emphasis on dynamical aspects, following my own main work on the subject. I also think that dynamical studies have so far been the best tool to learn about the origin and location of comet reservoirs. Despite the major emphasis on dynamics, I tried to also cover the physics and chemistry of comets, since this is a field in which we are rapidly gaining new insight thanks to the use of new ground-based and space-based observatories, as well as *in-situ* data gathered by spacecrafts during flyby missions. Yet, the physics of ion tails and their interaction with the solar wind and interplanetary magnetic field has not been covered - except for a brief introduction - since this is a very specific topic that somewhat departs from the main focus of the book.

The book contains some mathematical analysis of topics that I consider relevant, and that in general have not been treated in detail in other books. In particular, I develop in some extent the dynamics of Oort cloud comets, subject for which I do not know of any comprehensive treatment elsewhere. Whenever I deemed it convenient, I included an explanation of some technical terms, or demonstrated how a formula was obtained. I tried to cover the recent literature as much as possible, as well as some classical papers of several decades ago, or even earlier. Yet, given the huge production rate of papers on comets, it is possible that some relevant material could have been overlooked so, if that occurred, I present my apologies since now. The book also contains several tables with useful data. It may be suitable as a textbook for graduate students with some basic knowledge of celestial mechanics and astrophysics, as well as a consult book for comet researchers, or researchers from other related fields willing to start working on comets or get an updated view of the subject.

Last but not least, I like to thank several people who have contributed to this book either with comments, remarks, or material. In particular, I would like to mention Dominique Bockelée-Morvan, Adrián Brunini, Tabaré Gallardo, Walter Huebner, Dave Jewitt, Javier Licandro, John McFarland, James Scotti, Andrea Sosa, Gonzalo Tancredi, and Mario Wschebor. I especially thank Mark Bailey for his hospitality and stimulating discussions during a visit to the Armagh Observatory in which portions of this work were written. My due thanks also to Alejandro Crosa and Gabriel Santoro who were very helpful with the preparation of several figures.

Julio Angel Fernández
January 2005

EARLY IDEAS ABOUT COMETS

The ancient peoples paid special attention to whatever occurred in the heavens, noting on one side the regularity of several celestial phenomena, such as the rise and the setting of the Sun, Moon and stars, and the phases of the Moon, and on the other side the irruption of unexpected transient events, like eclipses, comets, novae and meteors, that broke such a regularity. Since the heavenly bodies were associated to divinities with influence on terrestrial affairs, the unexpected events caused concern and were regarded as portents of upcoming disasters.

Comets in particular were received with a mixture of woe and fascination, owing to their sudden and sometimes spectacular apparitions. The traditional use of the word "apparition", rather than a more sober one such as "appearance" or "passage", is itself a reminiscence of the old view that regarded comets as ghosts rather than natural objects. Very often the occasional witnesses looked at comets with fear, believing that they were forerunners of wars, pestilence and death. The ancient civilizations seem to have paid special attention to the observation of these bodies and other transient phenomena like fireballs and meteor showers, basically owing to their desire to predict future events rather than by mere scientific curiosity.

The ancient Greeks went beyond the mere contemplation to develop several theories about the nature of comets, though they presumably inspired on previous ideas held by the Chaldeans and the Egyptians. The word comet itself comes from the Greek word *kometes* that means "long-haired" star, alluding to their main distinctive features: a head or coma and a long tail or tails more or less directed in the antisolar direction (Fig. 1.1). The comet lore through history is a curious blend of scientific thought with superstitious tales that forms part of the rich cultural heritage of mankind. A detailed description of early ideas about comets can be found in the two excellent books by Bailey et al. (1990) and Yeomans (1991).

Figure 1.1. CCD image of comet C/2000 WM1 (LINEAR) observed with the 46-cm Centurion telescope of the Observatorio Astronómico Los Molinos (OALM) by Raúl Salvo and Santiago Roland. The field of view of the image is $1/2° \times 1/2°$, i.e. similar to that occupied by the Moon (Courtesy OALM).

1.1. Early records of comet apparitions

The Chinese were the most prolific observers whose meticulous observations of comets and other phenomena like novae, meteors, aurorae, eclipses and sunspots, have been preserved until now. Theirs is by far the most important source of reliable astronomical data covering a period from about 1100 BC to about 1700 AD. Besides the Chinese, the Koreans and Japanese also contributed with a significant number of observations during portions of the previously quoted time span (a comprehensive compilation of ancient observations from these Asian civilizations was done by Ho Peng Yoke (1962)). It is also possible that the Babylonians kept a good record of comet apparitions, though very little has reached us from them. The same can be said from the Mesoamerican civilizations (Mayans, Aztecs, and other peoples) from which very few documents have survived until present, though from the sketchy information available it seems that they also regarded comets as portents of impending calamities. Very few records of comet apparitions could also be recovered from the Hellenic civilization (which comprised Greece and surrounding areas around the Mediterranean Sea, including Alexandria), despite the attention paid by its philosophers to these bodies. Presumably, Greek philosophers were not so prolific and systematic observers as their Chinese counterparts, which explains the scarcity of references to observed comets. One of the few Hellenic sources of comet observations during the V-VI centuries BC is Aristotle's treatise *Meteorologica* written in 329 BC. There are some references to comet apparitions from Roman scholars, in particular Pliny the Elder and Lucius Annaeus Seneca (both from the first century AD). However, Roman accounts of comet apparitions are in general very vague, included incidentally within descriptions of historical events as bad omens. During the Middle Ages, Arab astronomers do not seem to have paid much attention to comets either, and their legacy in this matter is very scarce.

The number of recorded comet apparitions is very scant before the second century BC, and then it sharply raises to an average of above 25 comets per century, where it stays more or less constant until the eighth century AD (Fig. 1.2). In the following centuries the average rate raises somewhat to about 40-50 per century. The most ancient available references to comet apparitions date back to the 12th or 13th century BC, though they are extremely vague. The first reliable document describing a comet apparition in 674 BC was uncovered in a Babylonian stone tablet (Kronk 1999). Most comets cataloged until the fifteenth century rely heavily on Chinese and, in a lesser degree, Korean and Japanese reports. The Chinese referred to comets as *broom stars*, or *sparkling stars*, or later also as *long-tailed stars*. They also used the name *guest star*, but it is very likely that it usually referred to novae rather than comets. In some cases it is not clear that the recorded object is a comet, and it might have as well been a nova or a meteor.

Most of the recorded positions of ancient comets are unprecise and do not allow to compute a reliable orbit. In fact, most of the orbits computed before the 15th century correspond to periodic comets

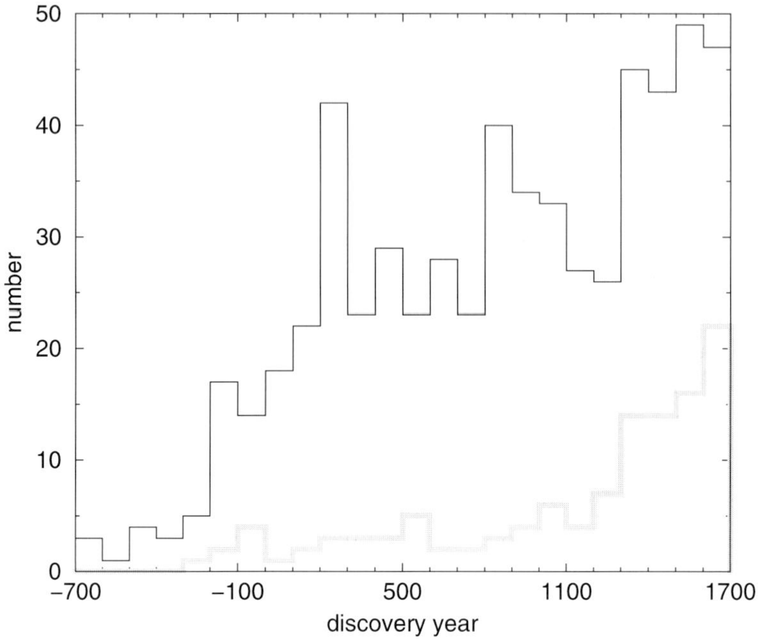

Figure 1.2. Record of comet discoveries per century before 1700 AD as presented by Kronk (1999) (thin histogram), and that corresponding to those apparitions whose observations allowed a reliable orbit determination as presented in Marsden and Williams (2003) Catalogue of Cometary Orbits (gray thick histogram).

1P/Halley, 109P/Swift-Tuttle and 55P/Tempel-Tuttle, for which the accurate computation of their orbits backward in time allowed to link the computed ephemerides of previous passages with ancient reports of their apparitions. From the fifteenth century, European observers began to play an ever increasing role in the discovery record. The Florentine physician and astronomer Paolo Toscanelli (1397 - 1482) observed accurately several comets, among them Halley in its 1456 apparition, and plotted their positions in the sky on charts which permitted the determination of their orbits by later workers. The fraction of recorded apparitions that allow an orbit determination shows a steady increase since the fourteenth century (Fig. 1.2). Stanislaus de Lubienietz produced one of the first comprehensive catalogues of Western observations of comets, titled *Historia Cometarum* (1666), which reports comet apparitions since the deluge time to the time of his book. It contains magnificent illustrations of comets on the heavens, one of them is shown below in Fig. 1.4.

As can be seen in Fig. 1.2, the number of recorded apparitions that led to reliable orbit determination, as presented in Marsden and Williams (2003) Catalogue of Cometary Orbits, constitute a small fraction of the total sample during most of the considered period. The fraction started to increase significantly in the fourteenth century following a better record of comet positions with better instrumentation like quadrants. The telescope was introduced for comet observation in 1618 by the Swiss Jesuit Johann Baptist Cysat (ca. 1586 - 1657) and the English astronomer John Bainbridge (1582 - 1643). But the first comet to be discovered telescopically was that of 1680 (now designed as C/1680 V1) by the German astronomer Gottfried Kirsch (1639 - 1710). It marks a turning point between the previous naked-eye discovery regime, and the posterior regime in which the telescope played an ever increasing role in the detection and follow-up of comets. As we shall see in the next chapter, one of the consequences of the telescope revolution was the dramatic growth of comet statistics in quantity as well as in quality. The time around 1700 also witnessed the rapid decline of the Chinese as predominant source of comet apparitions in favor of the Europeans.

1.2. Heavenly bodies or atmospheric phenomena?

The history of cometary thought began as a discussion on whether comets were celestial bodies or atmospheric phenomena. The Pythagoreans in the sixth century BC and Hippocrates of Chios (ca. 440 BC) are credited with the idea that comets were planets that appeared infrequently close to the horizon like Mercury. Anaxagoras of Clazomenae (ca. 500 - 428 BC) and the atomist Democritus of Abdera (ca. 460 - 370 BC) believed that each comet was produced by the close approach or conjunction of two planets giving the appearence of a single elongated object. Anaxagoras argued that both the Sun and comets were made up of burning stones. This peculiar interpretation of their physical nature was probably rooted in the observation of a bright comet followed by a meteorite fall in 467 BC whose bright trail in the sky was seen in daytime. This is the first reference to an association between comets and meteors.

Yet, other Greek thinkers regarded comets as phenomena much closer to the Earth. According to Xenophanes of Colophon (ca. 570 - 470 BC) comets were dry exhalations from the Earth in a similar manner that clouds were condensations of moisture raised from the sea. This idea was induced on early thinkers because comets bright enough to become observable with the unaided eye are generally close to the Sun. Therefore, they can only be observed in the early morning or early evening, and having the tail pointing toward the antisolar direction, they appear indeed to raise from the horizon. Aristotle (384 - 322 BC), one of the leading intellectuals of the ancient world, was to have a lasting influence on the ideas on comet's nature. He regarded comets, shooting stars and even the Milky Way as meteorological phenomena, and this is the reason why he included them in his treatise *Meteorologica* which dealt with the sublunar world. He ruled out the planetary nature of comets by asserting that they had been seen outside the zodiac. He also rejected the conjunction of planets or coalescence of stars, arguing that many comets had been observed to fade away without leaving behind one or more stars.

In Aristotle's view the sublunar world was composed of four concentric spheres ordered according to their density. The first densest sphere was the earth, followed by the watery, the airy and the fiery sphere on the top. The supralunar world populated by the heavenly bodies was composed by a fifth element or quintessence. He adopted the view, attributed to the Pythagoreans, that all celestial bodies moved in circles, considered to be the perfect curve. Irregular and vertical motions, like those attributed to comets, were only possible within the sublunar region. Following Xenophanes, Aristotle argued that comets formed from warm and dry exhalations that rose up from the earth when it was heated by the Sun. These exhalations ascended to the airy sphere and at the border with the fiery sphere they ignited by friction producing comets which were carried about the Earth by the circular motion of the heavens. Aristotle thus provided a physical explanation of why comets should foreshadow droughts, avoiding any kind of supranatural explanation of these bodies as portents or omens. Yet, supersticious fears were going to surround any comet apparition for another two thousand years.

Aristotle's view on comets as meteorological phenomena was mostly unchallenged for the following two millenia. A few dissenters, like Apollonius of Myndus (around the third century BC) still supported that comets were distinctive heavenly bodies, just as the Sun or the Moon, and attributed the changing brightness of a comet to its varying distance to the Earth, while Zeno of Citium on the island of Cyprus (circa 336 - 264 BC) considered that stars united their rays to create the image of an elongated star. Posidonius (135 - 51 BC) followed Aristotle's ideas about comets and added the interesting observation of a comet that became visible during a total solar eclipse, although it was previously concealed by the proximity of the Sun. This observation led him to conclude that comets should be much more numerous than usually observed, because some of them are lost in the glare of the Sun. In line with Aristotle's thought, Posidonius believed that comets burn as long as find nourishment in the aethereal region and that their appearance coincides with drought and their disappearance with heavy rains.

The Roman philosopher Lucius Annaeus Seneca (4 BC - 65 AD) noted in his *Quaestiones naturales* that comets could not be sudden fires that last at most for a few hours, but permanent creations of nature moving perhaps in close orbits. Seneca expressed genuine admiration for these bodies, being confident that "Men will some day be able to undestand their nature and paths in the heavens". Pliny the Elder (23 - 79 AD) discussed in his *Natural History* a classification of comets into 10 types according to their shapes and observed features. Pliny strongly supported the idea that comets were portents and that their shapes and the direction in which they dart their beams and what stars are nearby had influence on human affairs. Claudius Ptolemaeus or Ptolemy (ca. 100 - 175 AD) adopted Aristotle's view of comets as atmospheric phenomena, and for this reason they were not included in his masterwork the *Almagest* that dealt will all the heavenly bodies known at that time. Yet, Ptolemy described comets in his book *Tetrabiblos*, devoted to astrology, which shows that his main concern was to describe the ill effects brought by comet apparitions.

During the Middle Ages and in the Renaissance comets continued to be regarded in dual terms, as harbingers of disaster on one hand and as meteorological phenomena on the other, under the influence of the unquestioned authority of Aristotle and Ptolemy. Until the fifteenth century, no new original ideas or observations were added to the knowledge of comets, which were relegated to supertitious beliefs. Yet, the interest in these celestial bodies never subsided, which is illustrated in many drawings and paintings, either with the intention to describe their morphology or to reflect the awe their caused to the occasional witnesses (Fig. 1.3).

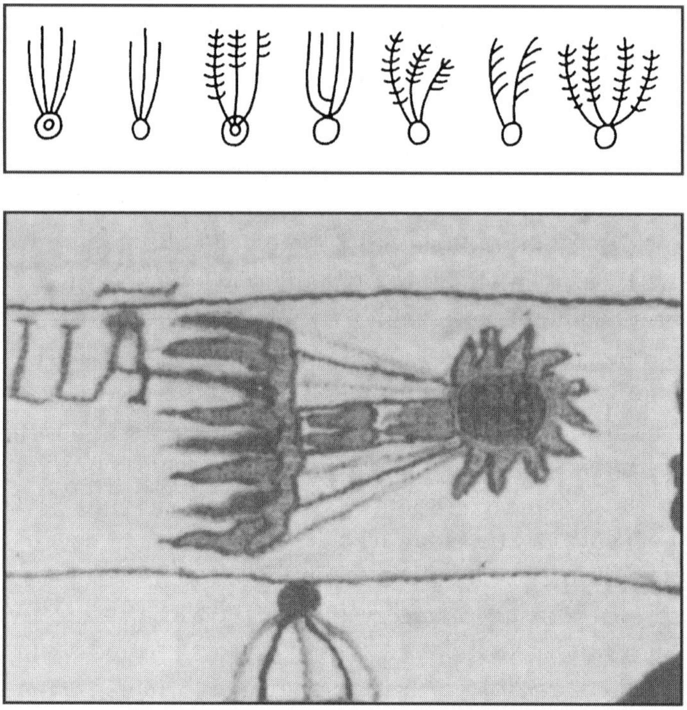

Figure 1.3. Comet drawings through history. (a) Comet types as they appeared in the Chinese Han tomb silk book (ca. 168 BC). (b) The 1066 AD apparition of Halley's comet as depicted in the Bayeux tapestry.

1.3. The confirmation of their celestial nature

From the eleventh century Western Europe started to recover very slowly from the state of ruin, disintegration and cultural darkness that followed the fall of the Roman Empire. There was a renewed interest in the works of the ancient Greek philosophers that reached Western Europe through the Arabs. Against this backdrop the interest in natural phenomena increased, in particular in comets. By the end of the middle ages, a new stimulating environment for scientific enquiry and discussion started to emerge, and with it the first attacks on the astrological signification of comets. Henry of Hesse (1325 - 1397) rejected the widely accepted thought that comets were prognosis of future events. Following Aristotle, he thought that comets were meteorological phenomena, and that pestilence often follows comets because they are produced by the exhalation from the Earth of pestilential vapor.

In 1456, the Viennese astronomer Georg von Peurbach (1423 - 1461) tried to determine the parallax of a comet (later known to correspond to one of the passages of Halley's comet). Some years later, Johannes Müller (1436 - 1476), known by his Latin name of Regiomontanus, attempted to measure the parallax of the great comet observed in 1472, though his derived value of 6° was highly erroneous. The most important legacy of Regiomontanus was to encourage scientific observations of comets with the aim to determine their distances to the Earth, their diameters and lengths of their tails. Girolamo Fracastoro (ca. 1478 - 1553) and Peter Apian (1495 - 1552) showed independently that comet tails always point away from the Sun, in fact a property already known by Chinese astronomers at least seven centuries before, and even Seneca wrote in his *Quaestiones naturales* that "the tails of comets fly from the Sun's rays", so in this point as in others the Renaissance scholars were just rediscovering phenomena already known by the ancients.

During the sixteenth century most astronomers were interested in determining the parallax of a comet in order to settle the debate on whether these bodies belonged to the sublunar world (and were thus atmospheric phenomena), or they belonged to the supralunar world and were thus celestial in nature. The mathematician Girolamo Cardano (1501 - 1576) noted that a comet seen in 1532 had an apparent speed smaller than that of the Moon, thus suggesting a greater distance which would place the comet in the supralunar world. But it was the bright comet of 1577 that gave astronomers all around Europe their great opportunity to measure its parallax (Fig. 1.4). Tycho Brahe (1546 - 1601) was among the observers that could successfully obtain a parallax, which placed the comet at least four times farther away than the Moon. Tycho also measured the apparent diameter of the comet's head and found it to be $8'$ which, according to its estimated distance, gave a diameter of nearly one fourth of the Earth's. Several other great astronomers of the time, as Michael Maestlin (1550-1631) and Helisaeus Roeslin (1544 - 1616) also find distances that put the comet in the supralunar world. As in many other cases in history of science, results can be controversial, mainly when the experiments or observations are pushed to the limits of the capabilities available at the moment. The comet of 1577 was one of these cases, and some respected scholars, among them the foremost astronomer of Eastern Europe Thaddaeus Hagecius (ca. 1525 - 1600), found for the comet a large parallax that placed it below the Moon (Hagecius rectified later his early estimate and recognized the comet to be supralunar). An interesting and well-

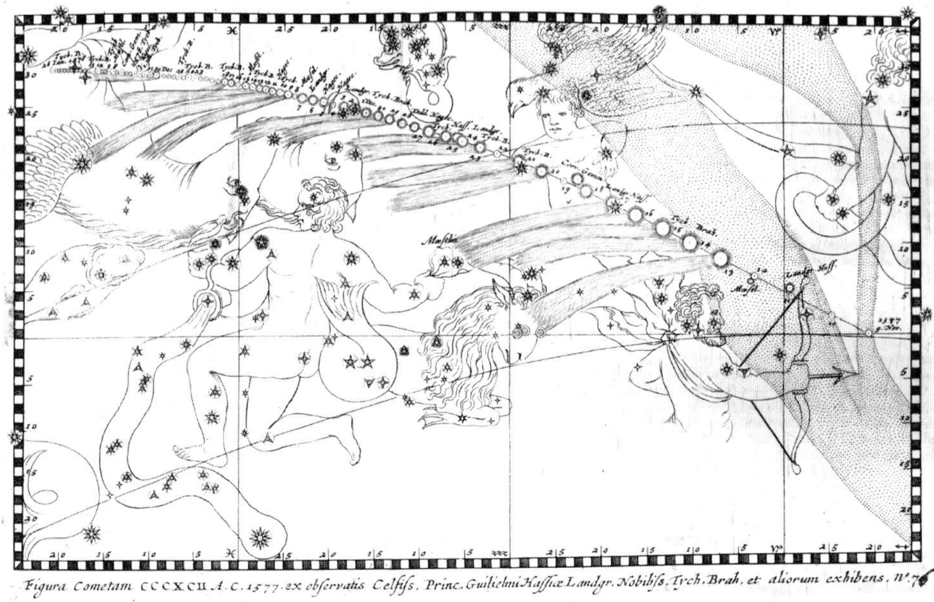

Figure 1.4. The motion on the heavens of the great comet of 1577 as illustrated by Stanislaus de Lubienietz in his book *Historia Cometarum* (1666) (courtesy of John McFarland, Armagh Observatory).

documented account about the comet of 1577 and the comet ideas about that time was presented by Hellman (1944).

Tycho suggested that comets moved around the Sun on circular orbits, like Venus and Mercury. In Tycho's system, the Sun itself, Mars, Jupiter, Saturn and the fixed stars moved around the Earth. Tycho even suggested that the orbit of the comet could be somewhat oblong, being the first time that somebody suggested that a celestial body might move on an orbit different from a circle. Johannes Kepler (1571 - 1630) believed that comets were ephemeral bodies that formed out of impurities in the celestial aether and moved along straight, rectilinear paths. He confirmed that tails pointed toward the antisolar direction and put forward the hypothesis that the sunlight passed through the comet's head and took with it some of the matter away from the Sun, leading eventually to its final consumption.

Even though the idea that comets were heavenly bodies had received a growing acceptance during the sixteenth and early seventeenth centuries, there was still a firm opposition from some highly respected scholars. Thus, Nicholas Copernicus (1473 - 1543) still believed that comets were terrestrial objects, and Galileo Galilei (1564 - 1642) went on to affirm that comets were vapors that move vertically upward and were made visible when sunlight reflected on the cloud of vapors. The absence of parallax was then explained by the reason that comets were insubstantial, as mere lights reflected on vapors.

The heavenly nature of comets reached finally wide acceptance by the end of the seventeenth century. Some of the main thinkers of the time discussed the place in heavens were comets originated. Thus, René Descartes (1596 - 1649) believed that comets formed together with planets around the Sun and other stars on vortices. For our own solar system comets were found at the outer edge, at the distance of Saturn.

Despite the advance in the understanding of the celestial nature of comets during the sixteenth and seventeenth centuries, and the discussion of their motion with scientific arguments, the supertitious fears unleashed by their apparitions did not subside. Such fears were shared by some of the most respected scholars of the time, like Tycho Brahe, Kepler and Michael Maestlin. Martin Luther (1483 - 1546) referred to comets as harlot stars and works of the devil.

1.4. The determination of their trajectories

Once the heavenly nature of comets was accepted by the majority of astronomers, the next step consisted in determining the kind of trajectory they followed. We have seen before that a wide range of opinions were compiting at that time, going from straight, rectilinear paths, as proposed by Kepler, to circular orbits as proposed by Tycho. The Italian-French astronomer Jean Dominique Cassini (1625 - 1712), one of the last great supporters of the geocentric system, considered that comets, like planets and the Sun, moved around the Earth but in highly eccentric orbits. The Polish astronomer Johannes Hevelius (1611 - 1687), author of the well-known treatise *Cometographia* (1668), after careful observations of several comets concluded that they moved on paths slightly curved toward the Sun, on either a hyperbola or a parabola. This was corroborated by the German astronomer Georg Dörffel (1643 - 1688), a student of Hevelius, who was able to fit a parabola, with the Sun at its focus, to the motion of the bright comet observed in 1680.

The debate on comet's trajectories could have continued for a long time, were Newton's theory of universal gravitation not ready. Fortunately, at the very same time as the first attempts to fit parabolas or hyperbolas to cometary paths were carried out, Isaac Newton (1642 - 1727) had almost completed his theory that predicted elliptic orbits for the planets moving around the Sun, which would occupy one of the focus. It was obvious that his theory should also apply to the case of comets. At the beginning Newton was reluctant with this possibility but, with time, he became convinced that planets and comets should obey the same laws, and developed a method to fit a parabola to the comet's motion given three observations more or less evenly spaced in time. This method was later included in his masterwork *Principia* (1687).

Newton's countryman Edmond Halley (1656 - 1742) was the first to fully exploit the new theory of gravitation to the case of comets. Halley computed parabolas for a sample of 24 well observed comets and noted that those comets observed in 1456, 1531 and 1607, shared parabolas of similar characteristics as those of the comet of 1682 observed by himself. This observation led Halley to conclude that these were different passages of the same comet and predicted that it would return again in 1758. The comet was recovered by the German farmer and amateur astronomer Georg Palitzsch (1723 - 1788) on Christmas evening of that year. Halley's prediction was corroborated and he had as a posthumous homage the comet named after him. The recovery of Halley's comet symbolizes the end of the era of discussion on the comet's motion: since then there was agreed that comets moved on parabolic, nearly parabolic, or slightly hyperbolic orbits, though a few of them, like Halley, had orbits elliptic enough to record several returns on historic times. Even though Halley was the first to successfully predict a comet return, he was not the first to look into this problem. Pierre Petit (ca. 1594 - 1677) and Adrien Auzout (1622 - 1691) firmly believed that comets were permanent celestial bodies moving in close paths, thus subject to return. Petit went further to claim that the comets observed in 1618 and 1664 were the same object, so its next return was due in 1710. Unfortunately, he was wrong.

The expected return of comet Halley for 1758-59 triggered a feverish computing activity aimed at predicting a more accurate date of perihelion passage, that involved some of the best mathematicians of the time. Leonhard Euler (1707 - 1783), noting the decrease in the period of comet Halley between 1531-1607 and 1607-1682, assumed that it was due to a drag force by the interplanetary aether. On the other hand, Alexis-Claude Clairaut (1713 - 1765) dismissed the drag effects of the aether but understood that a good orbital solution could be obtained only if the perturbations of Jupiter and Saturn were taken into account.

Newton's law of gravitation gave rise to the development of celestial mechanics whose goal is the study of the motion of celestial bodies under their mutual gravitational attraction. As the determination of cometary orbits became routine, it was deemed necessary to dispose of more manageable computing methods. Pierre-Simon Marquis de Laplace (1749 - 1827) developed a method that relaxed the stringent condition of Newton's method of having the observations more or less evenly spaced in time. Wilhelm Olbers (1758 - 1840) developed another simple method for determining the five elements needed for a parabolic orbit solution which was later widely used (description of these methods can be found in standard Celestial Mechanics textbooks like Roy (1982)).

1.5. Interstellar visitors or members of the solar system?

The motion of comets, that depart so markedly from that of planets, led to the idea that they might not be members of our solar system. Indeed, not only their orbits are quite different in shape and size but, while the planets moved all close to the ecliptic plane, most comets moved instead on orbits randomly oriented. Already Kepler believed that comets came from interstellar space, but it was Laplace who developed a complete theory of interstellar origin, becoming identified with it. Laplace argued that comets were condensations in an interstellar cloud, which attained their observed orbits as the result of the gravitational attraction of the Sun. He did not consider the motion of the Sun, but assumed that it was at rest immersed in an interstellar field of comets distributed uniformly and with all possible velocities between zero and infinity. Consequently, comets could be gravitationally attracted by the Sun from different directions which would explain the random orientation of their orbital planes. It is clear that bodies attracted from interstellar distances with very low relative velocities will move on paths very close to parabolas. Laplace explained the cases of comets in elliptical orbits, like Halley, as being captured by one of the planets after a close encounter.

Later, Giovanni Schiaparelli (1835 - 1910) pointed out the necessity to include the proper motion of the Sun relative to the cometary cloud. William Herschel had already found in 1783 that the Sun had a proper motion with respect to nearby stars, moving toward a point termed the *solar apex*. Therefore, if comets shared the interstellar space with stars, it was natural to think that the Sun would have likewise a proper motion with respect to the interstellar comet cloud. Being the velocity of the Sun toward the Apex of about 20 km s^{-1}, it would be extremely unlikely to find interstellar comets with relative velocities smaller than a few km s^{-1}. This would give an excess of hyperbolic comets far greater than Laplace had estimated. Furthermore, such comets would arrive preferentially from the direction towards which the Sun is moving. Since such a concentration of aphelion points in the apex direction was not observed, Schiaparelli concluded that the comet cloud should be comoving with the Sun.

To illustrate the previous situation, let us assume that the Sun moves with respect to a comet cloud with a velocity $\vec{u}$ (we neglect any random comet motion). A given comet will be attracted toward the Sun along a hyperbolic path of perihelion distance q and semimajor axis (negative) a (Fig. 1.5). If D is the "target radius" (i.e. the distance of closest

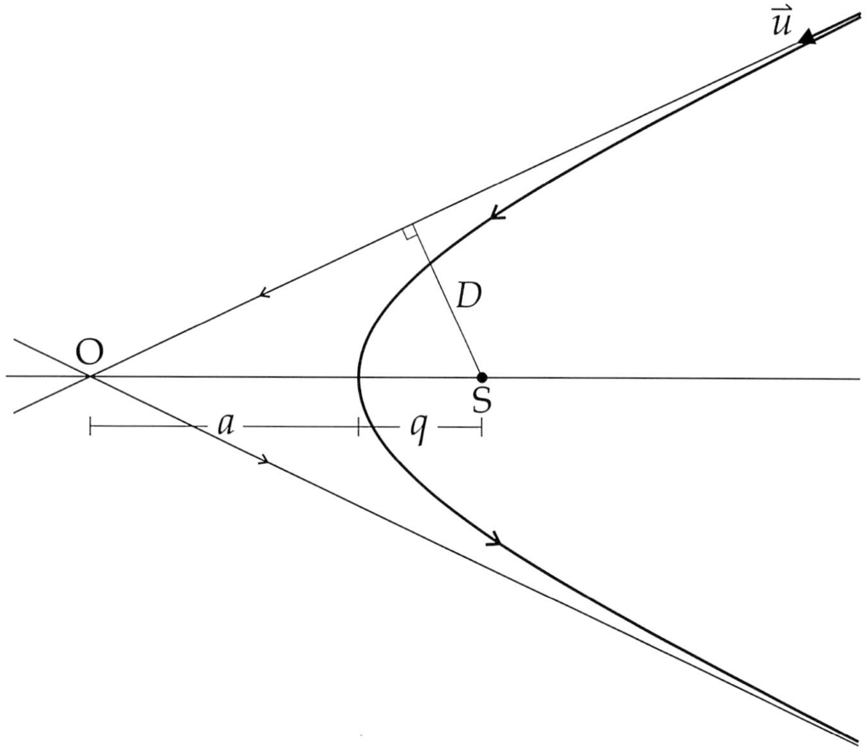

Figure 1.5. Geometry of the encounter of a comet with the Sun with a velocity at infinity $\vec{u}$ and a target radius D.

approach of the comet to the Sun, if it moved unperturbed along the asymptote), from conservation of angular momentum we have

$$Du = qv_q, \tag{1.1}$$

where v_q is the orbital velocity of the comet at perihelion. We should bear in mind that in a heliocentric system, $-\vec{u}$ will correspond to the velocity at infinity of the comet. For a Keplerian hyperbolic motion we have

$$u^2 = -\frac{\mu}{a}, \text{ and } v_q^2 = \mu\left(\frac{2}{q} - \frac{1}{a}\right),$$

where $\mu = GM_{\odot}$, G is the gravitational constant, and $M_{\odot}$ is the Sun's mass. By substituting these two expressions into eq. (1.1), we can obtain a relation between the perihelion distance and the velocity at infinity:

$$q = -\frac{\mu}{u^2} + \left(\frac{\mu^2}{u^4} + D^2\right)^{1/2}. \tag{1.2}$$

As shown in Fig. 1.6, for relative velocities greater than a few km/s it is impossible to get comets within a given "observable" region where they become potentially detectable (say, for perihelion distances $q < 2$ AU where most comets have so far been discovered), unless that the impact parameter $D \lesssim 10$ AU. An interstellar comet will enter the observable region if the encounter velocity $\vec{u}$ form an angle $\beta \lesssim D/r_{\infty}$ radians with the solar direction, where r_{∞} is the Sun-comet distance at the moment of capture. If before capture the comet was at 10^5 AU (that according to Laplace corresponded to the radius of the Sun's sphere of influence), the angle β should be at most $\sim 10^{-4}$ radians or about 20 arcsec. This shows that under the capture hypothesis, only those comets pointing toward or very near the Sun will enter the observable region and in all cases in clearly hyperbolic orbits. The probability that a comet with a randomly oriented velocity vector $\vec{u}$ will have it pointing to less than 10^{-4} radians to the Sun is $p \simeq 10^{-8}/4 = 2.5 \times 10^{-9}$, so only one comet in 4×10^8 could reach the observable region.

In 1929 Nicholas Bobrovnikoff concluded from the analysis of the lifetimes of 94 comets that these could not be older than one Myr, and that within this time the Sun must have therefore passed through an interstellar cloud from which it captured the comets. Nölke (1936) argued that condensations within the cloud could only become incorporated

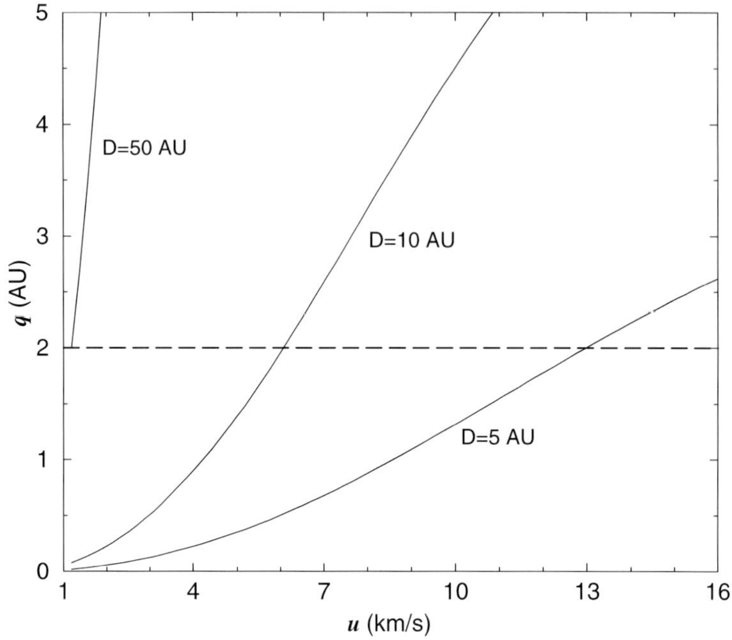

Figure 1.6. The perihelion distance reached by an interstellar comet captured along a hyperbolic path as a function of the velocity at infinity u and for the target radii shown beside each plot. The dashed line indicates a somewhat arbitrary "observable" boundary below which a comet would be potentially detected from Earth.

within the solar system if their motion was taking place in a resisting medium, which he associated to interstellar material composed of dust and gas (a more detailed description of these early works can be found in Richter 1963).

The theories postulating an interstellar origin for comets described before regarded the condensations within the interstellar cloud already formed when the Sun encountered them. On the other hand, in a series of papers published between 1948 - 1958, Raymond Lyttleton proposed that such condensations originated as a byproduct of the process of capture itself. Lyttleton considered Bondi and Hoyle's (1944) theory of accretion, according to which interstellar dust particles are gravitationally focused toward the antapex direction where they collide to each other. The collision zone will lie behind the Sun in a range of distances going from a few AU to about 10^3 AU, depending on the relative velocity of the dust particles at infinity. The dust particles lose kinetic energy via the inelastic collisions, so they will be transferred from the original hyperbolic orbits to elliptic orbits, moving around the Sun in discrete clouds that Lyttleton identified with comets (see Lyttleton 1951). Lyttleton's theory not only proposed an origin for comets, but also provided a physical model for its nucleus as an assemblage of interstellar dust particles. We will discuss further this model in Chapter 3.

At about the same time as Laplace presented his theory of interstellar comet origin, Joseph Louis Lagrange (1736 - 1813) proposed a theory of comet origin within the solar system. According to Lagrange, comets might originate from gigantic eruptions from one of the larger planets (like Jupiter or Saturn). Theories involving the occurrence of catastrophic events of gigantic proportions were not uncommon at that time. Olbers had previously proposed that the asteroids arose from the disruption of a parent planet located between Mars and Jupiter. Even though Lagrange's theory had the appeal of readily explaining the origin of the short-period comet family, whose concentration of aphelion points around Jupiter's orbit might suggest an origin in such a planet, it did not enjoy the favor of many astronomers. A variant of Lagrange's theory was presented in 1930 by the Soviet astronomer Sergei Vsekhsvyatskii (1905 - 1984) who argued that comets were formed from volcanic eruptions from the satellites of Jupiter and Saturn. This enabled him to overcome the difficulty of explaining the large velocities required to eject matter from Jupiter and Saturn (about 60 and 35 km s^{-1} respectively) to a mere few km/s, typical of the largest moons of Jupiter and Saturn. The interest in the planetary explosion theory for the origin of comets waned with time, given the formidable difficulties to explain the physics and the dynamics of the generated comets.

Around 1950 the debate on whether comets were interstellar objects or members of the solar system was still unsettled, as it was their physical and chemical nature. At that time there were a series of fundamental theoretical developments that were going to turn the tide of comet thought toward an origin in the solar system, and set the foundations of the "modern" comet science. So we shall stop here the review of the early ideas on comets and leave the "modern" ones for the following chapters.

STATISTICS OF COMETARY ORBITS AND MAGNITUDES

As we showed in the previous chapter, the earlier references of cometary apparitions are usually so vague that they do not allow to estimate fairly good orbits. Since about 1700 the record has steadily improved both in quantity and quality, adding to the naked-eye discoveries those made by telescopic means (Fig. 2.1). Ancient comet apparitions have been compiled in several catalogs, among the first ones we can mention the seventeenth-century tract *Historia Cometarum* by Stanislaus de Lubienietz, which we already referred to in the previous chapter. The most important reference source up to the eighteenth-century apparitions is the two-volume work *Cometographie ou Traité Historique et Théorique des Cometes* by Alexander Pingré written in 1783 and 1784. The most important modern source of cometary orbits is the *Catalogue of Cometary Orbits* by Brian Marsden and Gareth Williams, which brings all the comet apparitions whose observed positions in the sky were accurate enough to allow the determination of their orbits. It starts with the 239 BC apparition of comet Halley continuing up to the present, and it is annually updated.

Modern comet discoveries are usually followed by accurate astrometric measurements, which are essential for the computation of their orbits. The discovery rate showed a very low increase during the eighteenth and a large part of the nineteenth centuries. In 1892 a new breakthrough in the discovery techniques occurred when Edward Emerson Barnard (1857-1923) discovered by chance a comet in a photographic plate. Nowadays most comets are discovered in photographic plates or CCD cameras, and only a small fraction by visual searches carried out by amateurs. The extensive automated searches for near-earth objects (NEOs), like LINEAR, NEAT, LONEOS, Catalina, Siding Spring and Spacewatch, carried out during the last decade have serendipitously discovered many comets, which has produced a new big jump in the discovery rate of comets as shown in Fig. 2.1. A signi.cant fraction of the comets discovered in the last couple of decades are very faint, distant comets.

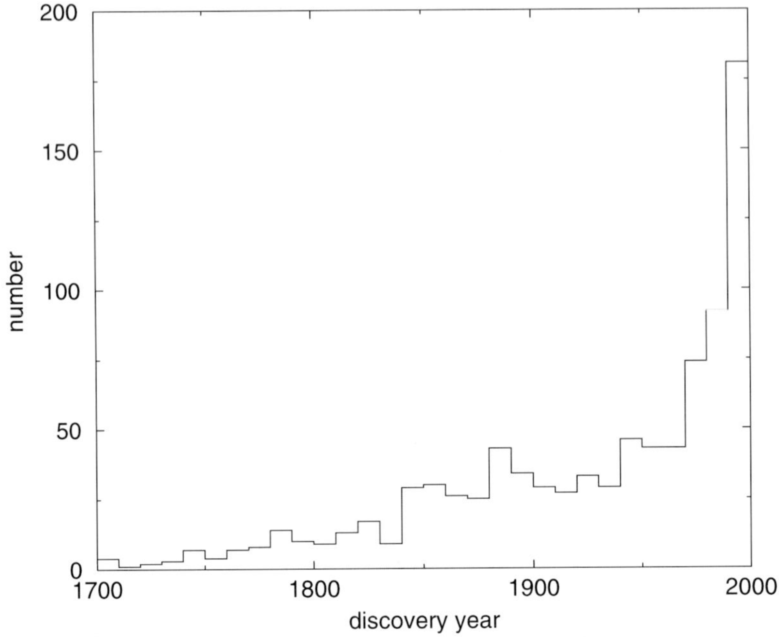

Figure 2.1. The discovery rate of comets during 1700-2000 (taking in 10-yr intervals). Both periodic and nonperiodic comets are included in the histogram. On the other hand, discoveries made from spacecrafts (IRAS, SMM, SOLWIND and SOHO) have been removed from the sample (data taken from Marsden and Williams's (2003) catalogue).

2.1. The orbital elements and Kepler's laws

A comet orbit is defined by a set of six orbital parameters of which two of them define the shape and size, they usually are: semimajor axis (a) and eccentricity (e), or sometimes instead of a and/or e, the perihelion distance (q) and/or the orbital period (P). There are in addition three angular parameters that define the orientation of the orbit in space, they are: inclination (i), argument of perihelion (ω), and the longitude of the ascending node (Ω). Alternatively, the longitude of perihelion defined as $\varpi = \omega + \Omega$ can be given instead of ω. The orbital elements are represented in Fig. 2.2. There is in addition a sixth element: the time of perihelion passage (τ) which is the particular epoch when the body is at perihelion. This epoch allows to fix the position of the body in the orbit at any other time t.

The position of the comet in its orbit is defined by the radius vector r and the true anomaly f (Fig. 2.2). In a two-body problem (where we

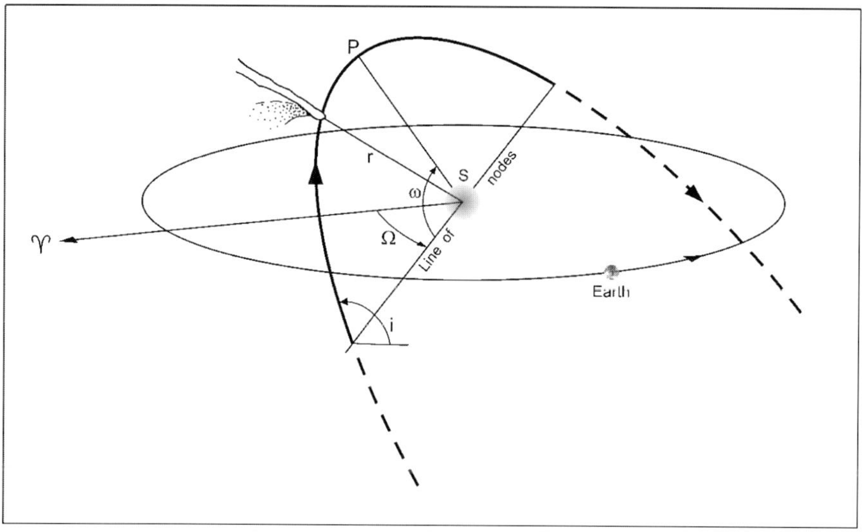

Figure 2.2. Orbital elements. The orbital plane is represented with respect to the ecliptic plane. P is the perihelion point and the angle P-S-comet is the true anomaly f. ♈ is the point of Aries.

neglect the perturbations from other bodies) we have

$$r = \frac{p}{1 + e\cos f} \tag{2.1}$$

where $p = a(1 - e^2)$ is the semilatus rectum. Equation (2.1) corresponds to a conic in which the Sun is at one of the focus (Kepler's first law). Actually, planetary perturbations will cause the comet's motion to depart from a purely Keplerian one, so the orbital elements will slightly change with time (this point will be analyzed in Chapter 4). If we neglect for the time being planetary perturbations, the total orbital energy of the comet will be conserved. The total orbital energy (per unit mass) is the sum of the kinetic and potential energies, namely

$$\frac{v^2}{2} + \frac{\mu}{r} = -\frac{\mu}{2a} \tag{2.2}$$

where $\mu = GM_{\odot}$, and $v^2 = \mu(2/r - 1/a)$ is the orbital velocity.

Instead of the semimajor axis a or the orbital period P, we can talk about the energy x of a comet. As eq. (2.2) shows, x is proportional to the reciprocal of the semimajor axis $(1/a)$, and from now on we shall refer to the latter (with opposite sign) as equivalent to x, namely $x \equiv -1/a$. Orbital energies x are negative for elliptic orbits and positive for hyperbolic ones, while $x = 0$ for a parabolic orbit. For elliptic orbits it is also possible to relate x to the orbital period P through Kepler's third law $P = a^{3/2} = (-x)^{-3/2}$, where P is given in years, a in AU and x in AU^{-1}.

It is noteworthy to point out that in a Keplerian motion the angular momentum is conserved (Kepler's second law), which can be expressed as

$$r^2 \dot{f} = h, \tag{2.3}$$

where the constant $h = [\mu a(1 - e^2)]^{1/2}$ defines the orbital angular momentum.

2.2. Periodic and nonperiodic orbits

Most comets are observed to move on highly eccentric orbits with orbital periods going from hundreds to several million years (Myr). A large fraction of these orbits are close to a parabola ($e \simeq 1$), and some comets even get hyperbolic orbits on their way out from the Sun, after gaining energy from the planets, so they will be lost to interstellar space. The rest of the comets move in less eccentric orbits (but nevertheless they are in general still more eccentric than those of the planets or asteroids) with orbital periods as small as a few years. Comets with the shortest periods are called periodic or short-period, whereas the rest are called non-periodic or long-period (LP). The limiting period is set rather arbitrarily at 200 years, based on the fact that comets with $P > 200$ yr have so far been observed only once (with the recent exception of comet 153P/Ikeya-Zhang with a period $P = 364$ yr, observed in 1661 and recovered in 2002). To a large extent this is an observational artifact due to the lack of systematic observations of comet apparitions, except for the last couple of centuries or so. A dynamical explanation for the boundary at $P = 200$ yr will be presented in Section 4.4, showing that, after all, it is not so arbitrary.

Periodic comets are usually divided into Halley-type (HT) and Jupiter-family (JF) or ecliptic comets with the boundary set at the rather conventional value of $P = 20$ yr. An additional criterion for distinguishing between HT and JF comets is the Tisserand parameter (to be defined in Chapter 7). Fortunately, the use of the orbital period or the Tisserand parameter to distinguish between HT and JF comets yields very similar samples. The distinction between HT and JF comets is not simply a matter of convention but it obeys to the reason that both populations may come from different source regions. This point will be analyzed in Chapters 4 and 7.

For LP comets we are particularly interested in computing their *original* semimajor axis a_{orig}, or orbital energy x_{orig}, since these parameters are related to their aphelion distances, and it is a measure of their dynamical age (i.e. the average number of revolutions they have performed by the inner planetary region), as we shall discuss in Chapter 4. The original orbital parameters are obtained by computing and removing the planetary perturbations that act on the comets during their incursion in the planetary region. The special group of LP comets moving on near-parabolic orbits with $a_{orig} > 10^4$ AU (or energies $x_{orig} > -10^{-4}$ AU^{-1}) are called "new" since they are believed to come to the inner planetary region for the first time.

The distributions of orbital inclinations of the different dynamical classes show striking differences. Thus, the observed i-distribution of LP comets fits approximately to a sine-law (Fig. 2.3a) indicating that the orbital planes of LP comets are more or less randomly oriented, though showing some excess of retrograde orbits (i.e. with $i > 90°$). This is not actually a new feature since it was already noted several decades ago (e.g. Porter 1963, Fernández 1981a). By contrast, HT comets move predominantly in direct orbits (Fig. 2.3b), whereas all JF comets so far discovered move in direct orbits with most of their orbital planes lying very close to the ecliptic plane (Fig. 2.3c).

Marsden and Williams's (2003) catalogue brings 1642 comets with computed orbits recorded up to the present, from which 1368 are LP comets and the rest are periodic. A significant fraction of the cataloged LP comets are sungrazers that probably come from the tidal disruption by the Sun of parent comets that get sungrazing orbits. The great majority of sungrazers (more than 560 by now) have been discovered from space-borne observatories, and in particular from the ESA/NASA *Solar and Heliospheric Observatory* (SOHO) spacecraft.

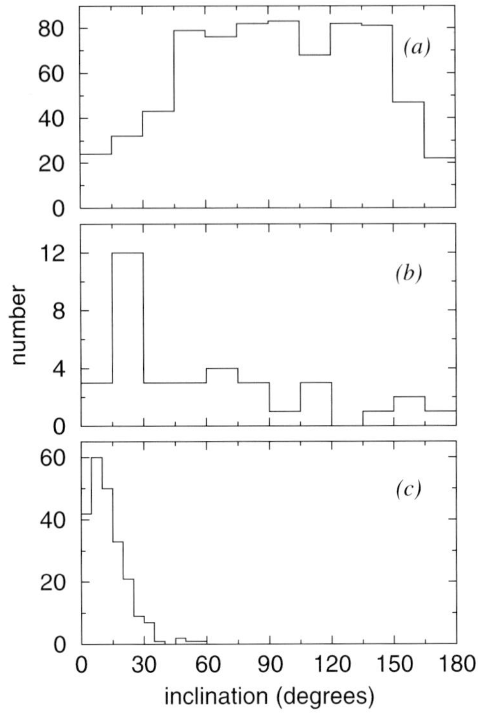

Figure 2.3. Inclination-distribution of: (a) long-period comets, (b) Halley-type comets, and (c) Jupiter-family comets. The families of sungrazers have been removed from the sample of LP comets. Samples taken from Marsden and Williams (2003).

2.3. Naming procedure

According to a resolution adopted by the International Astronomical Union in August 1994 (Minor Planet Circulars 23803-4), a designation is given to each discovered comet consisting of the year of discovery, an upper-case code letter identifying the halfmonth of discovery during that year (A = Jan. 1-15, B = Jan. 16-31, ... Y = Dec. 16-31, the letter I being omitted), and a consecutive numeral to indicate the order of discovery announcement during that halfmonth. This designation is prefixed, generally by "P/" or "C/", according as to whether the orbit shows the comet to be or not of "short" period. This means that most of the new discoveries of the so-called "intermediate-period" comets are now designated with the prefix "C/" since their orbits resemble *a priori* those of LP comets. The prefix "D/" is used for a few periodic comets that no longer exist or are deemed to have disappeared, and "X/" for a comet for which a meaningful orbit cannot be computed.

In addition to this official designation, the tradition of naming comets after their discoverer(s) is maintained. Comet discoveries are reported to the Central Bureau for Astronomical Telegrams (CBAT), which is the service in charge of receiving all the reports of observations of new astronomical phenomena. The 1994 resolution allows up to two independent discoverers, one less than what was previously admitted. For instance, the discovery of a comet was reported independently to the CBAT by the amateur astronomers Douglas Snyder in the U.S. and Shigeki Murakami in Japan on March 11, 2002, being the second in the first half of March. Follow-up observations showed that it was a LP comet. Its given name was accordingly C/2002 E2 (Snyder-Murakami). Turning to the comet discovered photographically by Barnard on October 12, 1892, later on it was shown to be periodic, but it has never been recovered in the following apparitions, so it is now considered to be extinct. Therefore, the official name for this comet is D/1892 T1 (Barnard). Many comets are discovered nowadays as part of automated sky surveys like LINEAR, so such comets are named after the sky survey program. If a periodic comet is observed to return, the prefix P/ is preceded by a sequential number. For instance, the first comet shown to be periodic by Edmond Halley in 1682 now receives the official designation 1P/Halley.

2.4. Total and nuclear magnitudes

The comet brightness or magnitude is of fundamental importance since it is closely related to the probability of detection. Furthermore, the brightness is also related to the size of the comet nucleus and its gaseous activity. However, the precise determination of comet magnitudes presents serious problems since comets do not appear as point sources like the stars but as nebulous sources. There have been devised several procedures to estimate the magnitude of the cometary coma that have rested on either de-focussing comparison stars to the size of the focal cometary coma, and then comparing the brightness of the extrafocal star images with that of the focal comet image, or to defocus both the comet and comparison stars by the same amount, or until they disappear in the sky background (see a review of the different estimation methods by Green 1996). One of more of these methods have been extensively used by the most prolific comet observers of the last century like Max Beyer, Sergej K. Vsekhsvyatskii and Nicholas T.

Bobrovnikoff. The measured magnitude will depend on the instrument employed. Thus, a telescope of small aperture will bring within the field of view the central condensation of the comet plus the extended coma, while a telescope of large aperture will only bring the central condensation. In the latter case the comet will thus look fainter than in the former case. Magnitudes measured by different observers with different instruments can differ by several orders of magnitude. To try to give more coherence to magnitude measurements, observers are requested to specify whether their measured magnitudes are 'total' (i.e. comprising the nuclear region and the extended coma), or 'nuclear' (comprising only the central condensation). It should be noted that in most cases the term 'nuclear' magnitude does not actually mean the magnitude of the comet nucleus, since it is very difficult to avoid contamination by coma light.

The apparent brightness B of a comet varies with its heliocentric distance r and its geocentric distance Δ, according to the law

$$B = B_o r^{-n} \Delta^{-2}. \qquad (2.4)$$

A pure reflection law would give an index $n = 2$ for the brightness variation with the heliocentric distance, but comets usually show indexes $n > 2$ which indicates that when approaching the Sun they brighten much more than expected for a bare solid body. When comets are close to the Sun, most of their light comes from sunlight scattered by dust particles in the coma and flurescent radiation from gaseous molecules, and not from sunlight reflected on a solid body, which greatly increases their photometric cross-section.

The total brightness B_T (nucleus plus surrounding coma) is usually expressed by the apparent total magnitude m_T, which is related to the former by: $m_T = C - 2.5 \log B_T$ (defined in a specific range of wavelengths), where C is a constant. Unless stated otherwise, we will deal with visual magnitudes. Taking logarithms in eq. (2.4) and making $B = B_T$ we obtain:

$$m_T = H_T + 2.5n \log r + 5 \log \Delta, \qquad (2.5)$$

where H_T is the absolute total magnitude, defined as the total magnitude the comet would show if it were ideally located at 1 AU from the Earth and the Sun. It is very often defined the absolute total magnitude H_{10}, which involves the additional assumption that the exponent $n = 4$ in eq. (2.5), which has an empirical justification as an average of the observed slopes of a large number of comet lightcurves.

For faint comets, detectors (e.g. CCDs) tend to record only the nuclear condensation while losing the broad (faint) coma the fades into the sky background. Therefore, the measured "total" magnitudes will be fainter than they actually are. To take into account this sky-fading effect, Kresák and Kresáková (1994) suggest the following empirical correction for the measured apparent total magnitudes of faint comets with $m_T \geq 9$

$$m_c = 0.5 m_T + 4.5,$$

where m_c is the corrected apparent magnitude. We note that $m_c = m_T$ for $m_T = 9$. No corrections are applied to comets brighter than $m_T = 9$.

Vsekhsvyatskii (1958, 1963, 1964, 1967) and Vsekhsvyatskii and Il'ichishina (1971) produced the most comprehensive catalog of absolute total magnitudes H_{10}. An updated catalog was later produced by Kresák and Kresáková (1989, 1994) for periodic comets.

We can also compute the magnitude of the nucleus, assumed to be spherical of radius R_N and (visual) geometric albedo p_v, at a distance r to the Sun. The geometric albedo gives a measure of the head-on reflectance (i.e. at phase angle $\alpha = 0$). The solar flux reflected on the comet's surface and received on the Earth at a phase angle α (Fig. 2.4) is given by

$$F(\Delta, \alpha) = \frac{F_{\odot}}{r_{AU}^2} \frac{R_N^2 p_v \phi(\alpha)}{\Delta^2} \tag{2.6}$$

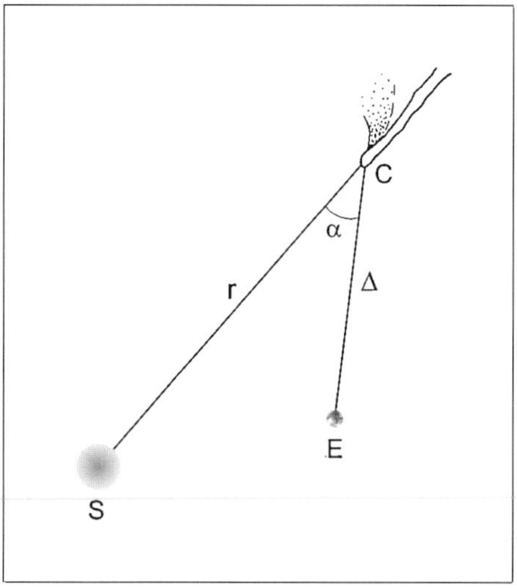

Figure 2.4. The geometry Sun-Earth-comet.

where $F_\odot$ is the solar constant, and $\phi(\alpha)$ is the phase function that gives the fraction of light scattered at phase angle α to that scattered at phase angle zero. We note that r_{AU} is given in AU, if we also want to express Δ in AU and R_N in km, we then have to multiply eq. (2.6) by 1.5×10^8. Taking logarithms in eq. (2.6), and bearing in mind that $m_N = C - 2.5 \log F(\Delta, \alpha)$ and $m_\odot = C - 2.5 \log F_\odot$ are the apparent (visual) magnitudes of the comet nucleus and the Sun respectively, we obtain

$$m_N = 14.11 - 5 \log R_N - 2.5 \log p_v \phi(\alpha) + 5 \log r\Delta, \tag{2.7}$$

where we have adopted $m_\odot = -26.77$.

The absolute magnitude of the nucleus H_N is defined as the magnitude the comet nucleus would have if it were at $r = \Delta = 1$ AU and at a phase angle $\alpha = 0$, in which case $\phi(0) = 1$. From eq. (2.7) we get

$$H_N = 14.11 - 5 \log R_N - 2.5 \log p_v. \tag{2.8}$$

We can adopt the empirical relation $-2.5 \log \phi(\alpha) = \beta_v \alpha$, where β_v is the phase coefficient in the visible. An empirical average value of $\beta_v = 0.04$ mag deg^{-1} has been derived for dark asteroids and some cometary nuclei (this point will be discussed in Section 7.6). By introducing this expression and eq. (2.8) into eq. (2.7) we obtain the following relation between the apparent and absolute nuclear magnitude

$$H_N = m_N - 5 \log r\Delta - 0.04\alpha. \tag{2.9}$$

We note that observers usually report 'nuclear' magnitudes that should not be confused with the true magnitude of the nucleus, since such magnitudes usually have some coma contamination. It is clear that as comets are observed at ever larger heliocentric distances, where they become inactive or almost inactive, the measured nuclear magnitudes approach the true magnitude of the nucleus. The improvement in nuclear magnitude determination with time for Jupiter family comets that now can be observed near their aphelia has been discussed by Fernández et al. (1999), Tancredi et al. (2000, 2005) and Lamy et al. (2004), who also analyze their relevance to estimate the true magnitude of the nucleus. A more detailed account will be presented in Chapter 7.

It is useful to have some expression that can relate total and nuclear magnitudes. We can say in advance that no accurate formula can be available, since the total magnitude depends not only on the comet's size but also on the fraction of surface area that is "active" (i.e. where volatile material is exposed to the Sun's radiation), and also on whether the comet is dusty or gaseous. Kresák (1978a), for instance, has used the following empirical relationship between the comet's diameter D (expressed in km) and the absolute total magnitude

$$\log D = 2.1 - 0.2 H_{10}, \qquad (2.10)$$

which is in reasonable agreement with analytical expressions derived by Fernández et al. (1999) for a model of comet nucleus surrounded by a coma composed of gas and dust.

The great majority of the absolute total magnitudes of LP comets are brighter than $H_{10} = 10$. Indeed, according to Everhart (1967a), the differential distribution of H_{10} shows an increase in the number of comets by a factor of about four for every one-magnitude increase up to $H_{10} = 7$, and then it still increases but at much lower rate, even when proper allowance for a greater incompleteness of comet discovery of fainter comets is made (Fig. 2.5). By studying the sample of LP comets

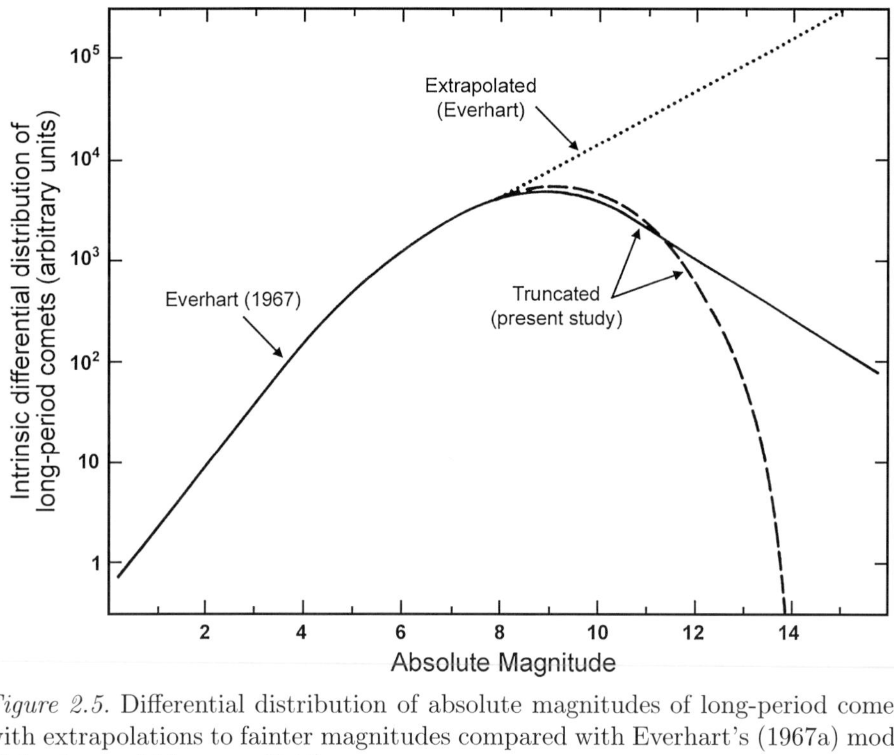

Figure 2.5. Differential distribution of absolute magnitudes of long-period comets with extrapolations to fainter magnitudes compared with Everhart's (1967a) model (Sekanina and Yeomans 1984).

that approached the Earth to less than 0.2 AU during the previous 300 yr, Kresák (1978 a,b) found a scarcity of faint comets, which he argued to be intrinsic because such Earth-approachers, even the faintest ones, can hardly pass undetected. Kresák concluded that there exists a definite cutoff in the number of active LP comets at an absolute total magnitude $H_{10} \sim 10.5$, or nucleus diameter $D \sim 1$ km according to eq. (2.10). A similar strategy was applied by Sekanina and Yeomans (1984) to study the comet population in the Earth's neighborhood. They considered all the encounters within 2500 Earth radii (0.1066 AU) and found that the intrinsic differential distribution of absolute magnitudes for LP comets levels off at $H_{10} \sim 8$. These results entirely confirmed Kresák's earlier conclusion about the scarcity of fainter LP comets, a much more radical conclusion than Everhart's (compare Everhart's and Sekanina and Yeomans's extrapolations in Fig. 2.5). Nevertheless, the problem of the faint end of the H_{10}-distribution is still far from solved. The fading of the coma brightness in the sky background of faint comets might strongly play against their discovery, even when they approach the Earth. In fact, several extremely faint comets, probably with $H_{10} \gtrsim 12$, have recently been discovered in the Sun and Earth's vicinity by the different sky surveys now in operation, in particular LINEAR.

The distribution of absolute total magnitudes of JF comets shows a clear shift toward fainter magnitudes (Fig. 2.6). There are two possible interpretations for this: (1) JF comets are on average smaller than LP comets, or (2) they are of comparable sizes but less active. We favor the latter interpretation since observations show that some JF comets are faint despite of being of large size, and this is because only a tiny fraction of their surfaces remain active (Fernández et al. 1999). If we assume that f represents the fraction of active area of a comet nucleus, the total brightness B_T of the comet (coma + nucleus) will approximately be proportional to $B_T \propto R_N^2 f$, so we get an expression of the type $H_T = C - 2.5 \log B_T - 2.5 \log f$. Instead of eq. (2.10) we can then use the more general expression

$$\log D = 2.1 - 0.2 H_{10} - 0.5 \log (f/f_{new}), \tag{2.11}$$

where f_{new} is the average fraction of active surface area of new comets. If all the surface is active, then $f_{new} = 1$, though this may be considered an upper limit since even new comets may have inactive zones covered by refractory material on their surfaces. As we shall see in Chapter 9, typical values of f for aged comets like 1P/Halley or JF comets (i.e.

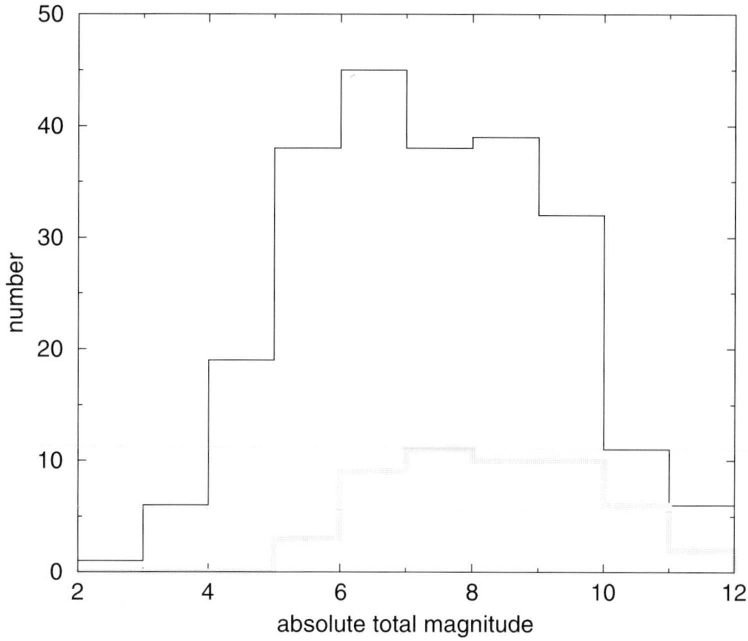

Figure 2.6. Distribution of absolute total magnitudes for the sample of LP comets taken from Everhart (1967a) (thin histogram), and for that of Jupiter family comets (thick histogram) taken from Kresák and Kresáková's (1994) catalogue. Both samples only include those comets with perihelion distances $q < 2$ AU.

that have already endured many passages by the Sun's vicinity) can be something in between 0.1- 20%. If new or young LP comets have, say, a fraction $f_{young} \sim 0.5$, and aged LP comets or JF comets have $f_{aged} \sim 0.05$, we get $-2.5 \log (f_{young}/f_{aged}) \sim 2.5$, i.e. a shift of 2.5 magnitudes for the same size distribution, which is more or less what is observed in Fig. 2.6.

2.5. Selection effects

By far the most important selection effect has to do with the comet's perihelion distance, by the simple reason that the larger q, the fainter the comet, as it is both more distant from the Earth and from the Sun. The rapid drop in the discovery rate for $q > 2$ AU is clearly shown in Fig. 2.7, though the discovery of distant LP comets is rapidly increasing in the last few years, in particular thanks to survey programs of near-Earth objects like LINEAR, NEAT and LONEOS. The observing geometry may also play a role in the potential discovery of

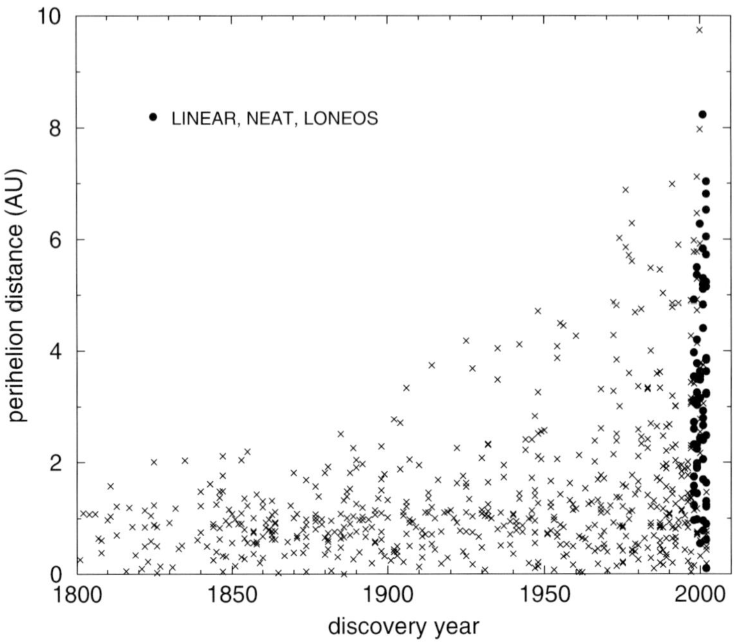

Figure 2.7. Discovery rate of LP comets since 1800. Filled circles are for comets discovered by the LINEAR, NEAT and LONEOS programs. We have removed all the families of sungrazers from the sample. Data taken from Marsden and Williams's (2003) catalogue.

a comet. The Holetschek effect is the best known one (e.g. Everhart 1967a,b, Kresák 1975). It was found by the Vienna astronomer Johann Holetschek more than a century ago, and it is associated with the fact that comets reaching perihelion on the opposite side of the Sun, as seen from the Earth, are less likely to be discovered. Figure 2.8 shows that indeed many more of the discovered LP comets reached perihelion near conjunction with the Sun (difference of longitudes of the comet at perihelion and the Earth close to 0° or 360°) than near opposition (difference close to 180°). The smaller discovery probability in the latter case is clearly understood since when these comets are close to perihelion, and thus have maximum brightness, happen to be far from the Earth. As seen, the Holetschek effect has virtually vanished in the last few decades (dashed histogram of Fig. 2.8), probably due to the influence of dedicated search programs that led to the discovery of many distant comets for which the geometric configuration with the Sun at the time of perihelion passage was unimportant.

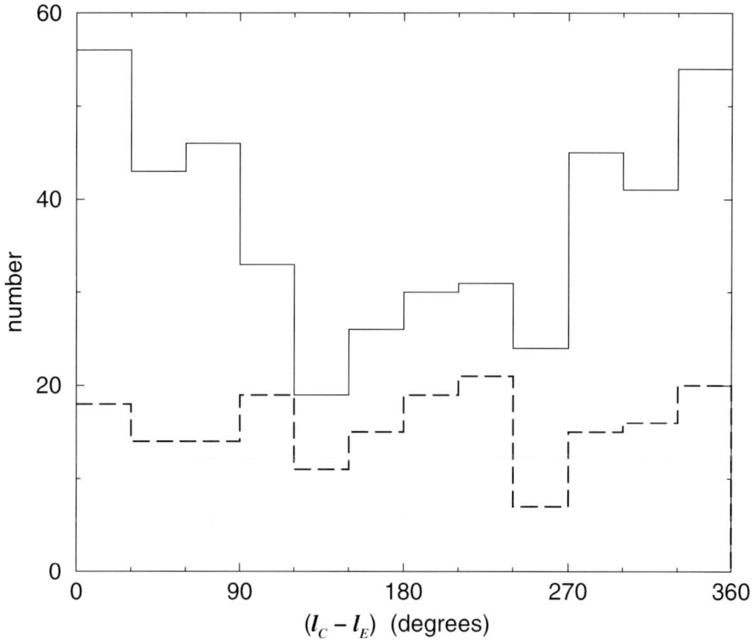

Figure 2.8. Frequency distribution of the angle ($l_C - l_E$) (difference between the heliocentric longitudes of the comet at the moment it reaches perihelion and the Earth) for the sample of LP comets discovered between 1801-1980 (solid histogram) and those discovered from 1980 to the end 2002 (dashed histogram). Sample taken from Marsden and Williams (2003) where the families of sungrazers were removed.

The unequal coverage of the Northern and Southern hemispheres may be another cause of bias in comet discovery. Yet, the distribution of declinations of LP comets at perihelion does not show a significant drop for high Southern declinations (south of $-30°$) (Fig. 2.9), which suggests that this effect has played only a minor role.

2.6. The distribution of perihelion distances

An important orbital parameter is the perihelion distance (q) that defines the distance of closest approach to the Sun. For LP comets, the frequency distribution of q shows a maximum near Earth's orbit ($q \simeq 1.0 - 1.1$ AU). It falls steeply closer to the Sun and farther away (Fig. 2.10). As discussed above, the latter can be explained in terms of an observational selection effect since most distant comets can pass undetected. The drop in the rate of passages of LP comets closer to the Sun may be explained in terms of much shorter physical lifetimes,

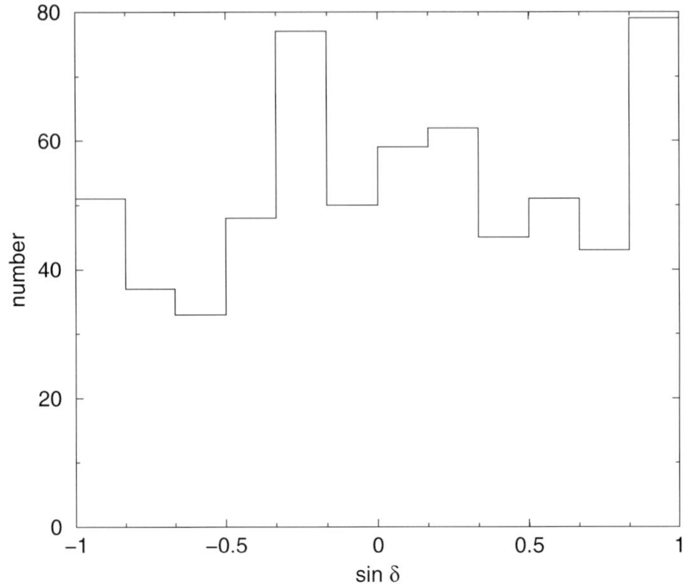

Figure 2.9. Distribution of declinations of LP comets at perihelion. Sample taken from Marsden and Williams (2003) where the families of sungrazers were removed.

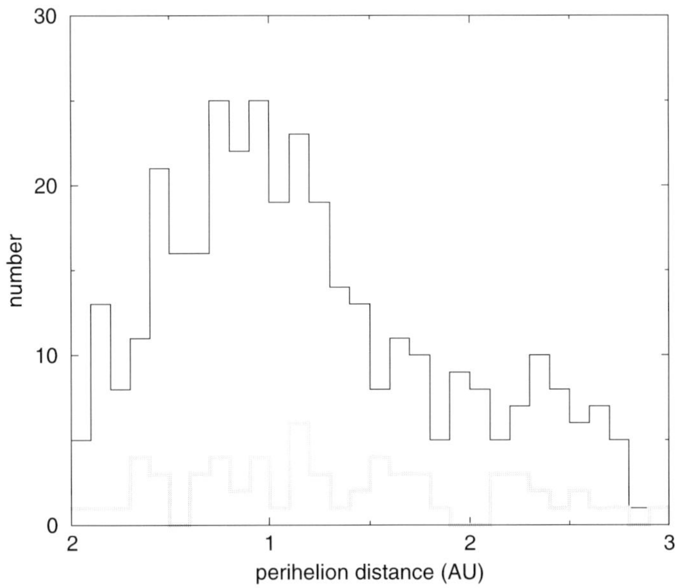

Figure 2.10. Distribution of perihelion distances of LP comets observed after 1900 (thin histogram), and of new comets (thick gray histogram). Data taken from Marsden and Williams's (2003) catalogue. All the sungrazers have been removed from the sample. For new comets, those classified as of class 2B (i.e. those whose original semimajor axes have the largest errors) have also been removed.

since they may disintegrate or fade away after a few passages. For instance, when we limit the sample of LP comets to the new comets, the q-distribution turns out to be roughly uniform, at least in the inner planetary region which, again, is an indication that such comets are entering this region for the first time, so they have still been little affected by the solar radiation.

It is noteworthy to analyze the distribution of LP comets in the parametric plane: perihelion distance (q) vs. original orbital energy (x_{orig}). As shown in Fig. 2.11, the concentration of comets along the line of zero energy (i.e. in near-parabolic orbits) is remarkable, and this effect is strengthened for the discovered LP comets with $q \gtrsim 2$ AU.

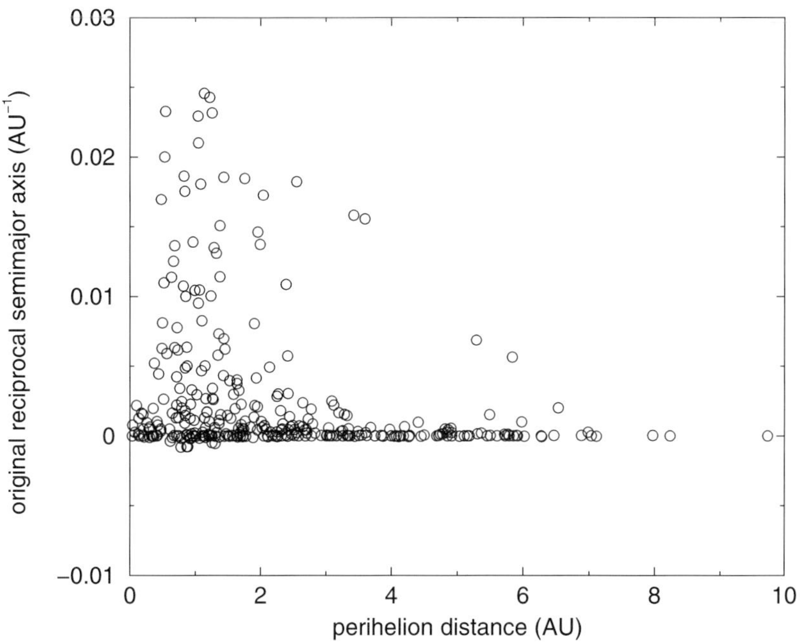

Figure 2.11. Original reciprocal semimajor axis versus perihelion distance of the long-period comets observed during the period 1850-2002 taken from Marsden and Williams's (2003) catalogue. The families of sungrazers are not included.

So far, the overwhelming majority of the discovered LP comets have perihelion distances smaller than a few AU. Very few have perihelia beyond Jupiter's orbit (> 5.2 AU), the farthest perihelion being so far for comet C/2003 A2 (Gleason) which has $q = 11.43$ AU.

2.7. The discovery rate and the flux of new and LP comets

The discovery rate of Earth-crossing or approaching LP comets ($q < 1.2$ AU) has remained more or less constant since 1850, although a moderate increase is noted during the last 20 - 30 yr. By contrast, the discovery rate of LP comets with $q \gtrsim 1.2$ AU has steadily increased since about 1900. The sample does not include the special group of more than 560 sungrazers discovered until now, mainly from space-borne observatories. As we shall see in Chapter 9, they come from the fragmentation of a few parent comets by tidal forces from the Sun. As mentioned, search programs like LINEAR, LONEOS, NEAT are playing a very active role in the discovery of distant LP comets (see Fig. 2.7). In particular, the *Lincoln Near Earth Asteroid Research* (LINEAR) project, that uses signal processing technology applied to large format, highly sensitive charge-couple devices (CCDs) attached to an 1-m size telescope, has been very successful in detecting near-Earth and main-belt asteroids, as well as comets. Since LINEAR began operations in March 1998, it has discovered 62 LP comets that represent about 60% of all the LP comets discovered until the end of 2002. Within a few years, the all-sky survey program Pan-STARRS consisting of an array of four 1.8-m telescopes, which will be able to reach an apparent red magnitude $m_R \sim 24$, promises to increase the discovery rate of comets by a factor of 10, mainly in large-q orbits (Jewitt 2003).

Since comets in Earth-crossing orbits have the greatest probability of being detected, we will concentrate on this sub-sample, though for new comets we will also compare the derived passage rate for $q < 1$ AU with that for $1 < q < 2$ AU. New comets have the advantage of being in general more active and therefore brighter, so they are more likely to be detected even at somewhat larger heliocentric distances (see Fig. 2.11). There are 23 new comets that crossed Earth's orbit during the last century (1901-2000), i.e. with computed $x_{orig} > -10^{-4}$ AU^{-1}. Actually, 8 out of the 23 comets have $x_{orig} > 0$ (i.e. hyperbolic orbits), which suggests large errors in their computation, perhaps due to the unaccounted effect of "non-gravitational" forces (to be analyzed in Chapter 4), so they may not pertain to the Oort cloud. On the other hand, some new comets may have escaped detection due to unfavorable observing geometries, and some comets classified generally as to be of "long-period", because their x_{orig} could not be determined, may be new. If we consider now the next interval $1 < q \leq 2$ AU, the number of observed passages for the period 1901-2000 is 26 (from which 8 have positive x_{orig}), which is in good agreement with the previous interval.

Another matter is the size distribution and the size limit of the observed new comets. As discussed in Section 2.4, the great majority of the discovered LP comets are brighter than $H_{10} = 10.5$, so we can set this as the upper magnitude limit for our comet sample. If we apply eq. (2.11) for $f = f_{new}$, we obtain a diameter $D = 1$ km for $H_{10} = 10.5$. Furthermore, the degree of completeness seems to be high for $H_{10} \lesssim 7$, which is suggested by the near constant rate of discoveries in the century. Vsekhsvyatskii's (1958) and Everhart's (1967a) tables of total magnitudes show that $\sim 60\%$ of the new comets with $q < 2$ AU has $H_{10} < 7$. If we make allowance for undetected new comets fainter than $H_{10} = 7$ (assuming that brighter new comets were all detected) by means of Sekanina and Yeomans's (1984) magnitude distribution (cf. Fig. 2.5), we find a ratio $N(H < 10.5)/N(H < 7) \sim 5$. By assuming now that 60% of the observed new comets during 1901-2000 had $H_{10} < 7$ and extrapolating the number to $H_{10} = 10.5$ we finally obtain

$$\dot{n}_{new}(H_{10} < 10.5) \sim 3 \text{ Earth-crossing new comets every 4 yr,}$$

from the different sources of uncertainty (degree of completeness of the sample, extrapolation to faint magnitudes) we can estimate the uncertainty of the above derived rate at about $\pm 30\%$.

The numbers of evolved LP comets ($a_{orig} < 10^4$ AU and $P > 200$ yr) that crossed Earth's orbit during 1901-1950 and 1951-2000 were 50 and 81 respectively. The ratio between the observed populations of evolved LP comets to new ones is about 6. Probably more LP comets were missed than new ones due to the decay of the gaseous activity after the first passage (cf. Figs. 2.10 and 2.11), so the ratio may raise somewhat to about 10 when this is taken into account. This ratio should be taken as an average for all comets with $q < 1$ AU. The ratio of evolved to new comets is a function of q because comets that get very close to the Sun have shorter physical lifetimes, so the number of evolved comets must decrease closer to the Sun.

Form the previous analysis we can estimate that about 7 LP comets with $D \gtrsim 1$ km pass perihelion within Earth's orbit per year. This theoretical estimate is getting close to the current discovery rate of a few LP comets with $q < 1$ AU yr^{-1} (which shows an increase with respect to previous years due to several LINEAR and NEAT discoveries). The previous derivation is in line with Kresák and Pittich's (1978) estimate that about 60% of all LP comets in Earth-crossing orbits are being discovered. Yet, at the time they published their results, such a percentage may have been overestimated by a factor of two. We can say that just recently such a degree of completeness may have been reached, thanks fundamentally to the above mentioned sky surveys. Our estimate is also in good agreement with that of Weissman (1990) who found a flux of 10.1 LP comets brighter than $H_{10} = 11$ that cross Earth's orbit per year. However, Weissman's conversion formula of absolute total magnitudes to masses (or diameters) leads to values more than one order of magnitude higher than those obtained from eq. (2.10), which illustrates the large disagreement among different authors regarding the relationship between H_{10} and comet size or mass.

THE UNDERSTANDING OF THEIR PHYSICAL NATURE

As we have seen (cf. Chapter 1), until the seventeenth century the most widely accepted idea was that comets were vapors or dry exhalations rising from the Earth. Once their heavenly nature was accepted, the exhalations that condensed into comets were assumed to rise from the atmospheres of the major planets (Hevelius), or from other stars (Cassini). On the other hand, Newton envisioned a comet as a body consisting of a solid nucleus shining by reflected sunlight, with a tail formed by vapors released from the nucleus. In the end, Newton's view proved to be correct, but the true physical nature of comets was going to be debated still for nearly three centuries.

Other problem was to determine the actual size of a comet. The extended comae seemed to correspond to huge bodies of size comparable to the Sun itself. We saw (cf. Section 1.3) how Tycho Brahe estimated a diameter of nearly one fourth of the Earth's for comet 1577. Of course, at that time there was no idea that the coma gases were very rarefied and thus very little matter was contained within the coma. Against this backdrop, it is not surprising that Georges-Louis Leclerc, Comte de Buffon (1707-1788) envisaged the formation of planets as the result of the collision of a comet with the Sun that produced the ejection of solar material that condensed into planets. Laplace was also of the idea that a close encounter of a comet with the Earth would raise the seas and produce a universal deluge. Kant believed that comets formed in the outer parts of the solar nebula and that bodies formed there tended to have larger masses with increasing distances to the Sun, so comets must have masses at least comparable to those of the largest planets.

3.1. Comet sizes and masses

The idea that comets had planetary size could not be sustained for too long. In 1805 Laplace could place an upper limit to the comet mass by noting that the close encounter of comet Lexell with the Earth in 1770 did not produce any perceptible perturbation in Earth's motion. He found that the mass of the comet could not be greater than 1/5000 of the Earth's mass. It was then becoming clear that comet heads and tails, despite their large volumes, were of extremely low density. This fact was confirmed by Friedrich Georg Wilhelm Struve (1793-1864) and Friedrich Wilhelm Bessel (1784-1846) who could not find any perceptible effect, either diminution or reflection, in the light of stars seen quite close to the apparent nucleus of Halley's comet during its apparition of 1835. The Great Comet of 1882 transited the Sun's disk and could not be seen, which allowed to place an upper limit of 70 km to its diameter. The same happened with Halley's comet that could not be observed during its transit before the Sun's disk on May 8, 1910, which allowed to set an upper limit of 100 km to its size.

There were many other different attempts to estimate comet masses based on different methods. On the basis of presumed mutual perturbations between two components of periodic comet 3D/Biela after its breakup in 1846, Josef von Hepperger estimated in 1906 a comet mass of 4.2×10^{-7} Earth masses. This value may be too large by three-four orders of magnitude, which is not surprising since the claimed gravitational perturbation between the fragments was presumably spurious. Another approach was tried by Vorontsov-Velyaminov (1946) who derived minimum initial masses of 10^{18} g for comet 109P/Swift-Tuttle and of 10^{16} g for 55P/Tempel-Tuttle, based on the mass estimates for the associated Perseids and Leonids meteor streams respectively, and assuming that the masses of the meteor streams were only a fraction of the masses still remaining in the respective nuclei.

The rare occasion of a close approach of a comet to the Earth provides a unique opportunity to estimate the comet's size in a rather straighforward manner. This was indeed the case of comet 7P/Pons-Winnecke that passed at a mere 0.04 AU from the Earth in 1927. The high resolution of the nuclear region attained during the close encounter allowed Vesto M. Slipher and Fernand Baldet to estimate an upper limit of 5 km for its diameter. As we will see below, this turned out to be a very good estimate but, unfortunately, almost unique since it depends on the - infrequent - occurrence of very close comet encounters with the Earth.

During the sixties and seventies Elizabeth Roemer carried out the most extensive and systematic determination of nuclear magnitudes in the pre-CCD camera era by means of photographic plates taken with long-focus reflectors. The large-scale comet images on the plates showed the nuclear condensation, usually of star-like appearance, with little trace of the coma. Her estimated comet sizes proved to be too large in most cases because her nuclear magnitudes were actually contaminated by light from the coma. Nevertheless, her effort was valuable at that time, setting meaningful constraints for the upper limits of comet sizes, from a few to some tens of kilometers (Roemer 1966).

3.2. 'Sand bank' versus 'dirty snowball' model

From his theory of interstellar capture of dust particles, Lyttleton (1951) envisaged the structure of a comet as a loosely gravitationally bound swarm of dust particles occupying a volume of the size of the coma (cf. Section 1.4). In actuality, he showed that the majority of particles simply described closed orbits round the Sun as a meteor stream. This is known as the 'gravel' or 'sand bank' model. Lyttleton argued that collisions among particles, especially towards the center of the comet, during the approach to the Sun, must lead to their pulverization and thereby furnished more finely divided material which, under the influence of the radiation pressure from the Sun, was forced into the tail causing its known forms.

However, Lyttleton's model soon met unsurmountable difficulties. Firstly, the observations showed that the nuclear region which shines by reflected sunlight is much smaller than the coma size, suggesting a much more compact structure than envisaged by Lyttleton. Spectroscopic observations showed that the proportion of mass in the coma in the form of dust is extremely small. Secondly, were comets constituted by swarms of dust particles, it was difficult to see how they could withstand close passages to the Sun without sublimating away.

Whipple's (1950) icy conglomerate model envisages the cometary nucleus as composed of a mixture of ices and dust particles. From the solar abundances and assuming a depletion of hydrogen, helium and neon, Delsemme (1977) derived a primordial dust (silicates) to gas (ices) mass ratio in comets ranging from 0.61 to 0.76. Whipple's model, also known as the 'dirty snowball' model, has been very successful in explaining several properties of comets as, for example, their ability to withstand close passages by the Sun and the nongravitational forces that make comet orbits to depart from those obtained from purely gravitational means. Whipple showed that these forces arise from the nonisotropic mass loss produced by the sublimation of the comet ices, giving rise to a recoil (rocket) effect (see more details in Section 4.2).

3.3. The solid nucleus

In general terms, Whipple's model was confirmed when an international flotilla of spacecrafts visited Halley's comet in its return of 1986 and could take images of its nucleus of irregular shape of dimensions 15 × 7.5 × 7.5 km. In September 2001 the NASA's Deep Space 1 spacecraft flew past the nucleus of periodic comet Borrelly at a close distance of 2200 km, showing again an 8-km long elongated body. Both nucleii are shown in Fig. 3.1. In essence, the main difference between comets and asteroids is that the former have gaseous activity owing to the presence of volatile material on or near the surface that sublimates by the Sun's radiation, while the latter are inactive rocky bodies (Fig. 3.2).

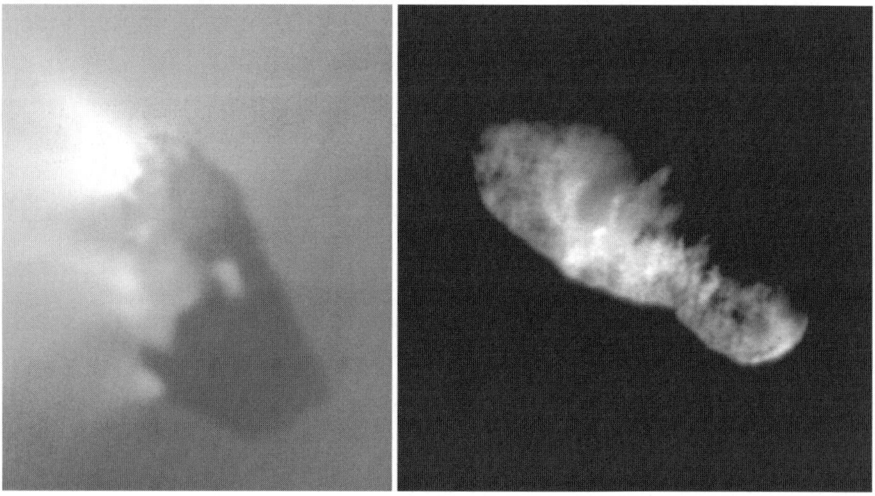

Figure 3.1. The nuclei of comets Halley (left) and Borrelly (right) (Image credits: ESA/Max Planck Institut für Aeronomie (Halley) and NASA/Jet Propulsion Laboratory (Borrelly)).

Comet 81P/Wild 2 has also been imaged from close distance by the *Stardust* spacecraft. Two images of the about 5-km diameter nucleus are shown in Fig. 3.3. The left image shows the rugged pock-marked surface of the bare nucleus, while the right image blots out the surface details while enhancing the surrounding coma and jets of gas and dust emanating from the nucleus. This marks again a clear difference with an asteroid in which no gas and dust are seen emanating from its surface. As regards the pock-marked surface of the comet nucleus, even though it resembles that of an asteroid (cf. Fig. 3.2), the origin of the features is probably quite different: asteroids pocks are impact craters caused by

Figure 3.2. Asteroid Ida with its small satellite Dactyl (at right) as it was imaged by the *Galileo* spacecraft from a distance of 10,900 km. Ida is about 56 km long (NASA/Jet Propulsion Laboratory).

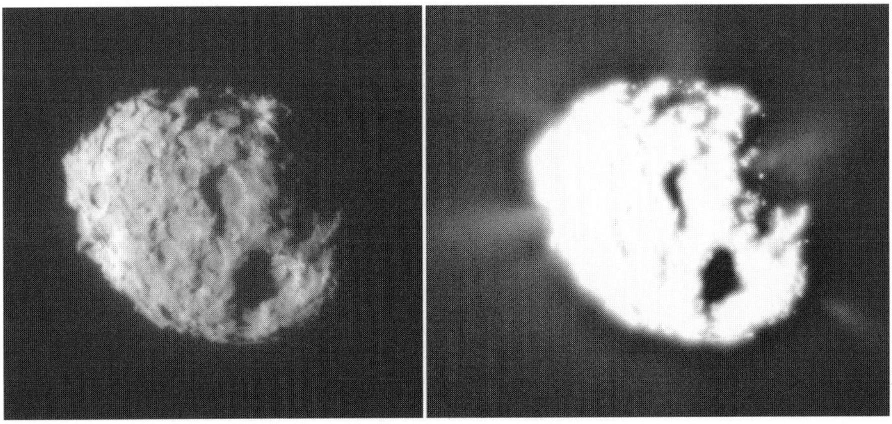

Figure 3.3. Comet 81P/Wild 2 imaged by the *Stardust* spacecraft from a distance of 500 km on January 2, 2004. The left image shows the surface details of the bare nucleus, while the right image enhances the dust and gas coma and jets surrounding the nucleus (NASA/JPL).

collisions with other asteroids and interplanetary boulders; some comet pocks on the other hand may be caused by sublimation of pockets of volatile material. The ice pockets sublimate when they are illuminated by the Sun and the gases produced escape into vacuum blasting out of the surface, leaving behind a big depression on the surface.

The comet nucleus is regarded as a very fragile and low-density structure. Donn (1963) estimated that the mixture of ices and meteoric matter would form an aggregate with a density of a few tenths g cm^{-3}. Weissman (1986) also proposed a nuclear structure consisting of aggregates of various sizes that he called the 'rubble-pile' model. Further details and other properties of the comet nucleus will be discussed in Chapter 7.

3.4. Sublimation of volatiles

We will analyze now the physics related to the sublimation of volatile material in the comet nucleus. The source of energy is solar radiation that falls on the rugged pitted surface of a comet nucleus, thus creating a complex pattern of shades and illuminated areas. To simplify the problem, we will assume in the following that the nucleus is of spherical shape with a smooth surface entirely covered by volatile material. Let us then consider a spherical cometary nucleus of radius R_N and visual Bond albedo A_v at a heliocentric distance r. The *Bond albedo* (A) is the fraction of incident light that is scattered in all directions. The Bond albedo is related to the geometric albedo p defined in Section 2.4 through the relation

$$A = pq, \tag{3.1}$$

where q is the phase integral given by

$$q = 2\int_0^{\pi} \phi(\alpha) \sin \alpha d\alpha, \tag{3.2}$$

where $\phi(\alpha)$ is the phase integral. Both albedos depend on the wavelength and we will usually use the albedos A_v, p_v for the visible part of the spectrum.

We will assume that the incident solar radiation evenly spreads across the nuclear surface, even though this is not completely true, even for a rapidly rotating nucleus. We can then write the energy balance equation (per unit surface area and time), between the input: solar radiation, and the following losses: (i) thermal infrared radiation, (ii) sublimation of ices, and (iii) heat conduction into the nucleus interior, namely

$$(1-A_v)\frac{F_\odot e^{-\tau}}{r_{AU}^2}\pi R_N^2 = 4\pi R_N^2(1-A_{IR})\sigma T^4 + \frac{QL_S}{N_A} + 4\pi R_N^2\kappa(T)\left|\frac{\partial T}{\partial z}\right|_{z=0}, \quad (3.3)$$

where $(1 - A_v)$ is the fraction of incident light that is absorbed; A_{IR} is the Bond albedo in the infrared, $A_{IR} \simeq 0$ for most ices; $F_\odot = 3.16\times10^{-2}$ cal cm^{-2} s^{-1} is the solar constant, namely the amount of incident solar energy per unit area and time at 1 AU from the Sun; r_{AU} is the heliocentric distance expressed in AU; $\sigma = 1.35 \times 10^{-12}$ cal cm^{-2} s^{-1} K^{-4} is the Stefan-Boltzmann constant; T is the equilibrium temperature; Q is the total sublimation rate in molecules s^{-1}; L_S is the latent heat of sublimation per mole, which is slightly dependent on the temperature; $N_A = 6.025 \times 10^{23}$ mol^{-1} is the Avogadro's number; $\kappa(T)$ is the thermal conductivity of the cometary material; and z is the depth. The exponent τ is the optical depth of the coma which is usually very small. Furthermore, it is not obvious what the overall effect of the coma dust should be on the energy budget. On one hand, the coma dust may absorb and scatter some light before reaching the nucleus but, on the other hand, it may even enhance the radiation received by the nucleus. This is because of the much larger cross section of the dust halo as compared to that of the comet nucleus. The solar radiation reaches the nuclear surface after multiple scattering by the dust particles and, indirectly, thermal energy re-radiated by the heated dust particles (Weissman and Kieffer 1981). We will thus neglect any extinction in the coma, since the overall effect, whether attenuation or enhancement, is very difficult to estimate and, in any case, it seems to be very small. Likewise, thermal conductity is also very small, so it can be neglected without losing too much accuracy.

In order to solve eq. (3.3) we must know the relation between Q and T. We can write $Q = 4\pi R_N^2 Z$, where $Z = Z(T)$ is the gas production rate in molecules cm^{-2} s^{-1}. By applying the kinetic theory of gases to a vaporization model in which there is a balance between the vaporizing flux and the condensing flux on the ice surface, the following expression is obtained (Delsemme 1981)

$$Z = P_v(2\pi\mu m_u kT)^{-1/2}, \tag{3.4}$$

where μ is the molecular weight, m_u is the unit atomic mass, k is the Boltzmann's constant, and P_v is the saturated vapor pressure. The latter is given by standard thermodynamical expressions such as the Clapeyron's equation applied to a phase change that takes place at constant temperature and pressure. For water molecules freely expanding into vacuum from the sublimation of water ice, the saturated vapor pressure is approximately given by (Rickman et al. 1990)

$$P_v = P_o \exp\left(-\frac{L_S}{kN_AT}\right) \simeq 1.2\times10^{13} \exp\left(-\frac{6000}{T}\right) \text{ dyn cm}^{-2}. \tag{3.5}$$

At large heliocentric distances, most of the solar radiation reaching the nucleus is spent in thermal re-radiation, so the equilibrium temperature can be determined from a balance between insolation and thermal re-radiation in a similar manner as done for rocky bodies like asteroids. By contrast, at small r most of the solar radiation is spent in the sublimation of the ices, which means that the thermal balance equation can be approximated by

$$(1 - A_v)\frac{F_\odot}{r_{AU}^2}\pi R_N^2 \approx \frac{QL_S}{N_A} \tag{3.6}$$

which, after introducing numerical values, leads to

$$Q \approx (1-A_v)\frac{F_\odot}{r_{AU}^2}\frac{N_A}{L_S}\pi R_N^2 \approx 5\times10^{28}\frac{R_N^2}{r_{AU}^2} \text{ molecules s}^{-1} \tag{3.7}$$

where R_N is expressed in km, and it is assumed that water ice controls the sublimation rate.

Equation (3.7) shows that at small r the sublimation rate changes in proportion with the solar insolation, i.e. like r^{-2}, no matter what the latent heat is. Note that eq. (3.7) implies that all the surface of the comet nucleus is freely sublimating. As we shall discuss in Chapter 9, in practice a fraction of the nucleus surface is usually covered by an insulating dust mantle, so the area under free sublimation may be substantially reduced. The gas production rate Q must then be multiplied by a factor $f(< 1)$ defined as the ratio of the free-sublimating area to the total surface area of the nucleus.

As discussed before, the temperature of the surface does not correspond to the equilibrium blackbody temperature, since the escaping gas carries away much of the heat as latent heat of sublimation. This is true at least for r smaller than a few AU. Under these circumstances, the temperature is essentially determined by the gas that controls the sublimation. For H_2O it varies between 210 K at 0.2 AU, 190 K at 1 AU, and 90 K at 10 AU. For a more volatile species like CO_2, it lies between 115 and 85 K for the range $0.2 < r < 10$ AU, while a highly volatile species like CO, the temperature range goes down to 30 - 45 K for the same range of r. The latent heats of these comet ices for the range of T indicated before, and the heliocentric distances where the onset of active sublimation occurs are shown in Table 3.1. Figure 3.4 shows the sublimation rate for several molecular species that may be present in comets. As seen, all of the candidate molecules are more volatile than water, so they are expected to be observed in cometary comae at heliocentric distances greater than that for which water sublimation starts to control the overall sublimation rate.

Table 3.1: Ices relevant to comets

Comet ice	L_S (cal mol^{-1})	r_{onset} (AU)$^{(*)}$
H_2O	11700	2.5
CO_2	6000	8.3
CO	1400	62.5

$^{(*)}$ from Delsemme (1981)

The sublimating gases are released at the thermal expansion velocity v_{th}, where

$$\frac{1}{2}\mu m_u v_{th}^2 = \frac{3}{2}kT$$

i.e.

$$v_{th} = \left(\frac{3kT}{\mu m_u}\right)^{1/2}, \tag{3.8}$$

where μ is the molecular weight of the gaseous species that controls sublimation. Typical expansion velocities are 0.5 km s^{-1} at $r \sim 1$ AU, which is much higher than the escape velocity from a comet nucleus.

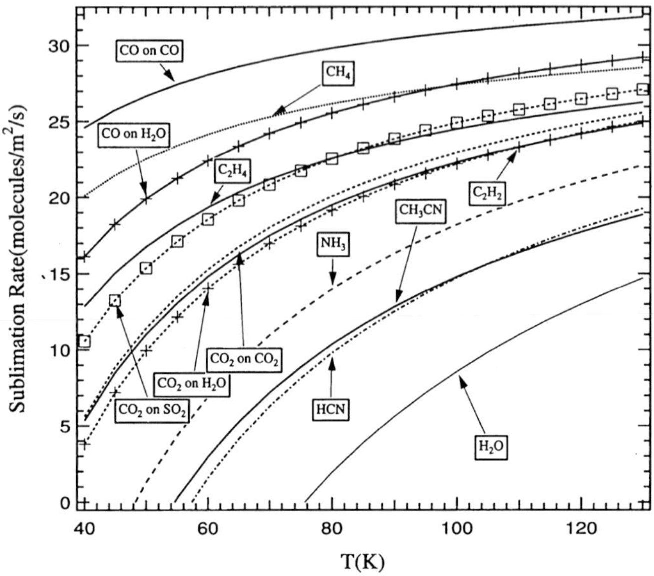

Figure 3.4. Sublimation rate of different volatile molecules as a function of the ice temperature. Sublimation rates are in a logarithmic scale (Mukai et al. 2001).

The coma is usually well developed for heliocentric distances $r \lesssim 3$ AU. Since the equilibrium temperature for a body at $r \approx 3$ AU is roughly equal to the sublimation temperature of water ice, the appearance of a well-developed coma at smaller r has been taken as evidence that the volatile component of most comets consists primarily of water ice. Yet, at distances $r > 3$ AU many comets show a substantial activity, which is mainly driven by the sublimation of CO. This can be seen in the case of comet C/1995 O1 (Hale-Bopp), one of the best studied comets in the last decades, for which the variation of the outgassing rate with the heliocentric distance can be fitted to the theoretical curve for the sublimation of CO for $r \gtrsim 2.5 - 3$ AU, and to that for H_2O for smaller r (Fig. 3.5).

Comet lightcurves usually show asymmetries between the branches pre- and post-perihelion which may be caused by the structural changes occurring in comet nuclei as they move near the Sun, as for instance, the purge of an insulating dust mantle that leaves exposed areas of fresh ices, or by thermal inertia. These two effects will tend to increase the gaseous activity after perihelion as is usually observed (see the cases of Hale-Bopp and of comet Halley shown in Fig. 3.6).

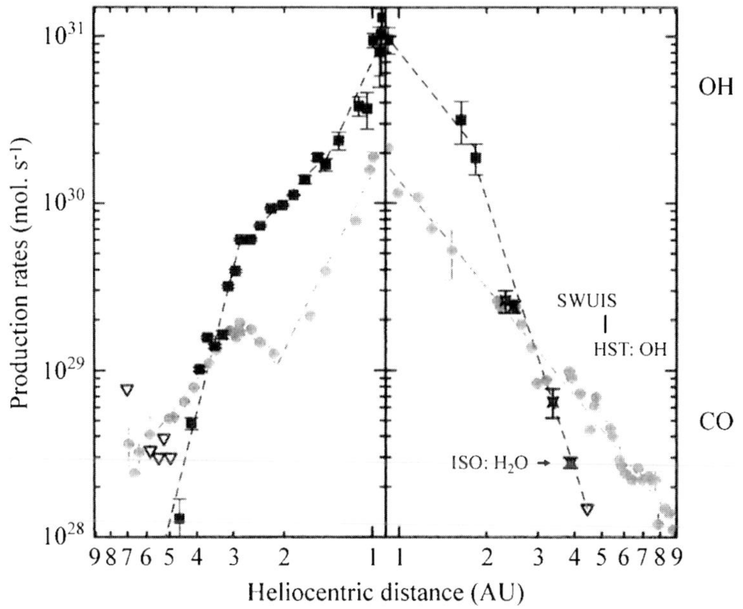

Figure 3.5. Observed CO and water outgassing rates of comet C/1995 O1 (Hale-Bopp) versus heliocentric distance (left: pre-perihelion; right: post-perihelion). Gray circles are for CO, and black squares for H_2O. Open triangles are upper limits for OH (or water production rate). These results were compiled from different observers (Biver 2001).

3.5. Spectroscopic observations

The first spectra of comets were obtained by the Florentine astronomer Giovanni Battista Donati (1826-1873) who observed and drew the spectrum of comet C/1864 N1 (Tempel). Donati's spectrum shows the three molecular emission Swan bands of C_2 that often dominate comet spectra in the visible region, though he was not able to identify their chemical nature. Afterwards Sir William Huggins (1824-1910) observed the spectrum of comet 55P/Tempel-Tuttle (1865 Y1) and noted that its coma had a broad continuum, indicating that most of its light was reflected sunlight, in agreement with earlier polarimetric observations by François Arago (1786-1859) who showed that at least part of the light coming from the tail of comet C/1819 N1 was polarized. Huggins also observed a bright molecular emission band in the comet's coma but he was not able to identify it. He observed spectroscopically other comets but it was not until his observations of comet C/1868 L1 (Winnecke) that he reached a turning point in the quest for uncovering

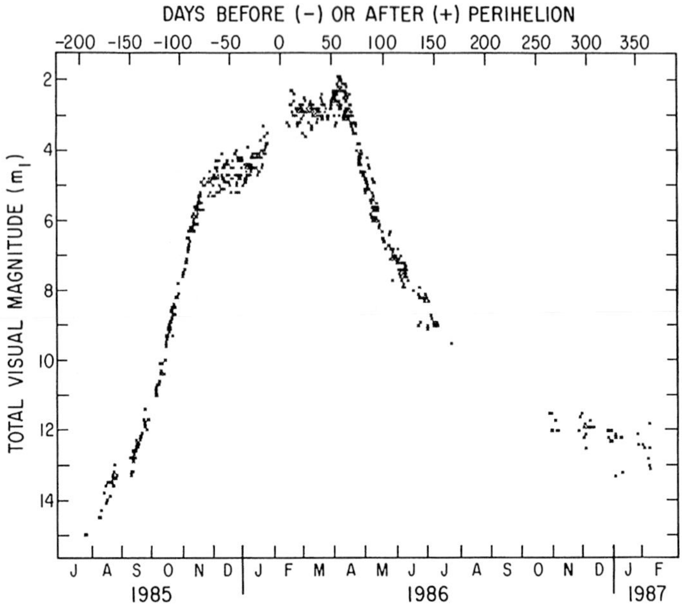

Figure 3.6. Total visual magnitudes of 1P/Halley versus time as compiled by Green and Morris (1987) (Larson et al. 1991).

the chemical composition of comets. He made a side-by-side comparison of the spectrum of comet Winnecke with the spark spectrum of ethilene vapor (C_2H_4). He found a good match which allowed him to conclude that carbon was present in the coma of the comet (in actuality, he observed the Swan bands of the radical C_2). In 1881 Huggins obtained the first photographic record of a comet spectrum (C/1881 K1). He found, besides the Swan bands of C_2, a strong emission band in the violet which he supposed to come from cyanogen (C_2N_2). It was later identified as the CN radical.

In 1882 two comets passed very close to the Sun: C/1882 F1 (Wells) and the Great September comet which allowed the identification of sodium (Na) in their active comae. A. de la Baume Pluvinel and Fernand Baldet used objective prisms in their comet observations with which they could observe emission bands in the tail of comet C/1908 R1 (Morehouse) which they identified as due to the CO molecule. These bands were later shown to be produced by ionized carbon monoxide (CO^+). de la Baume Pluvinel and Baldet also found a strong emission band at 3915 Ångstroms (Å) that they attributed to molecular nitrogen, and that it was later identified with its ion (N_2^+).

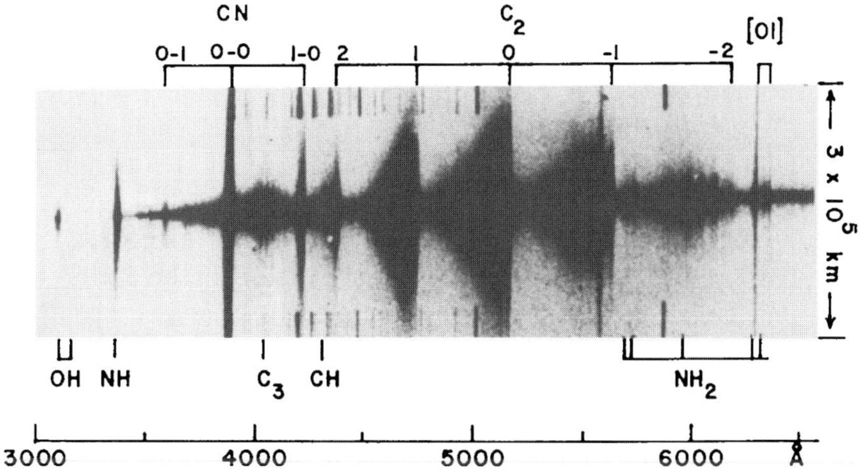

Figure 3.7. Spectrogram of comet C/1975 N1 (Kobayashi-Berger-Milon) showing the main molecular bands in the visible and near UV. The narrow strip along the central λ-axis is the continuum produced by scattering of sunlight in the dust particles of the inner coma ($\sim 10^4$ km radius) (Wyckoff 1982).

In summary, around 1900 there were several emission bands and lines identified in comets (Fig. 3.7). Yet, the physics behind it was not yet well understood. The basic question on whether comets shine by reflected sunlight, or whether they have some mechanism that produces luminosity, was still a matter of debate. The German astronomers Karl Schwarzschild and Erich Kron gave the correct answer in 1911, attributing the mechanism to the absorption and re-emission of solar radiation, i.e. the photon absorbed by the molecule is re-emitted as one photon of the same energy or in a cascade of photons of less energy (Schwarzschild and Kron 1911). This process is known as *fluorescence* (in the particular case that the re-emitted photon has the same energy as the absorbed one, it is called *resonant fluorescence*). The details of this mechanism were later elaborated by Hermann Zanstra in 1929.

In essence, atoms and molecules absorb and emit light at specified wavelenghts, i.e. the energy states are quantized. The energy levels that can be occupied by molecules are much more complex than for atoms, since molecules have energy levels for the electron envelope around the atomic nuclei, vibrational energy levels that correspond to stretching (i.e. variation of distance) and bending modes among the constituent atoms, and rotational energy levels that correspond to different rotational states of the molecule around its own axis. Transitions of electrons between different energy levels lead to emission (or absorption) of photons in the UV and visible part of the spectrum. Since electronic transitions are accompanied by changes in the vibrational and rotational states of the molecule, each electronic transition will give rise to a *band* consisting of multiple packed lines, each one of them arising from the combination of the given electronic transition plus a vibrational and a rotational transition. Vibrational transitions between different energy states carry less energy than the electronic transitions, so the emitted (or absorbed) photons fall in the infrared part of the spectrum. Again, since vibrational transitions are acompanied by rotational transitions, instead of lines we have bands. Finally, pure rotational transitions imply even less energy, so their corresponding lines fall in the radio region.

In 1941 a team led by the Belgian astronomer Polidore Swings (1906 - 1983) observed for the first time the spectrum of a comet in its ultraviolet part. In this way they were able to identify the ultraviolet bands of the hydroxyl radical OH at 3078 - 3100 Å in the nuclear region of comet C/1940 R2 (Cunningham). Swings suggested that the hydroxyl was probably produced by the photodissociation of water (H_2O) molecules. This was the first strong observational evidence of the presence of water in the comet's nucleus, that was to be confirmed later as the main constituent of comets. Since 1970, space-based UV observations with the second Orbiting Astronomical Observatory (OAO-2), the fifth Orbiting Geophysical Observatory (OGO-5), the International Ultraviolet Explorer and the Hubble Space Telescope allowed the detection of new species in the UV, as well as the observation and the determination of gas production rates of other species already observed in other spectral regions like the OH (Fig. 3.8).

The noble gases (He, Ne, Ar, Kr, and Xe) are both chemically inert and highly volatile, so they are particularly interesting as clues on the comet's thermal history. Unfortunately, their resonance transitions lie in the far UV ($\lambda \leq 1200$ Å), a wavelength region hard to access, so no convincing detection of any of these gases has so far been reported (Bockelée-Morvan et al. 2004), with the possible exception of argon (Stern et al. 2000).

Isotopic ratios can provide key information on the conditions of volatile formation in the solar nebula before its incorporation into comets. But, unfortunately, isotopic measurements require both very high sensitivity and very high resolution. The best determined isotopic ratios are for carbon $^{12}C/^{13}C$, and for the deuterium to hydrogen ratio D/H, which were obtained in more than one comet and in more

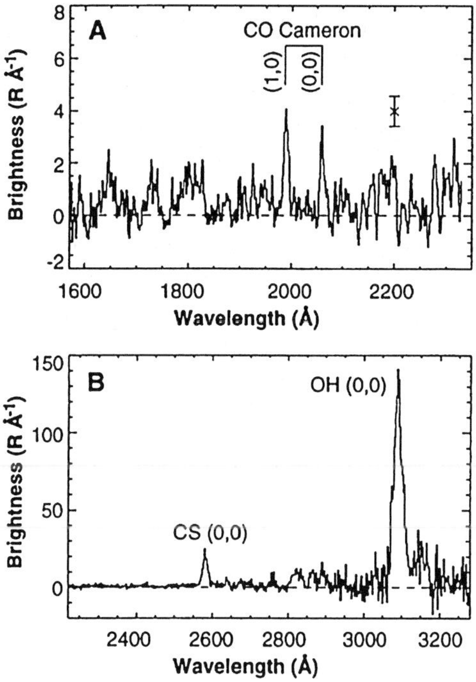

Figure 3.8. UV spectrum of comet Hale-Bopp taken with the Faint Object Spectrograph of the Hubble Space Telescope. The spectra show the Cameron bands of CO, a band of CS and the conspicuous band of OH (Weaver et al. 1997).

than one molecule containing the key atom. A carbon isotopic ratio of $^{12}C/^{13}C = 89$ was reported for comet Halley in agreement with the terrestrial value. However, the ratio D/H was found to be twice the value found for the Earth's oceans. Other isotopic ratios that have been measured are $^{14}N/^{15}N$, $^{32}S/^{34}S$, and $^{18}O/^{16}O$ (Bockelée-Morvan et al. 2004).

3.6. Inventory of parent molecules

The main cometary parent molecules, like CO, CO_2, H_2O do not have electronic transitions of some importance in the visible or UV, so they can only be detected by their vibrational transitions in the infrared, or by their rotational transitions in radio. An additional problem is that most molecules have short lifetimes against dissociation by UV sunlight, so most of them are confined to the inner coma. Their detection requires thus to explore a small region around the comet nucleus.

Water molecules have an important vibrational band at 2.7 μm, while CO_2 at 4.2 μm (see Fig. 3.9). The problem is that these molecules are also present in the Earth's atmosphere, so their detection can be done only from space or high-altitude observatories. Precisely the water molecule was detected for the first time in comet Halley from the NASA Kuiper Airborne Observatory flying at an altitude of 13,000 m. Later, methane (CH_4), acetylene (C_2H_2) and ethane (C_2H_6) were first identified in comet C/1996 B2 (Hyakutake) from the NASA Infrared Telescope Facility (IRTF) at Mauna Kea, Hawaii.

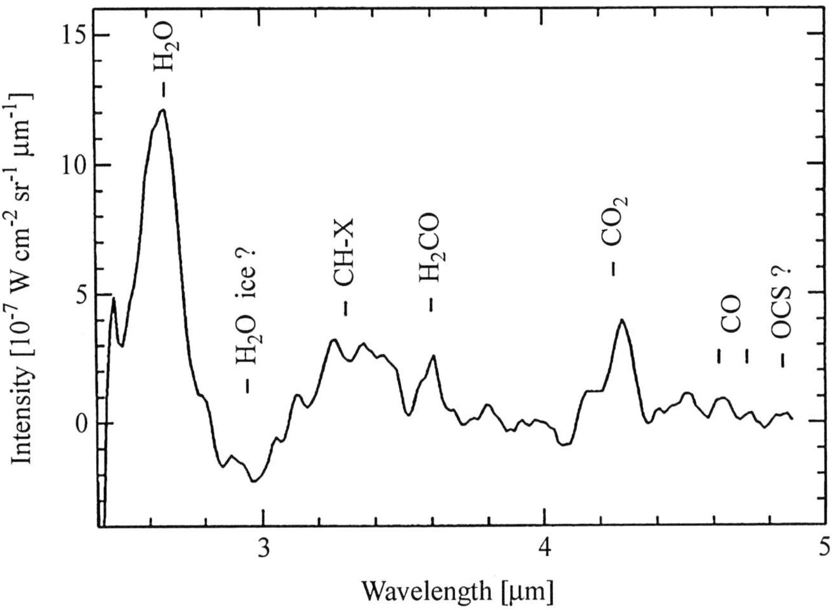

Figure 3.9. The 2.5- to 5-μm spectrum of comet Halley taken with the infrared spectrometer on board of the *Vega 1* Soviet spacecraft at a distance of $\sim$ 42,000 km. The molecular species identified are indicated in the figure; CH-X is for unidentified CH-bearing molecules (Combes et al. 1988).

The radio spectrum has shown to be very promising for the detection of parent molecules in comets (see Fig. 3.10). The first positive detection was the 18-cm line of the OH-radical in comet Kohoutek from the Nançay radiotelescope. Later, other molecular lines like those of the hydrogen cyanide (HCN) were detected in Halley, and methanol (CH_3OH) and formaldehyde (H_2CO) in comets C/1989 X1 (Austin) and C/1989 K1 (Levy). A list of the molecular species detected so far, the comets in which they were detected, and the method used for detection, are shown in Table 3.2.

Table 3.2: Parent molecules discovered in comets(*)

Molecule	Relative abundance (by number)	Comet	Technique
H_2O	100	1P/Halley	IR, mass spectroscopy
CO	$\sim 7-8$	C/1975 V1 (West)	UV
CO_2	3	1P/Halley	IR
H_2CO	0-5	1P/Halley	IR, radio
NH_3	1-2	1P/Halley	Giotto
CH_4	$< 0.2 - 4.5$	C/1986 P1 (Wilson)	IR
HCN	$< 0.02 - 0.1$	1P/Halley	radio
CH_3OH	1-5	C/1989 X1 (Austin)	IR, radio
H_2S	0.2	1P/Halley	Giotto
HNCO	-	C/1995 O1 (Hale-Bopp)	radio
HC_3N	-	C/1995 O1 (Hale-Bopp)	radio
OCS	-	C/1995 O1 (Hale-Bopp)	radio
SO_2	-	C/1995 O1 (Hale-Bopp)	radio
H_2CS	-	C/1995 O1 (Hale-Bopp)	radio
NH_2CHO	-	C/1995 O1 (Hale-Bopp)	radio
HCOOH	-	C/1995 O1 (Hale-Bopp)	radio
$HCOOCH_3$	-	C/1995 O1 (Hale-Bopp)	radio
CH_3CHO	-	C/1995 O1 (Hale-Bopp)	radio
$HC^{15}N$	-	C/1995 O1 (Hale-Bopp)	radio
DCN	-	C/1995 O1 (Hale-Bopp)	radio
HNC	-	C/1996 B2 (Hyakutake)	radio
HDO	-	C/1996 B2 (Hyakutake)	radio
$H^{13}CN$	-	C/1996 B2 (Hyakutake)	radio
C_2H_2	-	C/1996 B2 (Hyakutake)	IR
C_2H_6	-	C/1996 B2 (Hyakutake)	IR

(*) Sources: Irvine et al. (2000) and Bockelée-Morvan et al. (2005) and more references therein

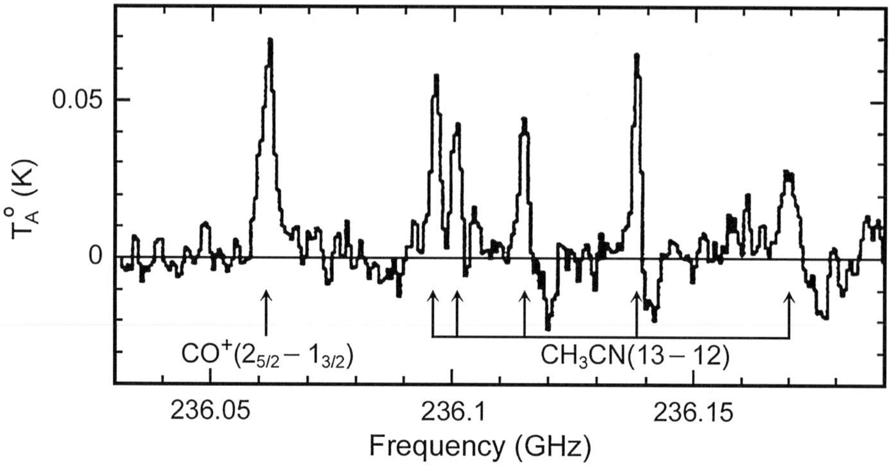

Figure 3.10. Radio spectrum of comet C/1995 O1 (Hale-Bopp) taken with the Caltech Submillimeter Observatory showing emission lines of CO^+ (identified for the first time in radio) and several lines of the methyl cyanide (CH_3CN) molecule (Lis et al. 1997).

3.7. Distribution of parent molecules and radicals in the coma

As most of the radicals observed in the visible radiation of comets cannot exist in the solid state, it was assumed since long ago that there must be invisible parent molecules from which the radicals originate in the coma by chemical (mainly photochemical) reactions. Gases are produced by the sublimation of cometary ices such as H_2O, CO, CO_2, NH_3 and CH_4. Once the parent molecules are vaporized, they are subject to the solar UV radiation that dissociates them into radicals, atoms and ions. The radicals and ions themselves enter into a complex chemistry that leads to the formation of new molecules. What is the final product of the photodissociation depends on the energy of the UV photon absorbed by the parent molecule. If the energy is high enough an ion can be produced. We can illustrate this process through the example of the most important parent molecule of cometary nuclei: water. Its photodissociation can lead to the following daughter products

$$H_2O + h\nu \quad \longrightarrow \left\{ \begin{array}{l} H + OH \\ H_2 + O \\ 2H + O \\ H_2O^+ \\ H + OH^+ \end{array} \right.$$

where $h\nu$ represents the energy of the photon, being h the Planck's constant and ν the photon's frequency. The photodissociation of the H_2O molecule in H and OH is the most important process (first line of the array) accounting for 85.5 percent of all water molecules that are photodissociated.

Carbon monoxide can be either a parent molecule or a product of the photodissociation of the CO_2 as follows

$$CO_2 + h\nu \rightarrow CO + O(^1D),$$

where ^1D is the excited state in which the oxygen atom is left, which leads by de-excitation to the emission of fluorescent radiation in the red line at 6300 Å. Therefore, the detection of this line in the comet's spectrum may be an indirect evidence of the presence of CO_2 in the comet nucleus.

The dust released with the sublimating gases slows down the gas outflow in the first few tens of meters above the nucleus through friction on the outflowing gas and energy exchange between the gas and dust. There is an increasing evidence that dust itself is a source of gas (e.g. CO, H_2CO, HCN, CN) and that the chemical composition of the gas may be changed by the presence of dust (e.g. recondensation of H_2O molecules onto dust grains). The gas released from the nucleus at the sonic speed rapidly expands and reaches supersonic velocities. The terminal velocity of the gas is reached far from the nucleus, but still in the inner coma (within $\sim$ 1000 km of the nucleus). Numerical calculations show that terminal velocities of parent molecules are usually between 0.5 and 2 km s^{-1} at a heliocentric distance $\sim$ 1 AU, in agreement with observed velocities.

Collisions between gas particles play an important role only in the near-nuclear region. The outer coma can be considered as a colisionless gas. The computation of the production rates of radicals in this region can be effected by the model proposed by Haser (1957). This model assumes isotropic radial outflow of the parent molecules and radicals and finite lifetimes for the different species. Let $n_p(s)$ be the number density of parent molecules at the distance s from the center of the nucleus, and v_p the outflow velocity. The production rate of parent molecules is then

$$Q_p = 4\pi R_N^2 n_p(R_N) v_p. \quad (3.9)$$

The equation of continuity gives

$$\frac{1}{s^2}\frac{d(s^2 n_p v_p)}{ds} = -\frac{n_p}{\tau_p}, \tag{3.10}$$

where τ_p is the lifetime of the parent molecule. Therefore we have

$$\int_{R_N}^{s} \frac{d(s^2 n_p v_p)}{(s^2 n_p v_p)} = -\frac{1}{v_p \tau_p}\int_{R_N}^{s} ds, \tag{3.11}$$

which, upon integration, leads to an expression for the coma density profile for parent molecules sublimating from the nucleus

$$n_p(s) = \frac{Q_p}{4\pi v_p s^2}\exp\left(-\frac{s}{v_p \tau_p}\right). \tag{3.12}$$

The product $v_p \tau_p = \gamma_p$ represents the spatial extent of the parent molecule, i.e. its scale length. For daughter species originating from a parent molecule, for instance by photodissociation, one must distinguish production rates Q_p and Q_d, outflow speeds v_p, v_d, lifetimes τ_p, τ_d and scale lengths $\gamma_p = v_p \tau_p$, $\gamma_d = v_d \tau_d$ for parent and daughter respectively. For the case in which parent and daughter species both move radially outward, Haser (1957) derived the equation for the density profile of the daughter species as follows. From eq. (3.12) we have that the rate of parent molecules at distance $X(< s)$ will decrease after traversing a distance dX by the amount

$$-d[4\pi X^2 n_p(X) v_p] = \frac{Q_p}{\gamma_p}\exp\left(-\frac{X}{\gamma_p}\right)dX. \tag{3.13}$$

Equation (3.13) will also correspond to the rate of formation of daughter molecules in the spherical shell between X and $X + dX$. When the daughter attain the distance s, their number will have decreased to

$$\frac{Q_p}{\gamma_p}\exp\left(-\frac{X}{\gamma_p}\right)\exp\left[-\frac{(s-X)}{\gamma_d}\right]dX. \tag{3.14}$$

The production rate of daughter molecules at s will result from the contribution of all daughter molecules produced at distances $X < s$, namely

$$Q_d = \int_{R_N}^{s} \frac{Q_p}{\gamma_p} \exp\left(-\frac{X}{\gamma_p}\right) \exp\left[-\frac{(s-X)}{\gamma_d}\right] dX$$
$$= \frac{Q_p}{\gamma_p} \exp\left(-\frac{s}{\gamma_d}\right) \int_{R_N}^{s} \exp\left[X\left(\frac{1}{\gamma_d} - \frac{1}{\gamma_p}\right)\right] dX. \quad (3.15)$$

After integrating eq. (3.15) and assuming that $\exp\left[-R_N\left(\frac{1}{\gamma_d} - \frac{1}{\gamma_p}\right)\right] \approx 1$ (since in general $R_N << \gamma_p$ and $R_N << \gamma_d$), and that $Q_d = 4\pi s^2 n_d(s) v_d$, we find that the number density of daughter molecules, $n_d(s)$, at the distance s is given by

$$n_d(s) = \frac{Q_p}{4\pi v_d s^2} \frac{\gamma_d}{\gamma_p - \gamma_d} \left[\exp\left(-\frac{s}{\gamma_p}\right) - \exp\left(-\frac{s}{\gamma_d}\right)\right]. \quad (3.16)$$

It is very likely that the daughter products do not move radially because the dissociation products from parent molecules are ejected isotropically in the parent's frame of reference. The assumption of a radial motion of the daughter molecules requires either $v_d << v_p$ which, because of the excess energy in the photodissociation process, is rarely satisfied, or that $v_p\tau_p << v_d\tau_d$, i.e. that the daughter product originates rather close to the nucleus. Models treating the case $v_d \simeq v_p$ have been studied by Combi and Delsemme (1980) and Festou (1981). In this case, significantly higher densities for the daughter product in the central part of the coma are obtained. The departure from a radial flow for the daughter species can be better addressed by a random walk (Monte Carlo or vectorial) model (Festou 1981).

3.8. The hydrogen corona

If water was indeed the main constituent of comets, it was expected to find an extended hydrogen coma produced by the photodissociation of water molecules. Unfortunately, the most conspicuous hydrogen emission line, Lyman-α, produced by resonant scattering of solar UV radiation, lies in the far UV at 1216 Å, and thus it is not possible to observe it from the ground. For this reason, it was not until 1970 that space-based observations with the *Orbiting Astronomical Observatory* (OAO-2) confirmed the presence of giant hydrogen coronas extending up to about 10^7 km in comets C/1969 T1 (Tago-Sato-Kosaka), C/1969 Y1 (Bennett) and 1971 II (Encke) (Bertaux et al. 1973). Lyman-α isophotes of comets Bennett and C/1973 E1 (Kohoutek) are shown in Figs. 3.11 and 3.12.

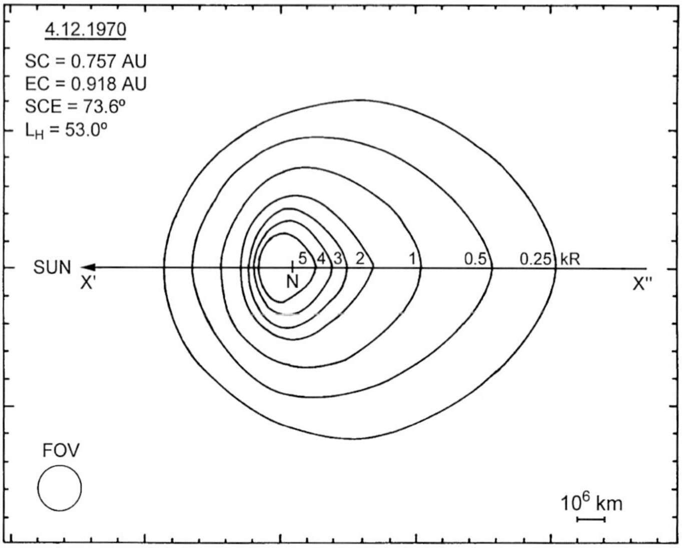

Figure 3.11. Lyman-α isophotes of comet C/1969 Y1 (Bennett) observed on April 12, 1970 with the Lyman-Alpha photometer onboard of the OGO-5 satellite. The isophotes are in kilo-Rayleight units (kR), where 1 kR $= 10^9$ photons cm^{-2} s^{-1}. N indicates the probable position of the comet nucleus. The comet's heliocentric and geocentric distances, elongation, and heliographic latitude are indicated at the upper left corner (Bertaux et al. 1973).

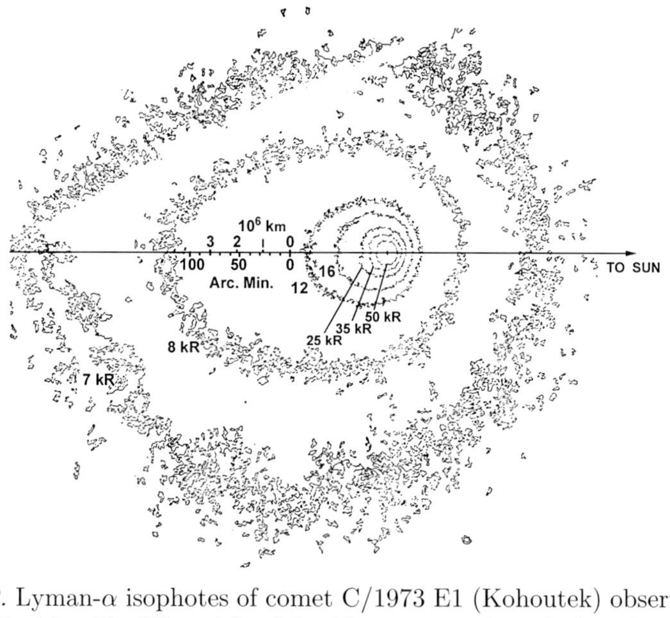

Figure 3.12. Lyman-α isophotes of comet C/1973 E1 (Kohoutek) observed on January 8, 1974 with a Far-Ultraviolet Schmidt camera onboard of an Aerobee rocket. The isophote for 35 kR has a radius of approximately 6×10^5 km (Carruthers et al. 1974).

Blamont and Festou (1974) observed in the UV comet C/1973 E1 (Kohoutek) from a high-altitude aeroplane. They considered expansion velocities for the sublimating water molecules and the radical OH produced in their dissociation: $v_{H2O} \simeq v_{OH} \simeq 1$ km s^{-1}. They derived for OH a lifetime (at $r = 0.6$ AU): $\tau_{OH} = (4.8 \pm 1) \times 10^4$ s. The H-atoms in the corona arise esentially from the photodissociation of two molecular species: H_2O and OH. Half of the H-atoms come from the H_2O and half from OH. The first process leaves the H-atoms thermalized with expansion velocities $v_H < 3$ km s^{-1}. The thermalization occurs because they are produced in the inner coma where collisions with other molecules are very frequent. On the other hand, the H-atoms produced by the photodissociation of OH are left with an excess energy that translates into fast velocities $v_H \simeq 8$ km s^{-1}. Because, the latter are produced at greater distances from the nucleus where the density is smaller, they suffer much less collisions and therefore keep their velocities (Blamont and Festou 1974). Since these fast H-atoms can travel longer distances before being destroyed, they determine the extent of the corona.

The hydrogen atoms are destroyed by photoionization by the EUV solar flux ($\lambda < 912$ Å), and by charge-exchange with solar wind protons. The lifetime τ_H at $r = 1$ AU is given by

$$\tau_H = \frac{1}{J + \sigma_{ex}\phi_p}, \qquad (3.17)$$

where $J = 7 \times 10^{-8}$ s^{-1} is the photoionization rate at 1 AU, $\sigma_{ex} = 2 \times 10^{-15}$ cm^2 is the charge-exchange cross-section, and $\phi_p = 1.5 \times 10^8$ protons cm^{-2} s^{-1} is the flux of solar protons at 1 AU (Bertaux et al. 1973). By substituting these numerical values into eq. (3.17) we obtain: $\tau_H(1AU) = 2.7 \times 10^6$ s. This lifetime can be scaled to 0.6 AU by multiplying by $(r/1AU)^2$, which yields: $\tau_H(0.6AU) \simeq 10^6$ s. The radius of the hydrogen corona will be approximately given by

$$\gamma_H = v_H \times \tau_H \simeq 8 \text{ km s}^{-1} \times 10^6 \text{ s} = 8 \times 10^6 \text{ km}, \qquad (3.18)$$

which agrees with the observations (cf. Figs. 3.11 and 3.12). The H-atoms generated by the photodissociation of H_2O molecules will reach less than 3/8 of that distance. Therefore, there will be an inner hydrogen corona composed of a mixture of slow and fast H-atoms, and an outer corona composed of only fast H-atoms.

The lifetime of the H_2O molecules is very short: $\tau_{H2O} \simeq 13$ hr ($\simeq 4.7 \times 10^4$ s) at 0.6 AU (Blamont and Festou 1974), so the coma of H_2O

molecules will have a radius of $\gamma_{H2O} = v_{H2O} \times \tau_{H2O} \simeq 4.7 \times 10^4$ km. Above this distance, the parent molecule is OH with a scale length $\gamma_{OH} = v_{OH} \times \tau_{OH} \simeq 4.8 \times 10^4$ km. Therefore, OH can reach distances of the order of 10^5 km, which is two orders of magnitude smaller than the distance reached by H atoms.

Under the assumption that the hydrogen envelope is optically thin, the Lyman-α intensity I_L emerging from the hydrogen envelope will be proportional to the integrated number density of hydrogen atoms, n_H, along the line of sight, namely

$$I_L = \frac{gF_L}{4\pi} n_H \text{ photons cm}^{-2} \text{ s}^{-1} \text{ sterad}^{-1}, \tag{3.19}$$

where gF_L is the excitation rate at the distance r or, in other words, the amount of energy that the hydrogen atom absorbs and re-emits every second. F_L is the flux at the center of the solar Lyman-α line and the factor g is given by

$$g = \frac{\pi e^2}{m_e c} f \tag{3.20}$$

where e and m_e and the charge and the mass of the electron, respectively, f is the oscillator strength of the Lyman-α transition, and c is the velocity of light. The oscillator strength is a dimensionless factor that enters into the precise quantum-mechanical formulae for deriving the probability of a certain transition between two energy levels by the absorption or emission of a photon of energy $h\nu$. For the Lyman-α transition we have $f = 0.42$. The Lyman-α solar flux at 1 AU is $F_{Lo} = 3.2 \times 10^{-11}$ photons cm^{-2} s^{-1} sterad^{-1}. The flux F_L at a heliocentric distance r will then be

$$F_L = F_{Lo} \left(\frac{r_o}{r}\right)^2,$$

where $r_o = 1$ AU.

From Haser's model we can make an estimate of the production rate of the parent species OH, $Q_{OH}(= Q_{H2O})$, and how the number density n_H varies with the distance s to the nucleus. Since $\gamma_H >> \gamma_{OH}$, eq. (3.16) can be simplified to

$$n_H(s) \simeq \frac{Q_{OH}}{4\pi v_H s^2} \exp\left(-\frac{s}{\gamma_H}\right). \tag{3.21}$$

Therefore, from measurements of the Lyman-α intensities in the corona, we can obtain n_H from eq. (3.19), and then the production rate of OH (equal to that of H_2O) by means of eq. (3.21).

The radiation pressure (per unit mass) due to the solar Lyman-α flux is

$$\beta_L = g/c = \frac{\pi e^2}{m_e c^2} f F_{Lo} \left(\frac{r_o}{r}\right)^2 = 0.57 \left(\frac{r_o}{r}\right)^2 \text{ cm s}^{-2} \tag{3.22}$$

which turns out to be about the same as the acceleration due to the solar gravity ($\simeq 0.6(r_o/r)^2$ cm s^{-2}). The action of the solar radiation pressure can thus explain the elongations of the hydrogen coronas in the antisolar direction (cf. Figs. 3.11 and 3.12).

3.9. The nature of dust particles

We have focused until now on the volatile component of cometary nuclei. We will turn now our attention to the refractory component, which can be observed as dust particles in the coma and tail. Dust particles are carried away by the sublimating gases, but while the gaseous parent molecules are rapidly dissociated and ionized outside the nucleus, the dust grains remain stable under the solar radiation. Even when they are warmed to temperatures of a few hundred Kelvin, they are still chemically and morphologically stable, thus preserving information on the conditions in which they formed, either in the interstellar medium or in the solar nebula. There is an agreement at present that silicate minerals formed in the interstellar medium are amorphous, while crystalline minerals may have formed in the solar nebula, either as gas phase condensates or amorphous silicates annealed into crystals when heated to temperatures above $\sim$ 1000 K (Wooden 2002). Dust grains may be a vehicle for the transportation of significant quantities of condensable elements from the circumstellar winds of evolved stars, novae and supernovae into the parent clouds of young stellar and planetary systems. Therefore, if some fraction of these grains survive intact, the study of their mineralogy and physical properties may shed light into stellar nucleosynthesis and evolution (Williams et al. 1997).

The spectrum of the dust coma is a continuum composed of a 5800 K blackbody curve due to scattered solar radiation and another blackbody curve due to thermal emission by the heated dust particles that reach equilibrium temperatures of a few hundred K (Fig. 3.13). The thermal

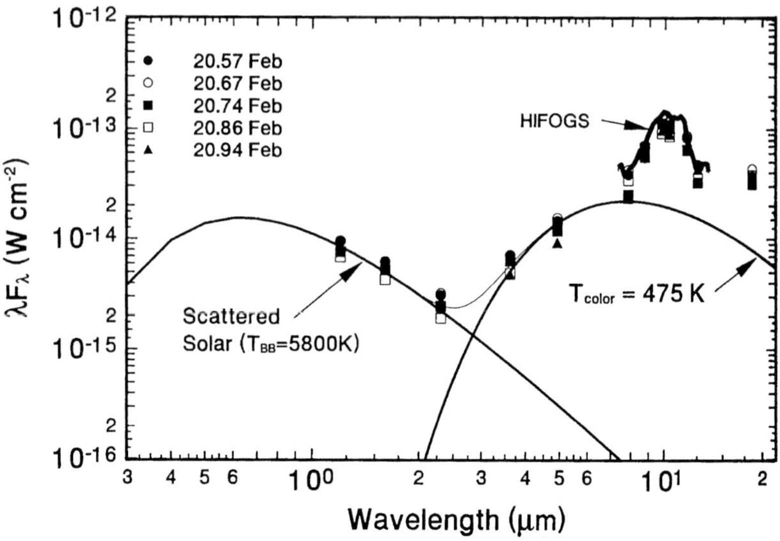

Figure 3.13. Spectral energy distribution of comet C/1995 O1 (Hale-Bopp). The observations were fitted to a 5800 K blackbody component due to scattered solar radiation on the dust particles, and a 475 K blackbody component due to thermal emission from small carbon grains. The color temperature T_{color} is defined by the ratio of the thermal fluxes at 3.6 μm and 4.9 μm and is a good measure of physical grain temperatures. The silicate emission feature centered around 10 μm is very prominent (Williams et al. 1997).

emission component shows in many cases a 10 μm emission feature atributted to radiation from Si-O bond stretching vibrations in the silicates contained in the dust grains. The exploration of comet 1P/Halley revealed the presence of a new population of grains composed of carbon, hydrogen, oxygen and nitrogen (called CHON), and the discovery of a new emission feature extending from 3.2 to 3.5 μm, attributed to the C-H stretch of organic compounds. Measurements with particle impact analyzer instruments onboard of Giotto and Vega 1 and 2 spacecrafts indicated that CHON material contributed about 30% of the total mass of the measured particles, with individual masses in the range 10^{-16} to 10^{-11} g, about one-third of the measured grains contained no significant organic component, the rest being a mixture of both silicate and organic components. The pure silicate grains were more heavily represented in the lower mass range (Mumma et al. 1993).

We note that the grain temperature of comet Hale-Bopp shown in Fig. 3.13 (475 K) is significantly higher than the equilibrium temperature of a blackbody at the same heliocentric distance r, which is given by (Gehrz and Ney 1992)

$$T_{BB} \simeq \frac{278}{r^{1/2}} \text{ K}, \tag{3.23}$$

where r is expressed in AU.

At the moment of the observations of Fig. 3.13, Hale-Bopp was at $r = 1.15$ AU, so that $T_{BB} = 259$ K. Why the grain temperature is much higher than T_{BB}? This is because dust grains absorb efficiently in the visible, where most of the solar radiation is concentrated, but their absorptivity drops in the infrared (i.e. their emissivity is also poor), so they are overheated. The absorptivity of a grain is described by the efficiency factor for absorption $Q_a(a, \lambda)$ which, for a given material, is a function of the grain radius a and the wavelength λ of the incident radiation. The grain temperature T_{gr} can be obtained from the equilibrium between the solar power absorbed by the grain and that emitted, namely

$$\frac{L_\odot}{4\pi r^2}\pi a^2 Q_a = 4\pi a^2 Q_e \sigma T_{gr}^4, \tag{3.24}$$

where $L_\odot = 3.827 \times 10^{33}$ erg s^{-1} is the solar luminosity, Q_a is the absorption efficiency of the grain averaged over the wavelength region where most of the solar radiation is concentrated (visible), and Q_e is the thermal emission efficiency for a grain of radius a at temperature T_{gr}. From eq. (3.24) we obtain

$$T_{gr} = \left(\frac{L_\odot Q_a}{16\pi\sigma r^2 Q_e}\right)^{1/4} = T_{BB}\left(\frac{Q_a}{Q_e}\right)^{1/4}. \tag{3.25}$$

Small absorbing grains (for instance, composed of graphite) have $Q_a > Q_e$, so $T_{gr} > T_{BB}$. However, larger grains, say with radii greater than a few microns, have $Q_a \sim Q_e \sim 1$, and thus $T_{gr} \rightarrow T_{BB}$. Grain temperatures measured in the comae of several comets are plotted in Fig. 3.14. We can see that in general grain temperatures are somewhat above the theoretical blackbody temperature at the same heliocentric distance. Figure 3.14 also includes comet dust trails detected by the *Infrared Astronomical Satellite* (IRAS) (Sykes and Walker 1992). The temperatures of the dust trails are very close to the theoretical blackbody temperatures, suggesting that they are composed of large grains. This is what should be expected anyway, since small grains are quickly dispersed by the forces associated to the Sun's radiation. Gehrz and

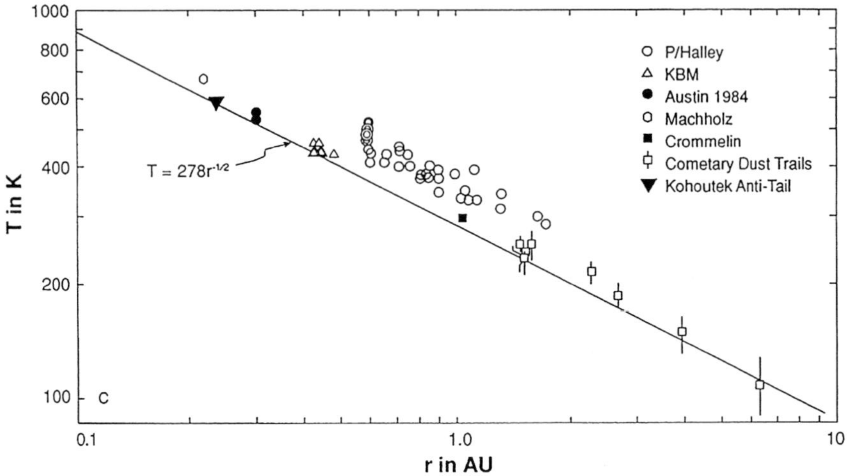

Figure 3.14. The observed coma grain temperature derived from the thermal infrared flux of several comets at different heliocentric distances. The solid line gives the theoretical blackbody temperature T_{BB} (Gehrz and Ney 1992).

Ney (1992) define the "superheat" of the grains as the ratio T_{gr}/T_{BB}. For the case of comet Halley, the grain "superheat" would suggest small radii of about 0.5 - 1 μm for the optically important grains of the coma. On the other hand, comets C/1975 N1 (Kobayashi-Berger-Milon) and C/1984 N1 (Austin) have $T_{gr} \sim T_{BB}$ suggesting that the optically dominant grains have sizes larger than several microns.

From 1.20 - 18.5 IR spectrophotometric studies of comet C/1995 O1 (Hale-Bopp), Williams et al. (1997) derive a temperature of 475 K for the thermal emission component (Fig. 3.13). Such a high grain temperature can be explained if the optically important grains dominating the visual scattering and near-IR emission from the coma have an average radius $\leq$ 0.4 μm. Therefore, the optically important grains in this comet seem to be somewhat smaller than those found in other comets. Yet, this interpretation has been challenged by Li and Greenberg (1998) who argue that fluffy aggregates have much higher absorption coefficients per unit mass than compact particles. Therefore, these authors interpret the thermal spectrum of comet Hale-Bopp as produced by fluffy aggregates, with typical sizes of several microns, composed of subunits of highly absorptive organic mantles on amorphous silicate cores.

3.10. The dust to gas mass ratio

We can estimate the production rate of several gaseous atomic and molecular species from the measurement of the intensity of their most important emission lines or bands in the ultraviolet, visible and radio. This is for instance the case of the conspicuous hydrogen emission line Lyman-α at 1216 Å discussed before. There has also been a wide use of narrow-band filter photometry, centered in some important emission bands of molecular species of cometary comae in the near UV and visible (e.g. A'Hearn et al. 1979, 1995). The chosen species have usually been OH, NH, CN, C_3 and C_2.

The intensity I_λ of light radiated by a species of atom or molecule in a line or band centered in λ is

$$I_\lambda \sim g F_{\odot,\lambda} N_{sp}, \tag{3.26}$$

where g is the fluorescence efficiency or luminosity per molecule (i.e., the g-factor defined in eq. (3.20) for the transition of interest), $F_{\odot,\lambda}$ is the solar flux in the wavelength λ, and N_{sp} is the number of atoms or molecules of the given species, which is given by

$$N_{sp} \sim Q_{sp}\tau, \tag{3.27}$$

where Q_{sp} is the production rate of the gaseous species and τ its mean lifetime. Values of the central wavelengths of the selected bands, the g-factors and τ are shown in Table 3.3 for the most commonly used molecular species. We note that the g-factors of the OH, NH and CN bands show variations within certain ranges. The reason is that they depend on the radial velocity of the Sun, as seen from the comet, that produces important variations in the intensity of the exciting solar radiation by Doppler effect. This is because the region of the solar spectrum around the wavelengths of the OH, NH and CN bands is heavily crowded with absorption lines that makes its intensity strongly dependent on λ. This is what is known as the *Swings effect* in honor to the astrophysicist Pol Swings who explained it.

Table 3.3: Molecular species selected for narrow-band photometry(*)

Species	λ	$g - factor$ (erg s^{-1} mol^{-1})	τ (s^{-1})
OH	3085	$1.4 - 8.3 \times 10^{-15}$	1.6×10^5
NH	3360	$4.9 - 7.6 \times 10^{-14}$	1.5×10^5
CN	3870	$2.4 - 5.0 \times 10^{-13}$	2.1×10^5
C_3	4060	1.0×10^{-12}	2.7×10^4
C_2	5115	4.5×10^{-13}	6.6×10^4

(*) from A'Hearn et al. (1995) and more references therein

By combining eqs.(3.26) and (3.27) we obtain for the production rate of the considered gaseous species

$$Q_{sp} \sim \frac{I_\lambda}{\tau g F_{\odot,\lambda}}, \tag{3.28}$$

By measuring the intensity I_λ, for instance by narrow-band photometry, it is then possible to estimate the production rate of several molecular and atomic species, so we can learn about their relative proportions and the total gas production rate Q_g. The estimate of Q_g requires in addition to know what are the parent molecules, for instance H_2O is the parent molecule of H and OH, and HCN may be the parent of the radical CN, although not exclusively. The Haser model can be used to describe the spatial distribution of the given species in the coma.

The mass production rate of dust Q_d (g s^{-1}) can be estimated by observations of the continuum flux in the visible (e.g. A'Hearn et al. 1995), by the thermal infrared (e.g. Hanner 1985), or by the study of the dust tails (e.g. Cremonese and Fulle 1990). All these methods are model-dependent since they rely on the emission velocities of dust particles from the nucleus surface, their typical size, size distribution, albedo and density. Therefore, we should bear in mind that large error bars should be attached to the estimated values of Q_d.

Let $I_{cont,\lambda}$ be the intensity of the continuum flux measured in a narrow wavelength range centered around λ, it can be expressed as

$$I_{cont,\lambda} \sim \frac{N(S)\pi s^2 p(\lambda)\phi(\alpha)}{\pi r^2 \Delta^2} F_{\odot,\lambda}, \tag{3.29}$$

where $N(S)$ is the number of dust particles of typical radius s within the nucleus-centered circular aperture of radius S, $p(\lambda)$ is the geometric albedo of the dust particles, $\phi(\alpha)$ is the phase function at phase angle α, r the heliocentric distance (in AU), and Δ the geocentric distance. $N(S)$ is related to the dust production rate by (Newburn and Spinrad 1985)

$$N(S) \sim \frac{\pi Q_d S}{2 m_d v_d}, \tag{3.30}$$

where $m_d = 4/3\pi s^3 \rho$ is the typical mass of a dust grain of density ρ, and v_d is the mean outflow velocity of the dust particles.

By measuring both Q_g and Q_d in a comet at more or less the same time, we can obtain its dust-to-gas mass ratio

$$\psi = \frac{Q_d}{Q_g}.$$

The estimated values of ψ for a small sample of well observed comets spread over two orders of magnitude. Some comets appear to be very dusty ($\psi > 1$), while others are found to be almost dust-free ($\psi < 0.1$). Even for the same comet, the dust loading of gas can greatly increase, as for instance during outbursts. Yet, A'Hearn et al. (1995) do not find any correlation of ψ with the heliocentric distance. Even though some comets appear to be very dusty, most comets are observed to be dust-poor or only moderately dusty with $\psi \sim 0.05$ to ~ 0.6 (Hanner 1985, Newburn and Spinrad 1985, Hanner and Campins 1986, Singh et al. 1992).

We can compare the measured values of ψ with the estimated mass ratio of dust (silicates) to gas (ices) corresponding to the primordial composition of the cometary material, which is found to be of about 0.7 (cf. Section 3.2). Yet we should note that the dust-to-gas mass ratio measured in the coma does not necessarily reflect the composition of the nucleus material. Dust particles are dragged away with the sublimating gases as long as they are not too heavy. Heavy particles will stay on the nucleus surface, so a "dust-free" comet may actually correspond to a comet whose typical grains are too heavy to be carried off with the gases. Furthermore, even if the coma contains large aggregates of dust grains (say, millimeter sized) they are inefficient light scatterers or emitters of thermal radiation, as compared to their submicron-sized counterparts. Indeed, studies of dust tails confirm that a large part of the dust mass released by a comet may be contained in millimeter or centimeter-sized particles (Cremonese and Fulle 1989, 1990).

Infrared Space Observatory (ISO) observations of the "gassy" comet 2P/Encke seem to confirm the previous conclusion. While optical observations lead to a low $\psi \sim 0.1$, the ISO observations permit to build a spectral energy distribution in the IR that is best fit to a model of grain size distribution in which most of the mass is contained in large grains $> 20\ \mu$m. This result greatly increases the dust-to-gas mass ratio of 2P/Encke to $\psi \sim 2.3$ (Lisse et al. 2004). Therefore, it might be possible that physically evolved - mostly JF? - comets that have developed dust mantles are losing most of their refractory material in large grain aggregates, thus giving dust-to-gas ratios well above unity.

3.11. Formation of the tails

Some comets develop extense tails, reaching lengths of the order of 10^8 km. One tail shows a curved shape and its spectrum corresponds to that of the Sun, showing that it is composed of dust particles that scatter sunlight. The other tail stretches to larger distances in the antisolar direction and is of bluish color (Fig. 3.15). The spectrum of this tail shows the emission bands of several ions, in particular CO^+, N_2^+, CH^+, CO_2^+, and H_2O^+. The ion tail shines by fluorescent radiation; in particular the CO^+ ion - the most abundant ion species in the tail - has a strong band in the blue part of the spectrum, giving the tail its characteristic color.

In 1836 Bessel developed a mechanical theory for the dust particle motion able to explain the cometary forms. He assumed that the diffuse appearance of a comet is due to an agglomeration of very fine dust particles which are expelled from the nucleus by repulsive forces originating at the Sun. In 1903 the Russian astronomer Fedor Bredikhin (1831-1904) assumed that the repulsive force varied as r^{-2} and it was defined in such a way that the solar mass (unity) was replaced by a hypothetical mass $\mu = 1-\beta$, where β is positive. The quantity $\beta = 1-\mu$ then represents the repulsion of the Sun, and since β can be greater than one the net force acting on the particle (Sun's gravitation minus repulsive force) can become positive (i.e. repulsive). It is thus possible to draw conclusions from the tail forms actually observed, by making assumptions about the original velocity of expulsion of the particles and about the strength of the repulsive forces. Repulsive forces of a factor greater than 100 the Sun's attraction were estimated for the dust particles in the tails of some comets.

Figure 3.15. Photograph of comet 153P/Ikeya-Zhang obtained with a 200/300 Schmidt camera on 2002 March 30.81. The field of view measures 6° by 3.5° (courtesy Michael Jäger).

The idea that the force of repulsion of the Sun, which was originally introduced purely as a hypothesis, may actually originate from radiation pressure of sunlight was first proposed by the Swedish chemist and physicist Svante Arrhenius (1859-1927). More detailed investigations by Schwarzschild showed in 1901 that it is indeed possible to explain repulsive forces on dielectric spheres of sub-micron to micron-size as large as 20 - 30 times the solar attraction but not more than that. This was a very important result since showed that another physical mechanism - besides radiation pressure - must be at work to explain much greater repulsive forces (for a discussion of early ideas on this topic, see Richter 1963 and Yeomans 1991).

The motion of a dust particle will thus be governed by two opposite forces: the solar gravity attraction F_G and the force associated to the solar radiation pressure F_R. For a spherical dust particle of radius a and density ρ at a heliocentric distance r, the two acting forces are expressed as

$$F_G = \frac{GM_\odot}{r^2}\left(\frac{4}{3}\pi a^3 \rho\right), \tag{3.31a}$$

$$F_R = \frac{Q_{pr}}{c}\left(\frac{L_\odot}{4\pi r^2}\right)\pi a^2, \tag{3.31b}$$

where c is the speed of light and Q_{pr} is the efficiency factor for radiation pressure, i.e. the ratio of the particle's cross section for radiation pressure to its geometric cross section πa^2. For homogeneous, isotropic, spherical grains of known refractive index, the efficiency factors can be computed using Mie's analytical solutions to Maxwell's equations (see, e.g. van de Hulst 1957). In general the value of Q_{pr} depends on whether the particle is a dielectric or an electrically conducting material.

Since both forces F_G and F_R are radial, opposite, and vary as r^{-2}, a dust particle will follow a Keplerian trajectory, corresponding to a reduced "effective" gravity field $(1-\beta)F_G$, where

$$\beta = \frac{F_R}{F_G} = 0.585 \times 10^{-4} \frac{Q_{pr}}{\rho a}, \tag{3.32}$$

where ρ and a are expressed in cgs units.

Absorbing materials like graphite reach values $\beta > 1$, i.e. the force $(1-\beta)F_G$ acting on these particles is repulsive. The same holds for metals because they are strong back scatterers. Dielectrics (e.g. silicates) are strong forward-scatterers and their values of β are much smaller ($\beta \sim 0.5$ for grains of a few tenths μm). Since $\beta \propto a^{-1}$ it is obvious that large particles will stay longer near the comet's nucleus. The same will hold for very small particles ($a \leq 0.1$ μm in which case β decreases because there is an increasing "transparency" of particles to radiation pressure (i.e. Q_{pr} decreases). Models of the density distribution of dust particles in the tail were developed by Finson and Probstein (1968a,b) based on three parameters: (i) the dust production rate; (ii) the size distribution of dust particles; and (iii) their emission velocity. In such models, the motion of the dust particles of a certain β are computed considering that they are subject to a central force $F = (1-\beta)F_G$. By comparing the model density distribution of dust particles with the observed density distribution from the photometric profile of the dust tail, it is possible to get information about size distribution of dust particles and the radius of the optically important grains.

Let us consider the position of a comet in its orbit at a certain time t_c (Fig. 3.16), not far away from the Sun, so it is very active releasing dust particles that go into the tail. As shown in the figure, the radiation pressure makes the particles to progressively depart from the comet's orbital motion. If all the particles of a given β released at time $t_c - t_k$ had zero velocity relative to the comet nucleus, then they would reach the same point after a time t_k. But dust particles have initial relative velocities $v_i \neq 0$, which will lead to a broadening of the locus occupied

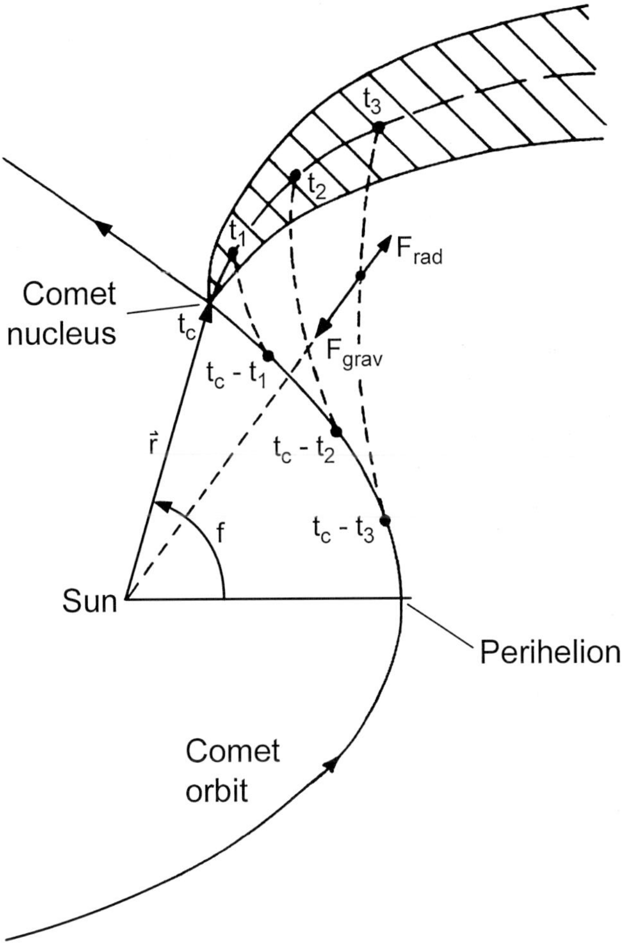

Figure 3.16. The differential motion of the dust particles with respect to the comet nucleus, under the action of the radiation pressure, generates de dust tail.

by the β particles at t_c of width $\sim 2v_i t_k (k = 1, 2, 3, ...)$. More complex patterns are obtained for the dust tail if we also consider different β values for the particles released at a certain time $t_c - t_k$. For the optically important dust grains of comet C/1956 R1 (Arend-Roland), Finson and Probstein (1968a,b) derived a radius of 2.8 μm if a mass density of $\rho = 1$ g cm^{-3} was adopted. This dynamical size is in general agreement with those derived from the thermal infrared spectrum.

The curved shape of the dust tails could thus be successfully explained by considering the motion of the dust particles under the action of the Sun's gravitational field and the radiation pressure. Yet, the puzzle of the straight tails in the antisolar direction still persisted around the middle of the last century. The straight tails contain atoms, molecules and ions - not dust particles - so the first step was to analyze the action of the radiation pressure from sunlight, as had been done before for dust particles In 1943 Karl Wurm evaluated the radiation-pressure effect on absorbing molecules of CN, C_2 and CO^+. The molecule was regarded in its behavior towards the incident light as replaceable by an isotropic and quasi-elastically bound electron of charge e, mass m_e and characteristic frequency ν. From the radiation field of the Sun, which at the position of the comet has a radiation-density denoted by u'_ν, the molecule absorbs and re-emits every second an amount of energy given by

$$\zeta = \frac{\pi e^2 f}{m_e} u'_\nu, \tag{3.33}$$

where, as in eq. (3.20), f is the oscillator strength. If u_ν denotes the density of the Planck radiation corresponding to the surface temperature of the Sun (~ 6000 K), then the radiation density at the distance of the comet is

$$u'_\nu = u_\nu \frac{R_\odot^2}{4r^2}. \tag{3.34}$$

The impulse on the molecule is then given by ζ/c. In addition to the radiation pressure, the molecule is also subject to the gravitational attraction of the Sun. The attraction of the cometary nucleus, however, can be neglected given its small mass. By applying eqs.(3.31a,b) and (3.32), Wurm computed the coefficient β as

$$\beta = \frac{\frac{\pi e^2 f}{m_e c} u'_\nu}{\frac{G M_\odot m}{r^2}} = \frac{\pi e^2 f}{4 g_\odot m_e m c} u_\nu, \tag{3.35}$$

where m is the mass of the molecule, and $g_\odot = G M_\odot / R_\odot^2$ the acceleration of gravity at the surface of the Sun. For the energy density we have the Planckian distribution

$$u_\nu = \frac{8\pi h \nu^3}{c^3} \frac{1}{\exp(h\nu/kT) - 1} \sim \frac{8\pi h \nu^3}{c^3} \exp(-h\nu/kT), \tag{3.36}$$

where the approximation is valid for ν in the visible range.

By substituting eq. (3.36) into eq. (3.35) we finally get

$$\beta = \frac{\pi e^2 f}{4 m_e m c g_\odot} \frac{8\pi h \nu^3}{c^3} \exp(-h\nu/kT). \tag{3.37}$$

Using values for f derived from quantum theory, Wurm (1943) found for the above mentioned molecules of CN, C_2 and CO^+ the following values of β: 0.69, 1.66 and 46.9 respectively. Since the radiation pressure and gravity more or less compensate each other for CN and C_2, these molecules would remain practically within the cometary head during their lifetimes, which is in fairly good agreement with observations. On the other hand, the value $\beta = 46.9$ for CO^+ would certainly explain why these molecules are driven into the tail, although it is still much too low in comparison with the repulsive forces that are actually observed (up to one thousand).

The discrepancy between the calculated radiation pressures and the observed forces of repulsion on molecules and ions stimulated the search for other possible physical processes that might explain the repulsion. In 1951 the German astronomer Ludwig Biermann proposed that the solar corpuscular radiation provided the accelerating force for the molecules in cometary tails. The free solar electrons, according to Biermann, are capable of transferring their momentum to the CO^+ ions of the cometary tail, whereby imparting accelerations to them of up to 10^4 cm s^{-2}, which are amply sufficient to explain the large repulsive forces. From this point of view the tail of a comet must therefore be regarded as a plasma composed of cometary ions and electrons mixed with solar protons and solar electrons, as well as neutral particles of cometary origin. A comet should therefore also show other characteristic properties of a plasma, in particular effects of magnetic fields and electromagnetic forces. Biermann also suggested that the particle streams from the Sun carry magnetic fields themselves. Therefore, the observed motions within the material of the tail, particularly movements perpendicular to the radius vector and curious helical structures may be traceable to magnetic fields. The corpuscular radiation predicted by Biermann was detected in 1959 by the plasma detector on board of the Soviet spacecraft *Lunik 3* and given the name of *solar wind*. At 1 AU, the solar wind has a velocity between 300 and 1000 km s^{-1} and a density between 1 and 100 protons and electrons cm^{-3}.

DYNAMICS OF LP COMETS ENTERING THE INNER PLANETARY REGION

A comet entering or approaching the planetary region will experience the action of planetary perturbations, so it will slowly depart from a Keplerian motion. The comet's orbit can in principle be determined from three accurate astrometric positions (in practice, several more) spanning through several days. First a parabolic solution is tried to fit the observations. Once good astrometric positions covering a period of several weeks are available, an elliptic (or sometimes hyperbolic) solution is tested as a better approximation to the actual orbit. Since the observed comet is continuously perturbed by the planets, its orbit will change with time. Therefore, what we obtain from the astrometric observations is the *osculating orbit*, that is the orbit the comet has at a particular instant. The epoch adopted for the osculating orbit is usually near perihelion where most comet observations are performed. In this chapter we will be mainly concerned with LP comets entering the inner planetary region, understood in a broad sense as those comets crossing or approaching Jupiter's orbit (i.e., almost all LP comets so far discovered), which are subject to the strong perturbations of Jupiter and Saturn and for which nongravitational forces may have some dynamical influence.

4.1. The original orbit

When we deal with a LP comet, we are interested in obtaining its *original orbit*, that is the orbit it had before entering the planetary region referred to the barycenter of the solar system. The knowledge of original orbits is vital to assess the place from where LP comets come and what is their dynamical age. To obtain the original orbit of an observed comet, the equations of motion are integrated from a given epoch (for which the osculating orbit has been obtained) backwards in time until the comet is so far away from the planets (say heliocentric distances $\gtrsim 50$ AU) that their perturbations on the comet can be assumed to be negligible, namely

$$\frac{d^2\vec{r}}{dt^2} = -\frac{GM_\odot\vec{r}}{r^3} + \nabla\mathcal{R} \tag{4.1}$$

where $\vec{r}$ is the Sun-comet radius vector expressed in the heliocentric frame of reference, and $\mathcal{R}$ is the disturbing function which describes the perturbations of the planets of masses m_i on the comet's orbit and is given by

$$\mathcal{R} = G\sum_i m_i \left(\frac{1}{d_i} - \frac{x_c x_i + y_c y_i + z_c z_i}{r_i^3}\right) \tag{4.2}$$

where d_i, r_i are the distances of planet i to the comet and the Sun respectively, and (x_c, y_c, z_c), (x_i, y_i, z_i) are the heliocentric coordinates of the comet and planet i, respectively. In many cases, a better solution can be obtained for the original orbit if an extra term that takes into account the nongravitational force (i.e., a jet reaction on the comet nucleus produced by the sublimating gases) is introduced in eq. (4.1). We will describe this force in more detail below.

The frequent classification of a comet orbit as a "parabola" (i.e., eccentricity $e = 1$ or energy $x = 0$), reflects our poor knowledge of the orbit. The energy (or reciprocal semimajor axis) of a LP comet is usually very close to zero, so the computation of its original value $(1/a)_{orig}$ requires a large number of accurate astrometric measurements spaced in time. Improvements in the observation of comets at large heliocentric distances, added to the use of fast computers that can handle the perturbations of all the planets, have facilitated the accurate determination of $(1/a)_{orig}$ for a large number of LP comets.

4.2. Change in the orbital energy

Comets coming into the inner planetary region will mainly be under the perturbing influence of Jupiter. For comets moving on near-parabolic orbits, their orbits are perturbed by Jupiter and the other planets, essentially when they approach the Sun to less than a few tens AU. The rest of the time they are so far away from the planets that their barycentric motion can be considered to be Keplerian without loss of accuracy. It is worth noting that a given perturbation on a near-parabolic orbit will be more efficient in changing the comet's reciprocal semimajor axis (or energy), $1/a$, when it is applied near perihelion than at large heliocentric distances. To show this, let us consider the comet's orbital velocity, v, at a heliocentric distance r

$$v^2 = \mu \left(\frac{2}{r} - \frac{1}{a} \right), \tag{4.3}$$

where again $\mu = GM_\odot$.

If at that instant a perturbation Δv occurs, the corresponding change $\Delta(1/a)$ will be

$$\Delta(1/a) = -\frac{2v}{\mu} \Delta v, \tag{4.4}$$

but the orbital velocity decreases with increasing r, so $\Delta(1/a)$ (in modulus) will be the greatest when r is the smallest (i.e. at perihelion).

On the other hand, the other orbital elements (q, i, ω and Ω) will experience negligible changes when the perturbation is imparted near perihelion (unless a close encounter with a planet occurs). Let us illustrate this with the case of the perihelion distance q. For a near-parabolic comet, q is related to the transverse component of the orbital velocity v_T by the equation: $v_T \approx (2\mu q)^{1/2}/r$, whereby a change Δv_T in the transverse velocity will correspond to a change Δq in the perihelion distance given by: $\Delta q/q \approx 2\Delta v_T/v_T$. Therefore, for a certain perturbation Δv_T the relative change $\Delta q/q$ will be the smallest when v_T is the greatest, namely around perihelion.

The average change in the comet's energy per orbital revolution (expressed in AU^{-1}) under the main perturbing influence of Jupiter is of the order $(M_J/M_\odot) \sim 10^{-3}$, where M_J is Jupiter's mass. This can be easily shown from eq. (4.4) where the perturbation in the comet's velocity Δv is approximately given by $\Delta v \sim F_J \times \Delta t$, where F_J is Jupiter's perturbing acceleration and Δt is the time during which Jupiter's gravitational action has the greatest effect. The modulus of F_J is given by (see, e.g., Roy 1982, p.158)

$$F_J = GM_J \left| \left(\frac{\vec{r}_J - \vec{r}}{d_J^3} - \frac{\vec{r}_J}{r_J^3} \right) \right|, \tag{4.5}$$

where d_J is the distance of Jupiter to the comet, and $\vec{r}$, $\vec{r}_J$ are the radius vectors of the comet and Jupiter respectively. Substituting Δv for $F_J \times \Delta t$ in eq. (4.4) and considering typical numerical values of a few AU for d_J, r_J and r, a comet's velocity $v \sim 40$ km s$^{-1} \sim 10$ AU yr^{-1}, and $\Delta t \sim 1$ yr, we obtain

$$|\Delta(1/a)| \sim 2v\frac{M_J}{M_\odot}\left|\left(\frac{\vec{r}_J - \vec{r}}{d_J^3} - \frac{\vec{r}_J}{r_J^3}\right)\right|\Delta t \sim \frac{M_J}{M_\odot}\ \mathrm{AU}^{-1}. \tag{4.6}$$

More accurate results of energy changes obtained from numerical integrations of parabolic orbits will be shown below, but eq. (4.6) gives essentially the correct order of magnitude.

Equation (4.6) only gives, within the order of magnitude, an average value of the energy change for comets crossing or approaching Jupiter's orbit. The energy change ϵ per orbital revolution will actually be a complex function of the comet's orbital elements and of the planetary configuration the comet meets during its perihelion passage. Let us then define $\Psi(\epsilon)$ as the probability distribution function of energy changes ϵ per perihelion passage for a population of near-parabolic comets with perihelion distances and inclinations within certain ranges. We should be now more specific about what we understand as a 'near-parabolic' orbit. For instance, a LP comet with $q \sim 2$ AU and $P = 200$ yr (the shortest orbital period for this dynamical class of comets) will have $e \simeq 0.94$. Therefore, broadly speaking, we can consider as near-parabolic comets those with eccentricities $e \gtrsim 0.9$. Numerical integrations of fictitious comets (e.g. Kerr 1961) has shown that $\Psi(\epsilon)$ can be approximated either by a Gaussian distribution

$$\Psi(\epsilon) = \frac{1}{\sigma_\epsilon\sqrt{2\pi}}\exp\left(-\epsilon^2/2\sigma_\epsilon^2\right), \tag{4.7a}$$

or by a double-exponential distribution

$$\Psi(\epsilon) = \frac{1}{\sqrt{2}\sigma_\epsilon}\exp\left(-\sqrt{2}|\epsilon|/\sigma_\epsilon\right), \tag{4.7b}$$

where σ_ϵ is the standard deviation. Equations (4.7a) and (4.7b) do not quite match the empirical ϵ-distributions, since the latter ones show additional long tails of large values of $|\epsilon|$ due to strong planetary perturbations in close encounters (Everhart 1968).

We will define the *typical energy change* ϵ_t per perihelion passage as equivalent to the standard deviation of $\Psi(\epsilon)$, namely $\epsilon_t \equiv \sigma_\epsilon$. Values of ϵ_t are plotted in Fig. 4.1 as a function of q and for different ranges of i. They were obtained from the computation of energy changes of samples of fictitious comets on initial parabolic orbits within different ranges of

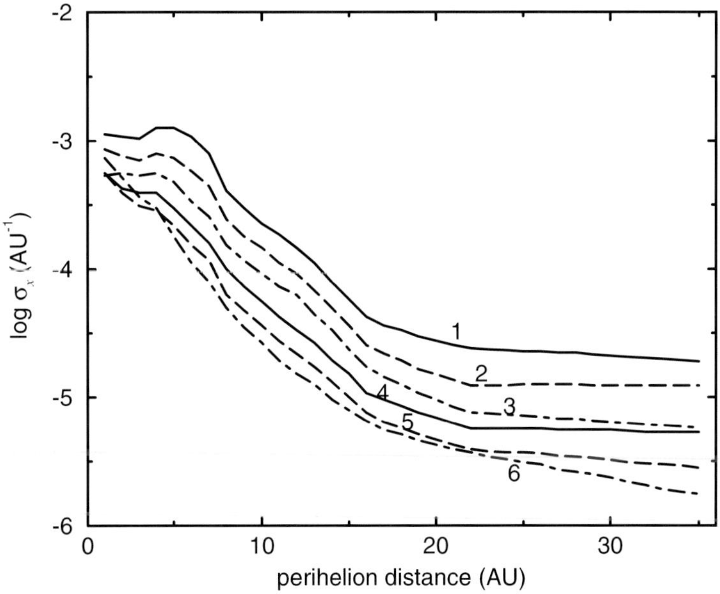

Figure 4.1. Typical energy changes per perihelion passage of bodies in near-parabolic orbits, as given by the standard deviation of the ϵ-distribution of samples of test bodies with perihelion distances and inclinations within certain ranges, as a function of the perihelion distance and for the inclination ranges: $0 < i < 30°$ (curve 1), ... $150° < i < 180°$ (curve 6) (Fernández and Brunini 2000).

q and i, and taking random values for the other two orbital parameters (ω, Ω) within the range $(0, 2\pi)$. We see that the values derived for $q \lesssim 5$ AU are of the order of $(M_J/M_\odot)$, in agreement with eq. (4.6). ϵ_t decreases very quickly with q, as comets pass at increasing distances from Jupiter, the greatest perturber. There is also a dependence on the comet's inclination. It is in general larger for direct orbits $(i < 90°)$ than for retrograde ones $(i > 90°)$. Comets in retrograde orbits will meet a planet at larger relative velocities, so they will stay less time in the planet's neighborhood, being then on the average less perturbed.

4.3. Nongravitational forces

In 1819 the German astronomer Johann Franz Encke discovered that the comet with the shortest orbital period so far known (3.3 years) departed significantly from a purely gravitational motion. Thus, comet 2P/Encke as it is now named, showed a shortening in its orbital period of about 2.5 hours per orbital revolution. Secular decreases of orbital periods were also found for other periodic comets like 3D/Biela and 16P/Brooks. The first idea was that a resisting medium affected the orbital motion of small light bodies like comets. However this explanation did not apply to comets like 8P/Tuttle showing a secular increase of their periods, or 21P/Giacobini-Zinner that initially showed an increase and later a decrease. In 1836 Friedrich Wilhelm Bessel suggested that the observed material expelled predominantly in the sunward direction must exert a recoil force on the comet nucleus. Though essentially correct, this idea remained largely ignored for more than a century, mainly because there was not agreement about the true nature of the comet nucleus. The American astronomer Fred Whipple returned to Bessel's idea explaining nongravitational forces in terms of the jet reaction produced by the non-isotropic, sublimating gases from the comet nucleus (Whipple 1950). According to Whipple, the thermal inertia in a rotating nucleus will cause the region of maximum outgassing to shift towards the nucleus "afternoon", giving rise to a net force $\vec{J}$ deviated from the radial direction to the Sun (Fig. 4.2). Therefore, there will be in general, radial, transverse, and normal nongravitational components (J_r, J_t, J_n) acting on the comet nucleus. The transverse component will cause an acceleration or deceleration of the comet's motion depending on whether it is directed along the motion or in the opposite sense. In a more realistic situation of a comet nucleus with a few active areas scattered on a dust-mantled surface, the sublimating gases may follow a rather more complex pattern than the one depicted in Fig. 4.2 but, nevertheless, the physical principles are the same.

The main nongravitational effect that can be detected in a periodic comet observed at previous apparitions is a delay or advance in the time of the perihelion passage, with respect to that derived from purely gravitational theory, that corresponds to a change ΔP in its orbital period P. For instance, for the last few apparitions 1P/Halley has arrived at its perihelion with an average delay of $\Delta P \simeq 4.1$ days. The change ΔP can be expressed in terms of the radial and transverse nongravitational components by means of the planetary equations under the Gauss form, which leads to

$$\Delta P = \frac{6\pi}{n^2 a} \int_0^P \left[\frac{e \sin f}{(1-e^2)^{1/2}} J_r + \frac{a(1-e^2)^{1/2}}{r} J_t \right] dt, \tag{4.8}$$

where $n = 2\pi/P$ is the mean motion, e the eccentricity and f the true anomaly.

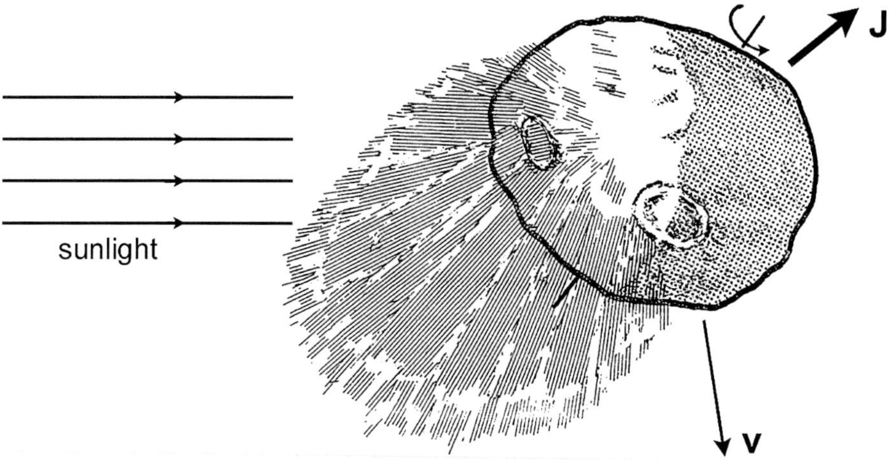

Figure 4.2. The action of nongravitational forces on comets. The sublimating gases from the nucleus give rise to a net jet force $\vec{J}$ in the opposite direction to the maximum outgassing. $\vec{V}$ is the orbital velocity of the comet. Because of the thermal inertia, the zone of maximum outgassing on a rotating nucleus will lag with respect to the subsolar point.

Nongravitational accelerations have been evaluated for most periodic comets observed in more than one apparition, since it is possible to link several perihelion passages and thus obtain reliable estimates of changes ΔP attributable to nongravitational effects. Marsden et al. (1973) found a ratio 10:1 between the radial and transverse components, implying an average lag angle $\lambda \approx 6°$. They also found a negligible value for the normal component. Therefore, the standard model for the nongravitational force provides values for the components

$$J_r = \Gamma g(r) \cos\lambda = A_1 g(r), \tag{4.9a}$$

$$J_t = \Gamma g(r) \sin\lambda = A_2 g(r), \tag{4.9b}$$

$$J_n = 0, \tag{4.9c}$$

where Γ is the magnitude of the nongravitational force at 1 AU from the Sun and $g(r)$ is the variation law with the heliocentric distance. It is customary to give the nongravitational acceleration acting on a comet through the parameters A_1 and A_2 (expressed in units of 10^{-8} AU day^{-2}), such that the equation of motion given by eq. (4.1) becomes

$$\frac{d^2\vec{r}}{dt^2} = -\frac{GM_{\odot}\vec{r}}{r^3} + \nabla\mathcal{R} + A_1 g(r)\hat{r} + A_2 g(r)\hat{T}. \tag{4.10}$$

Marsden et al. (1973) found that the nongravitational effects on the motion of several comets are better explained if water snow is assumed to be a major component in cometary composition, and they suggested the expression

$$g(r) = \alpha \left(\frac{r}{r_o}\right)^{-m} \left[1 + \left(\frac{r}{r_o}\right)^{n}\right]^{-k} \tag{4.11}$$

where $\alpha = 0.1113$ is a normalization factor such that $g(1) = 1$, $m = 2.15$, $n = 5.093$, $k = 4.6142$ and $r_o = 2.808$ AU. This expression is an empirical fit to Delsemme and Miller's (1971) curve for the sublimation rate of water snow.

For a comet with a symmetric lightcurve (outgassing) with respect to perihelion, the change ΔP will depend only on the transverse component J_t. In this case, the term of eq. (4.8) containing J_r integrated over the whole orbital period P will vanish, and only the term containing J_t will remain. Yet, as pointed out by Rickman (1986), most comet lightcurves are moderately or highly asymmetric (cf. Figs. 3.5 and 3.6), so the integral of the term containing J_r will not vanish; indeed, it may become dominant. Yeomans and Chodas (1989) developed a nongravitational acceleration model that takes into account the lightcurve asymmetry with respect to perihelion. The approach is simply a slight modification of the standard symmetric model, and consists in the offset of the time when the water vaporization curve reaches its maximum value by an interval ΔT that can be either positive (maximum after perihelion), or negative (maximum before perihelon). Thus, we have simply to substitute $g(r)$ in eq. (4.10) by $g(r')$, where $r' = r(t + \Delta T)$, and then proceed as in the standard case.

The evaluation of nongravitational forces is more difficult for LP comets, since these have not been observed in a second apparition to check for an advance or delay in the time of perihelion passage. Nevertheless, nongravitational terms have been fitted to the equations of motion of several LP comets leading to more satisfactory orbital solutions. Most of the computed values of A_1 fall in the range $1 - 10 \times 10^{-8}$ AU day^{-2}, while $A_2/A_1 \sim 0.1$. For an assumed $A_1 = 10^{-8}$ AU day^{-2} and $A_2/A_1 \sim 0.1$, Bolatto et al. (1995) found energy changes $\Delta x >> 10^{-4}$ AU^{-1} for samples of fictitious LP comets with perihelion distances $q \lesssim 0.25$ AU, i.e. the energy change due to nongravitational forces becomes comparable to that produced by planetary perturbations. On the other hand, periodic comets show typical values of A_1, A_2 one-two orders of magnitude smaller than for LP comets. Thus, pertubations on the energy x will be very small for periodic comets, though the change in the orbital period $\Delta P/P \sim \Delta x/x$, observed as a delay or advance of the time of the perihelion passage, can be measurable.

4.4. Resonant and non-resonant motion

We can ask now the following question: what is the main difference(s) between the dynamics of periodic comets, and other bodies of the solar system like asteroids, and the dynamics of LP comets? The answer is: the dynamics of the former populations is ruled to a lesser or greater degree by resonances, while for the latter bodies resonances play a very minor or no role at all. To better understand this point, let us introduce very briefly the concept of *resonance*, leaving a more thorough description for Chapter 8. A resonance implies the existence of commensurabilities in the motion of two or more bodies, for instance between their orbital periods. As a consequence, perturbations will add coherently because certain planetary configurations will repeat themselves periodically. A mean-motion resonance (MMR) involves commensurabilities between two or more orbital periods. We have many examples of MMRs in the solar system. For instance, Neptune and Pluto are locked in the 2:3 MMR (namely, three revolutions of Neptune correspond to two revolutions of Pluto), which has the effect that when Pluto is at perihelion Neptune is always far away, thus preventing Pluto from suffering strong perturbations.

Let us illustrate the concept of resonance with an example. Let us consider the Sun, Jupiter on a circular orbit, and a massless comet on an eccentric orbit with an orbital period twice that of Jupiter. If, for instance, the comet at perihelion aligns with Jupiter and the Sun, at the following perihelion passage (that corresponds to two revolutions of Jupiter), the comet will align again with Jupiter and the Sun. Actually, the commensurability does not have necessarily to be exact in mathematical terms (a condition hard to get in nature), but if the bodies are close enough to the commensurability, in some configurations they can oscillate, or *librate*, back and forth around the exact value. Such librations are said to be stable. The usual mechanical analogy is the pendulum librating around the equilibrium (bottom) point.

The problem with very eccentric orbits, as the case of LP comets, is that their binding orbital energies are very small, so the perturbations in the energy per orbital revolution Δx becomes significant as compared to the binding energy x of the comet itself. Imagine that a LP comet has an orbital period P commensurable with that of Jupiter $P_J = 11.9$ yr, namely $jP = kP_J$, where j and k are integers such that $k >> j$. The perturbation in the orbital period $P = a^{3/2} = (-x)^{-3/2}$ is

$$\frac{\Delta P}{P} = \frac{3}{2}\frac{\Delta x}{x}, \tag{4.12}$$

where Δx is the energy change per orbital revolution, namely $\Delta x \sim \epsilon_t$ (cf. Section 4.2). As shown in Fig. 4.1, $\epsilon_t \sim 10^{-3}$ AU^{-1} for a LP comet with $q \sim 1$ AU. If the perturbation on the comet's orbital period were such that the comet meets Jupiter displaced, say half Jupiter's orbit, from its location at the previous comet passage (Fig. 4.3), we could

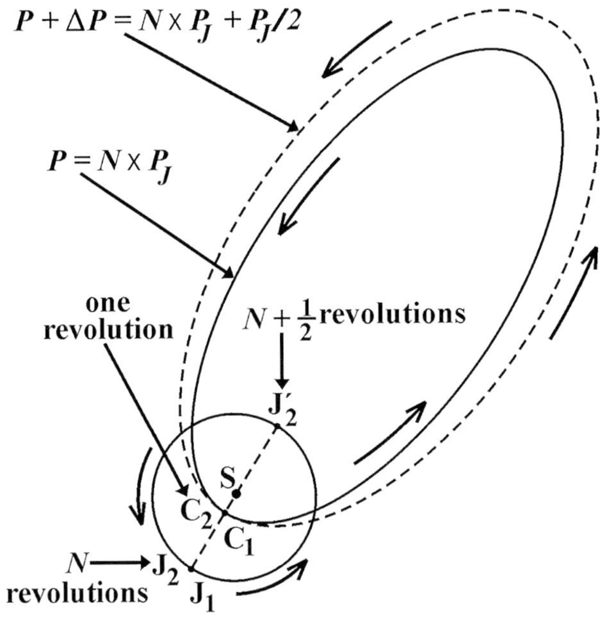

Figure 4.3. Sketch depicting the ideal case of Jupiter on a circular orbit of period P_J, and a comet on an eccentric orbit of period $P = N \times P_J$, where N is an integer number. Let Jupiter be at a given time at opposition as seen from the comet (configuration $J_1 - C_1 - S$). If the comet's orbit were not perturbed, the same configuration would repeat itself after one revolution of the comet ($J_2 - C_2 - S$). However, in a more realistic situation in which the comet's orbit is perturbed, the configuration will change. In the case of the figure in which the orbital period of the comet experiences a perturbation $\Delta P = P_J/2$, Jupiter will be in conjunction as seen from the comet after one comet's revolution ($J_2' - S - C_2$).

confidently say that the equilibrium configuration is destroyed, and no libration can restore it, in other words, the comet leaves the resonance. In the case of a pendulum, it would be as if we push it forcefully in such a way that it passes from librations around the bottom point to circulation in one direction around the fixed point.

The condition for breaking the resonance will thus be $\Delta P \sim P_J/2$. If we introduce this value in eq. (4.12), we can derive a limiting semimajor axis a_R, such that for $a > a_R$ resonances are not longer possible, namely when

$$a_R \sim \left(\frac{2}{3}\frac{\Delta P}{\Delta x}\right)^{2/5} \sim \left(\frac{P_J}{3\epsilon_t}\right)^{2/5} \simeq \left(\frac{11.9}{3 \times 10^{-3}}\right)^{2/5} \simeq 27.5 \text{ AU}, \tag{4.13}$$

and the limiting period is $P_R = a_R^{3/2} \simeq 144$ yr.

From the study of small-amplitude librations in the restricted three-body problem Sun-Jupiter-comet, Chambers (1997) found that stable librations can occur for up to the 1:9 MMR for low-inclination comets (i.e., up to $P = 11.9 \times 9 \simeq 107$ yr), whereas stable librations can extend up to the 1:21 MMR ($P \simeq 250$ yr) for retrograde orbits. From eq. (4.13) we can find a similar dependence on the comet's inclination, since ϵ_t is a function of i. Again, from Fig. 4.1 we can see that typical energy changes go from: $\epsilon_t \sim 1.3 \times 10^{-3}$ AU^{-1} for $i \sim 0$ to $\epsilon_t \sim 5 \times 10^{-4}$ AU^{-1} for $i \sim 180°$. By introducing these numerical values in eq. (4.13) we obtain a range of a_R values between ~ 25 AU ($P \sim 124$ yr) and ~ 36 AU ($P \sim 216$ yr), in close agreement with Chambers's results. It is interesting to point out that several Halley-type comets are indeed librating around MMRs. These are the cases of 23P/Brorsen-Metcalf, 12P/Pons-Brooks and 13P/Olbers in the 1:6 MMR ($P = 71$ yr), C/1921 H1 (Dubiago) in the 1:5 MMR ($P = 59$ yr), C/1942 EA (Väisälä) in the 1:7 MMR ($P = 83$ yr) (Carusi et al. 1987), and 109P/Swift-Tuttle in the 1:11 MMR ($P = 130$ yr), so far the comet with the longest period found in resonance (Chambers 1997). Therefore, the conventional limiting value of $P = 200$ yr between periodic and long-period comets acquires now a dynamical support, since it roughly corresponds to the average P_R for different inclinations.

4.5. Random-walk in the energy space

According to our previous discussion, we can conceive the dynamical evolution of a LP comet through successive passages by the planetary region as a stochastic process in which in every perihelion passage the comet meets a planetary configuration completely unrelated to the previous one. As shown above, the essentially non-resonant character of the dynamics of LP comets is what establishes a clear distinction with the dynamics of periodic comets or some other populations of the solar system. The LP comet evolution can then be described as a random-walk in the energy space, where each step $\epsilon = \delta(1/a) = -\delta x$ corresponds to the energy change in a passage, and where the other orbital elements: q, i, ω and Ω remain essentially unchanged. We note that since comets can either gain or lose energy after a perihelion passage, ϵ can be positive or negative.

Comets injected in the planetary region in initial near-parabolic orbits ($x \simeq 0$) will then follow a random-walk in the energy space between two "absorbing" walls: hyperbolic ejection ($x > 0$), and transfer to a periodic orbit $P < P_R$), where $P_R = 200$ yr, or an orbital energy $x_R = -0.0292$ AU^{-1}. Figure 4.4 illustrates the fate of two fictitious comets starting on near-parabolic orbits in which one of them is ejected after 314 revolutions, while the other is transferred to a periodic orbit after 734 revolutions. Since such comets start near one of the absorbing walls (hyperbolic ejection), the great majority will hit it before they have the chance to random-walk to the other more distant absorbing wall. In actuality, unless comets disintegrate after several passages by the Sun's neighborhood or collide with the Sun or one of the planets, the only exit door is hyperbolic ejection, even for those reaching periodic orbits.

If comets start their dynamical evolution in near-parabolic orbits, then those having energies close to zero will be dynamically "young", in the sense that they will have on average no more than a few passages by the planetary region (see the two fictitious comets of Fig. 4.3). In particular, comets whose original orbital energies fall in the narrow range $0 > x > -10^{-4}$ AU^{-1} are called "new". As we saw in Section 2.2, they are presumed to be in their first passage through the inner planetary region, bearing in mind that the typical energy change per orbital revolution ϵ_t largely exceeds their binding energies. The more passages comets have, the more they will sink in the Sun's potential well, so we should expect for such dynamically "old" comets to have on average larger binding energies (i.e. small semimajor axes). We stress that this only applies in statistical terms. As Fig. 4.4 shows, one of the comets returns to a near-parabolic orbit, i.e. it looks "young" again, prior to its ejection. The average number of revolutions required for a

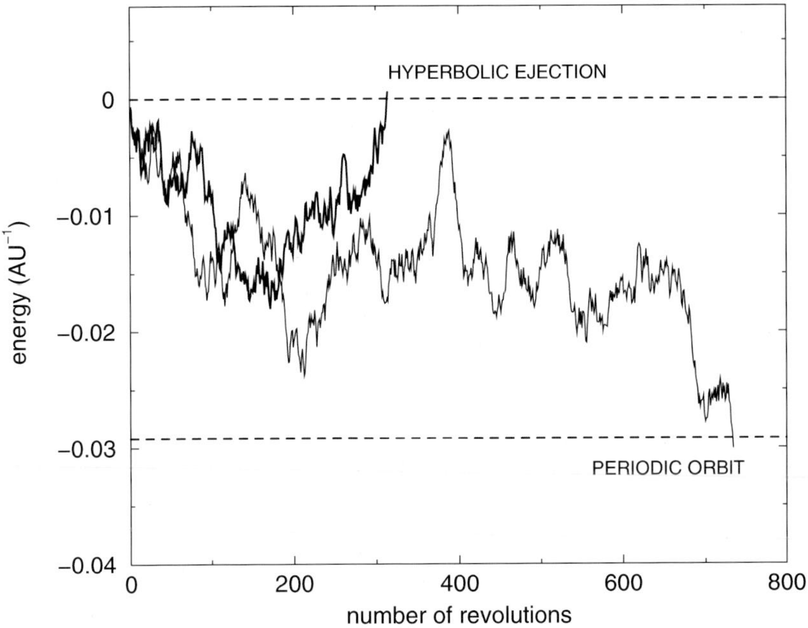

Figure 4.4. Random-walk of two fictitious comets in the energy space. They start their dynamical evolution in parabolic orbits (energy = 0). One of them is finally ejected and the other ends up in a periodic orbit.

comet to pass from an initial parabolic orbit to an orbit with a certain energy x_P will be of the order

$$< N >= \left(\frac{x_P}{\epsilon_t}\right)^2. \qquad (4.14)$$

For instance, if $\epsilon_t \sim 10^{-3}$ AU^{-1}, the average number of revolutions required for a comet to pass from a parabolic orbit to one with an energy $x = -10^{-2}$ AU^{-1} (semimajor axis $a = 100$ AU) is one hundred.

The action of nongravitational forces may change the simple picture of a random-walk in the energy space, since such forces may have a systematic effect, either as a loss or as a gain of orbital energy Δx per perihelion passage. The overall effect will be to speed up the dynamical evolution with respect to that given by eq. (4.14). As described in the previous section, the nongravitational contribution might be significant only for those LP comets approaching the Sun to less than a few tenths AU.

4.6. The diffusion equation

Let $\nu(x,t)dx$ be the number of LP comets passing perihelion per year with energies in the range $(x, x+dx)$ at a certain time t. van Woerkom (1948) derived the shape of the distribution $\nu(x,t)$ starting from an initial population of comets in parabolic orbits. The total number of comets $N(x,t)dx$ in this range of energies will be related to the flux $\nu(x,t)dx$ by means of

$$N(x,t)dx = a^{3/2}\nu(x,t)dx, \tag{4.15}$$

where $a^{3/2}$ is the comet's orbital period in years and a is expressed in AU. The increase in the number of comets of a certain x per year will be given by $\partial N(x,t)/\partial t$. The diffusion equation will be expressed as

$$\frac{\partial N(x,t)}{\partial t} = \int_{-\infty}^{+\infty} \nu(x-\epsilon,t)\Psi(\epsilon)d\epsilon - \nu(x,t), \tag{4.16}$$

where the first term on the right hand side represents the number of comets acquiring the energy x per year, and the second term those changing from x to other values.

We can develop $\nu(x-\epsilon,t)$ in ascending powers of ϵ

$$\nu(x-\epsilon,t) = \nu(x,t) - \frac{\partial \nu}{\partial x}\epsilon + \frac{1}{2}\frac{\partial^2 \nu}{\partial x^2}\epsilon^2 ..... \tag{4.17}$$

If we neglect terms of order three or superior in eq. (4.16) and bear in mind that

$$\int_{-\infty}^{+\infty} \Psi(\epsilon)d\epsilon = 1,$$

$$\int_{-\infty}^{+\infty} \epsilon\Psi(\epsilon)d\epsilon = 0,$$

and

$$\int_{-\infty}^{+\infty} \epsilon^2\Psi(\epsilon)d\epsilon = \sigma_\epsilon^2,$$

after substituting eq. (4.17) into eq. (4.16), and having in addition from eq. (4.15) that $\partial N/\partial t = -x^{-3/2}\partial \nu/\partial t$, we can write the diffusion equation under the partial differential form

$$\frac{\partial \nu(x,t)}{\partial t} = \frac{1}{2}\sigma_\epsilon^2 x^{3/2}\frac{\partial^2 \nu(x,t)}{\partial x^2}. \tag{4.18}$$

An initial population of parabolic comets implies that $\nu(x,0) = 0$ for all values of x except for $x = 0$. From the resolution of eq. (4.18) with this initial condition, van Woerkom (1948) found the following solution at time t

$$\nu(x,t) = A\left(1 + \frac{8x^{1/2}}{\sigma_\epsilon^2 t}\right)\exp\left(-\frac{8x^{1/2}}{\sigma_\epsilon^2 t}\right), \tag{4.19}$$

where A is a normalization constant. It is obvious that $\nu(x,t) \to A$ for $t \to \infty$, namely the passage rate of the surviving comets will approach a uniform distribution of x. As we will see in the next chapter, the x-distribution shows a strong concentration of comets at $x \simeq 0$ in an otherwise rather uniform x-distribution as expected from the previous analysis. Explanations for such a spike of comets at $x \simeq 0$ based on either a very young comet population, or that comets do not last long enough to diffuse to larger binding energies (or smaller semimajor axes) turned out to be unsatisfactory, since a recent capture from interstellar space would leave an excess of direct orbits (cf. Chapter 1). As we will analyze in the next chapter, Oort (1950) could find a satisfactory explanation for such a comet concentration postulating the existence of a comet reservoir at nearly interstellar distances.

Solutions to the diffusion equation were later found by Shteins and Riekstyn'sh (1961) who added a condition for disintegration of comets as a function of time and perihelion distance. As expected, their solution showed a rapid falloff in the number of passages $\nu(x,t)$ with increasing binding energies for large t, in contrast with the uniform x-distribution derived from eq. (4.19).

4.7. Dynamical time scale for capture into a periodic orbit

A LP comet random-walking between two absorbing walls will essentially have the same probability to occupy any energy range $(x, x+dx)$ between the boundaries $x_o \simeq 0$ and $x_R = -0.0292$ AU^{-1}. Therefore, we can compute the average orbital period $\overline{P}$ as

$$\overline{P} \simeq \int_{P_R}^{P_o} P(x)\frac{dx}{(x_o - x_R)} \tag{4.20}$$

and bearing in mind that $P = (-x)^{-3/2}$, so that $dx = 2/3P^{-5/3}dP$, we get

$$\overline{P} = \frac{1}{(-x_R)} \int_{P_R}^{P_o} P \times \frac{2}{3} P^{-5/3} dP = \frac{2}{3(-x_R)} \int_{P_R}^{P_o} P^{-2/3} dP,$$

which upon integration, and bearing in mind that $(-x_R) = P_R^{-2/3}$, leads to

$$\overline{P} \simeq 2P_o^{1/3} P_R^{2/3}. \tag{4.21}$$

Equation (4.21) has a meaning only if we place the outer absorbing wall not at infinity (since in that case $P_o \to \infty$), but at a distance at which we can assume that external perturbers (e.g. galactic forces or passing stars) can remove comets from the planetary region, say $a_o \sim 10^4$ AU or $P_o \sim 10^6$ yr (we will analyze this point more properly in the next chapter). Furthermore, if we take $P_R = 200$ yr, we obtain $\overline{P} \simeq 6.8 \times 10^3$ yr.

For a LP comet crossing Jupiter's orbit, the dynamical time scale, t_{per}, for capture into a periodic orbit from a near-parabolic orbit will be obtained as the product of the average period $\overline{P}$ by the average number of revolutions to reach the energy x_R given by eq. (4.14), namely

$$t_{per} = \overline{P} \times < N >_R \simeq \overline{P} \times \left(\frac{x_R}{\epsilon_t} \right)^2 \simeq 5.8 \times 10^6 \text{ yr}, \tag{4.22}$$

where, as in eq. (4.14), we adopted $\epsilon_t = 10^{-3}$ AU^{-1}.

4.8. Dynamical losses

From an initial population of near-parabolic comets injected in the inner planetary region, a few of them will finally reach periodic orbits (orbital energies $\lesssim -0.03$ AU^{-1}) through the random walk in the energy space. The rest will be ejected to interstellar space (i.e. they acquire positive orbital energies). We can estimate the number of comets, $n(N)$, that will remain gravitationally bound to the solar system after N perihelion passages. To this end, let us assume that the comet's energy changes either by $+\epsilon_t$ or $-\epsilon_t$ in each perihelion passage, where ϵ_t is the typical energy change per orbital revolution as defined in Section 4.2. The condition for the comet to remain bound to the solar system after k passages, is that the sum of energy kicks Δx_i received by the comet be negative, namely

$$S_k = \Delta x_1 + ... + \Delta x_k < 0, \tag{4.23}$$

where $\Delta x_i = \pm\epsilon_t$. There are many possible combinations of $+\epsilon_t$ and $-\epsilon_t$ in k steps. Each one of these combinations defines a *path* in the parametric plane (N, S). We can use combinatorial analysis to compute the total number of possible paths with a varying number of positive and negative elements (see, e. g. Feder 1968, Chapter III). We have

$$\binom{k}{0} + \binom{k}{1} + ... + \binom{k}{k} = 2^k, \tag{4.24}$$

where each one of the binomial coefficients gives the number of possible arrangements of k elements containing $p \leq k$ positive ones. We have

$$\binom{k}{p} = \frac{k!}{p!(k-p)!}. \tag{4.25}$$

From eq. (4.24) we deduce that the probability $p_{path,k}$ that the comet will follow a particular path (or arrangement of the k steps) is

$$p_{path,k} = 2^{-k}. \tag{4.26}$$

The comet will be ejected after N revolutions if $S_N = 0$ (i.e. it becomes a parabolic comet). The condition $S_N = 0$ means that $N/2$ steps are positive and $N/2$ negative (Fig. 4.5). We have several possible paths (from the negative side of gravitationally bound orbits) that can connect the starting point $S_0 = 0$ with the return to zero: $S_N = 0$. It should be noted that a return to the level $S = 0$ implies that N is necessarily even. The number of paths, $n_{path,N}$, that fulfill this requirement is expressed as the number of possible arrangements of N elements in which $N/2$ are positive, namely

$$n_{path,N} = \frac{1}{2} \times \binom{N}{N/2} = \frac{1}{2} \times \frac{N!}{(N/2)!(N/2)!}, \tag{4.27}$$

where the factor $1/2$ arises from the fact that we disregard all the paths on the positive side of the $S = 0$ axis (hyperbolic ejection).

When N is large, we can use the approximation given by Stirling's formula

$$N! \sim \sqrt{2\pi} N^{(N+1/2)} \exp(-N), \tag{4.28}$$

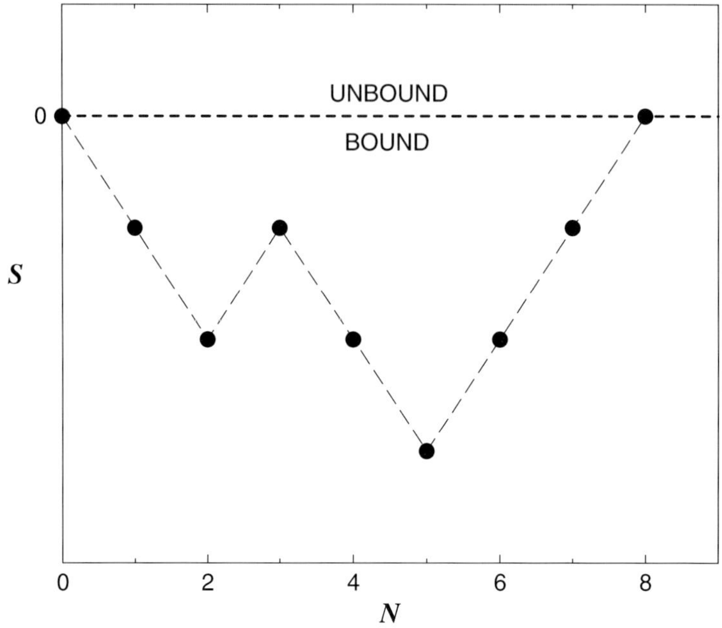

Figure 4.5. Sketch showing a path on the negative side of the S-axis, starting and returning to $S = 0$.

which substituted in eq. (4.27) leads to

$$n_{path,N} \sim \frac{\sqrt{2\pi}N^{(N+1/2)}\exp(-N)}{[\sqrt{2\pi}(N/2)^{(N/2+1/2)}\exp(-N/2)]^2} = \frac{1}{2}\frac{2^{(N+1)}}{\sqrt{2\pi N}}. \tag{4.29}$$

Therefore, the probability that a gravitationally bound comet reaches $S_N = 0$ after N revolutions is

$$p_N = n_{path,N} \times p_{path,N} \sim \frac{1}{2} \times \frac{2^{(N+1)}}{\sqrt{2\pi N}} \times 2^{-N} = \sqrt{\frac{1}{2\pi}}N^{-1/2}. \tag{4.30}$$

This result was found empirically by Everhart (1976) and Fernández (1981b). Strictly speaking, eq. (4.30) does not apply for small values of N. Nevertheless, we can can adopt as a good approximation for all values of N

$$p_N \approx \frac{1}{2}N^{-1/2}. \tag{4.31}$$

As expected, for $N = 1$ we get $p_1 = 0.5$, i.e. half the initial population of near-parabolic comets remains and the other half is ejected; in other words, for a population at the edge of the cliff, the half receiving a positive kick will fall into the precipice, while the other half receiving a negative kick will retreat one step from the edge into the space of bound orbits.

4.9. The capture problem

A comet starting in a near parabolic orbit will random walk in the energy space reaching an energy x_f after a number of passages (cf. eq. (4.14))

$$N_{xf} \sim (x_f/\epsilon_t)^2 \tag{4.32}$$

where, as before, ϵ_t is the typical energy change per orbital revolution. Now, if we substitute eq. (4.32) into eq. (4.31), we get the probability p_{xf} that the comet reaches an energy x_f, i.e.

$$p_{xf} \sim \frac{1}{2}\frac{\epsilon_t}{x_f}. \tag{4.33}$$

Equation (4.33) seems to provide a result in contradiction with diffusion theory. According to this theory, a particle starting with an energy x_i and diffusing between two absorbing walls at energies x_o and x_f would have a probability of capture to the energy level x_f given by

$$p_{xf} = \frac{x_i - x_o}{x_f - x_o}, \tag{4.34}$$

which is independent of ϵ_t in contradiction with the previous expression. This was pointed out by Stagg and Bailey (1989) as a paradox. Yet there is an explanation for this seemingly paradox: we should bear in mind that in our case $x_o \sim 0$, and if $\epsilon_t >> |x_i|$ (as is the case for comets on initial near-parabolic orbits), eq. (4.34) has really any meaning after one perihelion passage when about half of the comets remains and the other half is ejected. After one perihelion passage, the comets that remain bound have an average (negative) energy $|x_i| \sim \epsilon_t$, so if we substitute this value and $x_o \sim 0$ into eq. (4.34) and take one half we get eq. (4.33).

The energy of a comet reaching a short-period orbit ($P = 20$ yr) is $x_f = -0.136$ AU^{-1}. If we adopt a typical change $\epsilon_t = 7 \times 10^{-4}$ AU^{-1} for a Jupiter-crossing LP comet on a randomly-oriented orbit, the probability that it will survive to reach a SP orbit is $p_{xf} \simeq (1/2)(7\times 10^{-4}/0.136) \simeq 2.6\times10^{-3}$. If the comet moves on a low-inclination orbit, the typical energy change raises to $\sim 1.5\times10^{-3}$ AU^{-1} (cf. Fig. 4.1) and $p_{xf} \sim 5.5 \times 10^{-3}$, namely a comet in about 180 will reach a SP orbit, which is in fairly good agreement with Everhart's (1972) numerical experiments. This is of course a very rough approximation since, strictly speaking, random-walk techniques cannot be applied to periodic orbits, as explained in Section 4.4.

4.10. From "new" comets to "old" comets. The problem of physical decay

Let us consider an initial population of n_{new} "new" comets injected into the inner planetary region (by definition, their initial orbits are near-parabolic with energies $0 > x > -10^{-4}$ AU^{-1}). The number of comets that will remain after N passages is

$$n_N \sim n_{new} \times p_N = \frac{1}{2} n_{new} N^{-1/2}, \tag{4.35}$$

where p_N is given by eq. (4.31).

A number n_{new} of new comets will give rise to a number of returns as evolved LP comets given by

$$n_{ev} = \sum_{N=1}^{N_{max}} n_N \sim \frac{1}{2} n_{new} \sum_{N=1}^{N_{max}} N^{-1/2} \sim n_{new} N_{max}^{1/2}, \tag{4.36}$$

where N_{max} is the maximum number of returns allowed for a comet during its evolution. In theory, N_{max} is only limited by the solar system age but, in practice, for small-q comets the limit will be imposed by physical decay. In Section 2.7 we saw that the ratio of new to evolved comets that cross Earth's orbit is about 1:10. By substituting this ratio in eq. (4.36) we get $N_{max} \approx 10^2$; in other words, in order to observe 10 evolved LP comets for every new one in an Earth-crossing orbit, its average lifetime should be of $\sim 10^2$ revolutions. Physical causes will set this limit rather than dynamical ones. We will come back to this point below.

After N revolutions an Oort cloud comet will diffuse to an orbital energy $x(< 0)$ given by

$$x \sim -N^{1/2} \epsilon_t. \tag{4.37}$$

Therefore, the larger x (in modulus), the older the comet is from a dynamical point of view. By substituting eq. (4.37) into eq. (4.35) we obtain

$$n_N \sim \frac{1}{2} n_{new} \frac{\epsilon_t}{(-x)}. \tag{4.38}$$

Therefore, the number of comets n_x per unit energy is

$$n_x = \frac{n_N}{\epsilon_t} \sim \frac{1}{2} n_{new} \frac{1}{(-x)}, \tag{4.39}$$

where we assume for the time being that ϵ_t is independent of the inclination.

Actually, eq. (4.39) only considers dynamical losses. After many passages comets will also be lost due to physical effects such as sublimation, splitting, or collision with the Sun or one of the planets. Therefore, only a fraction of the n_x comets that remain bound to the solar system will actually survive as active comets, the rest will disintegrate or become inactive. This fraction can be expressed as

$$n'_x = n_x e^{-\tau_{ph} N}, \tag{4.40}$$

where $\tau_{ph} = 1/N_{ph}$, and N_{ph} is the mean physical lifetime in number of revolutions. Of course, we could use other laws to describe physical losses but, given our ignorance in this respect, we consider eq. (4.40) to be a suitable approximation to gain insight into the problem of physical decay.

If we substitute eq. (4.39) and eq. (4.37) into eq. (4.40) and divide by the corresponding orbital period $P = (-x)^{-3/2}$, we finally obtain the rate of passages per unit energy, ν_x, yielded by the initial injection of n_{new} new comets, i.e.

$$\nu_x = \frac{1}{2} n_{new} (-x)^{1/2} \exp\left(-\tau_{ph} \frac{x^2}{\epsilon_t^2}\right). \tag{4.41}$$

The energy distribution of the evolved LP comets with $q < 1.5$ AU discovered after 1850 is shown as a histogram in Fig. 4.6 (new comets are not included in the histogram). Three theoretical curves derived from eq. (4.41) for mean physical lifetimes of 10, 100 and 1000 revolutions are superimposed on the histogram. As seen, the fit of the histogram to the theoretical curves is not good: there is a large number of young comets with small values of x (in modulus) and a long tail of

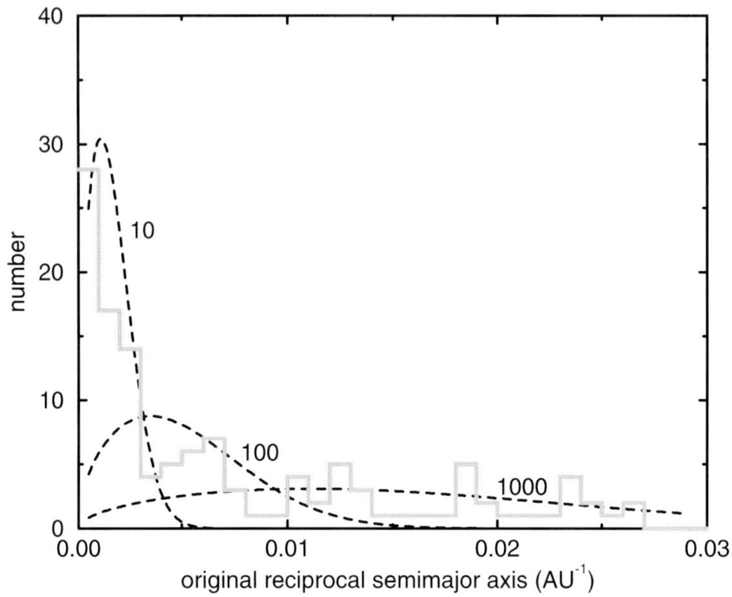

Figure 4.6. Distribution of the original reciprocal semimajor axes of long-period comets with binding energies $x < -10^{-4}$ AU^{-1} (or $a_{orig} < 10^4$ AU) and perihelion distances $q < 1.5$ AU. Three theoretical distribution functions, derived for physical lifetimes of 10, 100 and 1000 revolutions as given by eq. (4.41), are superimposed to the histogram (Fernández and Gallardo 1999).

older comets with larger x. The sharp drop in the number of observed comets at $x \sim -3 \times 10^{-3}$ AU^{-1} may be explained as due to a large population of short-lived comets (of about 10 revolutions or so) that decay fast due to their small size or fragile structure. The long tail may represent larger comets able to withstand hundreds or thousands or revolutions before disintegration or sublimation.

Most of the large-q LP comets so far discovered have original x close to zero (near-parabolic orbits), i.e, they are dynamically new or young. The almost complete absence of evolved LP comets with $q \gtrsim 2$ AU and large binding energies (cf. Fig. 2.11) suggests that: (1) they already disappeared by physical or dynamical causes, and/or (2) they exist but are too faint to be observed because of their larger distances to the Sun and the Earth.

4.11. Changes in the inclination-distribution of LP comets with the dynamical age

The random character of the orbit inclinations of LP comets is approximately verified only for the sub-sample of new comets ($0 > x > -10^{-4}$ AU^{-1}) (Fig. 4.7a). Randomness implies that the i-distribution follows a sine-law. However, we note clear departures from the sine-law as comets age. The trend is rather complex: the sample of "young" LP comets ($-10^{-4} > x > -10^{-3}$ AU^{-1}) shows an excess of retrograde orbits (Fig.4.7b), while middle-aged ($-10^{-3} > x > -5 \times 10^{-3}$ AU^{-1}) and old LP comets ($5 \times -10^{-3} > x > 2.92 \times -10^{-2}$ AU^{-1}) show a growing excess of direct orbits as we go toward more tightly bound orbits (Fig. 4.7c,d).

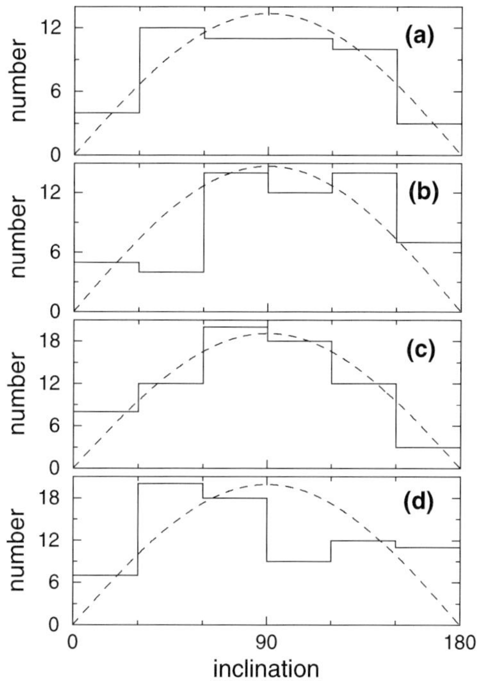

Figure 4.7. Inclination distribution of LP comets with perihelion distances $q < 2.5$ AU for different dynamical states of evolution: (a) "new" ($0 > x > -10^{-4}$ AU^{-1}), (b) "young" ($-10^{-4} > x > -10^{-3}$ AU^{-1}), (c) "middle-age" ($-10^{-3} > x > -5 \times 10^{-3}$ AU^{-1}), and (d) "old" ($-5 \times 10^{-3} > x > -2.92 \times -10^{-2}$ AU^{-1}). A sinusoidal curve has been fitted to each one of the histograms. Data taken from Marsden and Williams (2003) catalogue, leaving aside the families of sungrazers.

To understand this behavior, let us note that the typical energy change ϵ_t depends on the inclination. Young LP comets cover a range of energies $\sim (\Delta x)_{young}$ comparable to the typical energy change. Therefore, the transit of LP comets through the young age is very fast: it takes no more than one or a few passages. Since the typical energy change for direct orbits, $\epsilon_{t,D}$, is somewhat larger than that for retrograde orbits, $\epsilon_{t,R}$, retrograde comets will require on average more passages to transit the young state. This effect then explains why we should find more young LP comets in retrograde orbits than in direct ones.

After the fast transit through the young age, the surviving LP comets really enter into the diffusion regime for which the equations shown in the previous section are applicable. We can re-write eq. (4.41) taking now into account that ϵ_t depends on the inclination obtaining

$$\nu_{x,i} = \frac{1}{2}\frac{\epsilon_{t,i}}{\epsilon_t} n_{new}(-x)^{1/2} \exp\left(-\tau_{ph}\frac{x^2}{\epsilon_{t,i}^2}\right), \tag{4.42}$$

where ϵ_t now represents the typical energy change averaged over the whole inclination range $(0, \pi)$, and $\epsilon_{t,i}$ is the typical energy change over a certain range $(i, i + \Delta i)$. If we now consider direct and retrograde comets with the typical energy changes $\epsilon_{t,D} \simeq 7.6 \times 10^{-4}$ AU^{-1} and $\epsilon_{t,R} \simeq 5.4 \times 10^{-4}$ AU^{-1} averaged over the i-ranges $(0, \pi/2)$ and $(\pi/2, \pi)$ respectively, we can obtain the ratio of retrograde to direct orbits from eq. (4.42) as

$$\frac{\nu_{x,R}}{\nu_{x,D}} = \frac{\epsilon_{t,R}}{\epsilon_{t,D}} \exp\left[-\tau_{ph}x^2\left(\frac{1}{\epsilon_{t,R}^2} - \frac{1}{\epsilon_{t,D}^2}\right)\right]. \tag{4.43}$$

Values of the ratio R/D are shown in Fig 4.8 for three physical lifetimes N_{ph} = 10, 100 and 1000. We find that the excess of direct orbits among aged LP comets can be understood as a combination of two causes: (1) dynamical, since the probability of capture to a given energy x is proportional to ϵ_t (eq. (4.33)) that is smaller for retrograde orbits, and (2) physical, since the evolution of comets in the energy space is slower for retrograde orbits, these are more likely to fade before reaching the old state.

We note in the i-distribution of old LP comets an excess of orbital planes toward the ecliptic plane, both in direct and retrograde orbits. Even though the excess of small-i comets can be understood in terms of the dynamical and physical effects described before, the excess of retrograde orbits cannot be explained by these effects and is even

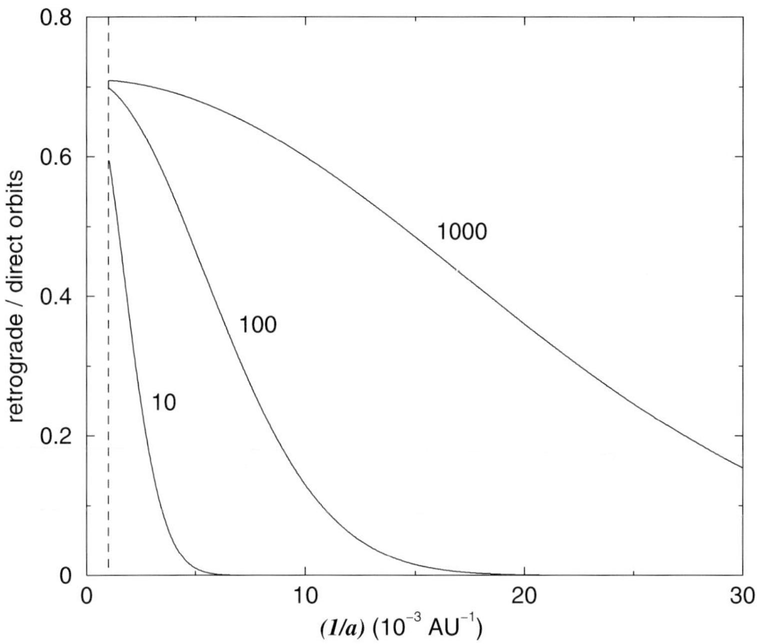

Figure 4.8. The ratio of direct to retrograde aged LP comets, as derived from eq. (4.43), as a function of the reciprocal of the semimajor axis.

in conflict with them. Fernández and Gallardo (1994) explained this feature as a dynamical effect by which near-perpendicular comets (i close to 90°) tend to shift to retrograde orbits. This can be understood from the Tisserand parameter T defined as

$$T = \frac{a_J}{a} + 2\cos i \left[\frac{q}{a_J}\left(2 - \frac{q}{a}\right)\right]^{1/2}, \tag{4.44}$$

where q, a, and i are the perihelion distance, semimajor axis and inclination of the comet, and a_J is the radius of Jupiter's orbit assumed to be circular. In the restricted circular three-body problem, T is a constant of the motion. In our case: Sun-Jupiter-comet, T will vary because of the perturbations of the other planets, and because Jupiter's orbit is not circular. Nevertheless, T will vary very slowly as compared with the comet's orbital parameters q, a and i, so it can be taken as a constant (more details will be discussed in Section 7.1). If we introduce $x = -1/a$ in eq. (4.44), differentiate it, and then assume that $x \sim 0$ for LP comets, we get

$$(-a_J^{3/2} + 2^{-1/2}q^{3/2}\cos i)\Delta x - (2^{3/2}q^{1/2}\sin i)\Delta i + (2^{1/2}q^{-1/2}\cos i)\Delta q = 0. \tag{4.45}$$

For near-perpendicular orbits we have $i \sim 90°$, so eq. (4.45) reduces to

$$-a_J^{3/2}\Delta x - 2^{3/2}q^{1/2}\Delta i \simeq 0,$$

i.e.

$$\frac{\Delta x}{\Delta i} \simeq -\frac{2^{3/2}q^{1/2}}{a_J^{3/2}} < 0. \tag{4.46}$$

Therefore, as the LP comet ages (i.e., increases on average its binding energy), $\Delta x < 0$ and thus $\Delta i > 0$, i.e. the comet increases its inclination. This effect can then explain the anomalous high number of LP comets with $i \gtrsim 120°$ among the middle-aged and old populations. This effect can somewhat counteract the other two effects seen before, but not entirely, so the excess of direct orbits appears as a clear feature of the old LP comet population.

We have so far analyzed the dynamics of LP comets subject to the action of planetary perturbations. These comets enter the inner planetary region for the first time as new comets from the Oort cloud. Before their injection into the inner planetary region, they may have already spent a long time in the Oort cloud, in general in orbits entirely outside the planetary region. Their dynamics is thus quite different from what we have studied in this chapter, and it will be the main scope of the next two chapters.

THE OORT CLOUD

As discussed in Chapter 1, at the middle of the last century the debate on whether comets were members of the solar system or interstellar bodies was still unsettled. The main hurdle against an origin in the solar system was to explain how cometary orbits, that presumably shared at the beginning the coplanarity of the planets and asteroids, could have later acquired a random distribution. In 1950 the Dutch astronomer Jan Hendrik Oort (1900-1992) found that the distribution of original reciprocal semimajor axes of the sample of LP comets known at that moment showed a strong excess within the narrow range $0 < (1/a)_{orig} < 10^{-4}$ AU^{-1}. As we discussed before, these are the so-called "new" comets. It is very likely that after a single passage most new comets will be either ejected to interstellar space or transferred to more tightly bound orbits. This finding led Oort (1950) to the conclusion that a huge swarm of $\sim 10^{11}$ comets surrounds the solar system at distances of a few 10^4 AU. This structure, called the *Oort cloud*, is generally supposed to be the source of LP comets. According to Oort, comets originally formed in the planetary region, the asteroid belt being the most likely source region, which was the only substantial population of minor bodies known at that time. He further argued that planetary perturbations were responsible for scattering comets to near interstellar distances where perturbations by passing stars randomized their orbital planes and re-injected some of these comets in the inner planetary region, thus becoming potentially observable. Oort depicted what later became a standard model of the Oort cloud, with the exception of the source region, The asteroid belt does not seem to be a suitable source, basically because of the very different compositions: asteroids are rocky bodies while comets are ice-rich bodies.

In actuality, Oort was not the first author to refer to the perturbing action of passing stars on near-parabolic comets. There are at least two earlier treatments by Fesenkov (1922) and Öpik (1932). Yet, none of these authors could develop a model of the comet cloud. We will review in this chapter the progress made in our understanding of the Oort cloud and its dynamics in the last half a century. Let us only note that Oort had available for his study a sample of only 16 LP comets with computed original orbits, whereas now it has increased to 386 LP comets (Marsden and Williams 2003).

Besides passing stars, Oort cloud comets are also subject to the perturbing action of molecular clouds and galactic tides. The quasi-steady supply of new comets is due to the action of galactic tides and distant passing stars (say, distances greater than $\sim 3 \times 10^4$ AU), whose average effect is nearly steady on time scales comparable to the orbital periods of Oort cloud comets. By contrast, stars passing at closer distances and penetrating encounters with molecular clouds may cause sudden enhancements in the flux of new comets, which are called "comet showers" (Hills 1981, Fernández 1992). This point will be analyzed in the next chapter. In this chapter we will concentrate on the dynamics of Oort cloud comets subject to external perturbers.

5.1. The distribution of original energies

We found in the previous chapter that the passage rate of comets scattered by planetary perturbations will show a uniform distribution of their orbital energies. By contrast, the observed distribution shows a spike at near-zero energies ($0 < (1/a)_{orig} \lesssim 10^{-4}$ AU^{-1}) in an otherwise smooth $(1/a)_{orig}$-distribution (Fig. 5.1). Comets in the spike come from the Oort cloud. Since for Oort cloud comets crossing Jupiter's orbit the average perturbation by Jupiter is well above their original energy, they are more likely to be in their first visit, i.e. they are "new" (though they may have passed on average many times by the outer planetary region where planetary perturbations are much weaker as shown in Fig. 4.1).

As shown in Fig. 5.1, a few comets have original slightly hyperbolic orbits, but these may be due to observational errors and/or nongravitational forces that are not accounted for in the computations. Marsden and Williams classify the computed original semimajor axes $(1/a)_{orig}$ in the quality classes 1A, 1B, 2A, and 2B according to their accuracy. Kresák (1992a) estimates mean errors of ± 5 for 1A, ± 12 for 1B, ± 50 for 2A, and ± 250 for 2B (in units of 10^{-6} AU^{-1}). It is obvious that the Oort cloud spike will be blurred for comets of class 2B, since their mean error largely exceeds the width of the spike of ~ 100 (10^{-6} AU^{-1}). The intrinsic errors can thus explain the presence of a small fraction of comets with original hyperbolic orbits. In this regard, Kresák (1992b) showed that for the comet with the largest positive energy so far known, C/1975 X1 (Sato) with $(1/a)_{orig} = -734 \times 10^{-6}$ AU^{-1}, an alternative original

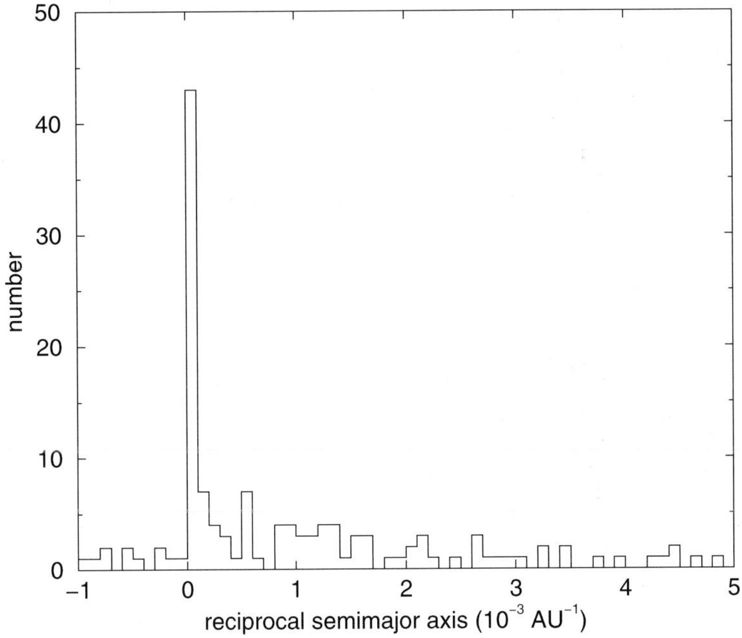

Figure 5.1. Distribution of original reciprocal semimajor axis of LP comets with $(1/a)_{orig} < 5 \times 10^{-3}$ AU^{-1} and $q < 2$ AU taken from Marsden and Williams's (2003) catalogue. Comets of the poorest quality class 2B were removed from the sample.

elliptic orbit with $(1/a)_{orig} = +15\times10^{-6}$ AU^{-1} can be fitted, giving O-C residuals only 10% to 15% larger. Furthermore, nongravitational forces may introduce additional errors in the computation of $(1/a)_{orig}$. As we showed in Section 4.3, the energy changes caused by nongravitational forces may be comparable to planetary perturbations for $q \lesssim 0.25$ AU, in other words, nongravitational forces may introduce errors > 100 (10^{-6} AU^{-1}). This may explain that some positive $(1/a)_{orig}$ of small-q comets become negative (hyperbolic) because of unaccounted nongravitational forces. Confirming this presumption, the sample of dynamically young comets with large perihelion distances - where nongravitational forces are probably less effective - shows very few "hyperbolic" members, whereas these are more frequent for smaller q (Fig. 5.2). We showed in Chapter 1 that comets captured from the interstellar space would come in strongly hyperbolic orbits, so their lack is a firm evidence that comets belong to the solar system.

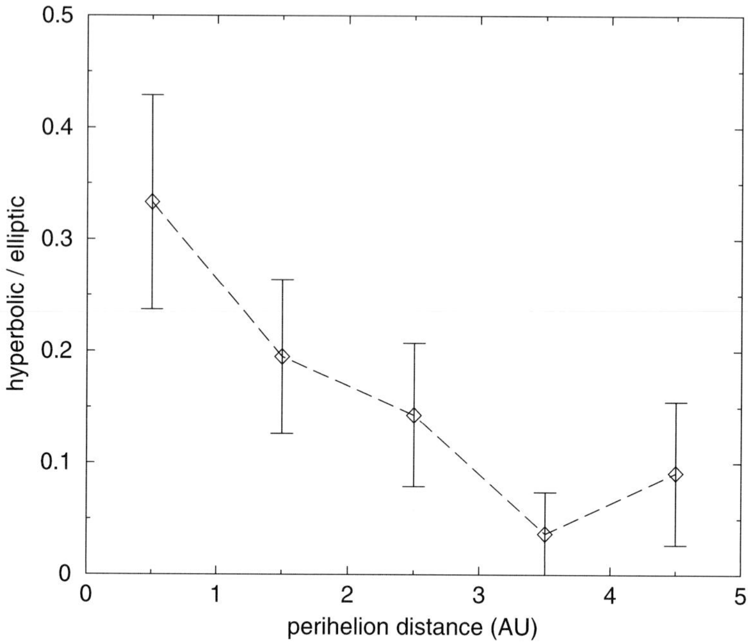

Figure 5.2. Ratio of comets with original hyperbolic orbits to comets with $0 < (1/a)_{orig} < 10^{-3}$ AU^{-1} within ranges of perihelion distances $\Delta q = 1$ AU. The error bars for the computed sample are also indicated.

5.2. Stellar perturbations

Let us now evaluate the magnitude of the perturbation caused by a passing star of mass M and relative velocity V with respect to the Sun on a comet located at a heliocentric distance r (Fig. 5.3). The usual approach is to assume that the comet is at rest in a heliocentric frame of reference during the stellar encounter. This is justified bearing in mind that typical orbital velocities of comets in the Oort cloud are about 0.1 km s^{-1}, while typical velocities of passing stars are about 30 km s^{-1}. If we assume that the greatest dynamical influence on the comet's motion occurs when the star is within 10^5 AU from the point of closest approach to the Sun, then during the time it takes the star to cross 2×10^5 AU, the comet will have moved only $\sim 10^3$ AU, i.e. a small fraction of its tour around the Sun. Furthermore, it is also assumed that the star's path is only slightly perturbed by the Sun's gravity, so it can be taken as a straight line without loss of accuracy (Öpik 1932, Oort 1950).

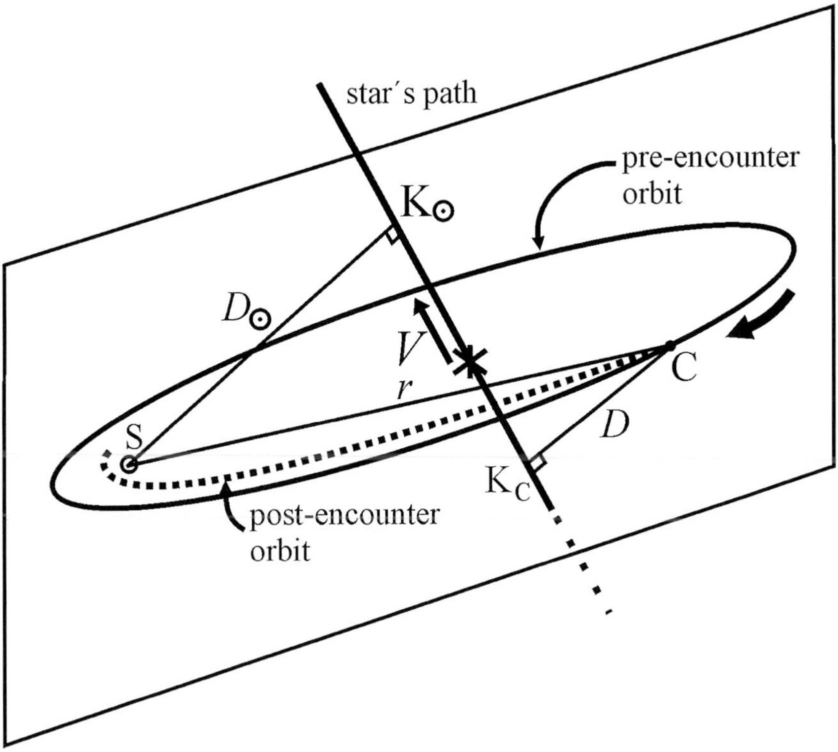

Figure 5.3. Geometry of a stellar encounter.

Under the above simplifying assumptions, the change in the comet's velocity v_c due to the star passage can be computed as

$$\Delta v_c = \int_{-\infty}^{+\infty} F_* \times dt = \int_{-\infty}^{+\infty} \frac{GM}{(D^2 + u^2)^{1/2}} \frac{du}{V} = \frac{2GM}{VD}, \tag{5.1}$$

where u is the distance along the straight path to the point K_c of closest approach to the comet, and D is the distance of closest approach of the star to the comet (Fig. 5.3). Furthermore, we can express the star's velocity $V = du/dt$. It is to be noted that the only component of the star's gravitational force that really matters is that along $\vec{D}$, because the sum of the impulse component along the star's path before K_c cancels out with the sum after passage by K_c. The vector $\vec{\Delta v_c}$ will thus have the same direction as $\vec{D}$. The star will also impart an impulse to the Sun that can be computed in the same manner as eq. (5.1). In this case the action of the star on the Sun can be reduced to an impulse imparted at the point of closest approach to the Sun, $K_\odot$, and that follows the direction $\vec{D}_\odot$, where $D_\odot$ is the distance of closest approach of the star to the Sun. Therefore, the impulsive change in the comet's velocity with respect to the Sun is

$$\vec{\Delta v} = \vec{\Delta v_c} - \vec{\Delta v_\odot} = \frac{2GM}{VD}\frac{\vec{D}}{D} - \frac{2GM}{VD_\odot}\frac{\vec{D_\odot}}{D_\odot} \tag{5.2}$$

We can make an analytical approach to estimate the change in the comet's velocity by stellar perturbations if we divide the stellar encounters in "close" and "distant", following the approach developed by Rickman (1976). For the case of a star coming very close to the Sun, we have $D_\odot << D$, so eq. (5.2) becomes approximately

$$|\Delta v| \simeq |\Delta v_\odot| = \frac{2GM}{VD_\odot} \tag{5.3a}$$

It is also possible that the star passes very close to a comet becoming $D << D_\odot$, so the change $|\Delta v|$ as given by eq. (5.3a) will hold changing $D_\odot$ by D.

For distant stellar encounters $r << D_\odot$ and $r << D$, so $\vec{D}$ and $\vec{D}_\odot$ become nearly parallel and eq. (5.2) reduces to

$$|\Delta v| \simeq \frac{2GMr\cos\beta}{VD_\odot^2} \tag{5.3b}$$

where β is the angle between the vectors $\vec{r}$ and $\vec{D}_\odot$.

During an orbital revolution of period P a comet will be perturbed by many stars. Let $s(D_\odot)dD_\odot = 2\Phi_* D_\odot dD_\odot$ be the rate of stellar passages with impact parameters in the range $(D_\odot, D_\odot + dD_\odot)$, where Φ_* is the stellar flux in the Sun's neighborhood. In Table 5.1 we show the mass density $(\rho_{*,i})$, number density $(n_{*,i})$, and the velocity dispersion $(\sigma_{*,i})$ of different types of stars in the Sun's neighborhood. If we combine all the stellar types of Table 5.1 (leaving aside the brown dwarfs), we obtain a star's density in the Sun's neighborhood $n_* = 0.073$ pc^{-3} and a mass density $\rho_* = 0.036$ M$_\odot$ pc^{-3}, which gives an average mass per star ~ 0.5 M$_\odot$. Since the Sun's velocity with respect to the Local Standard of Rest is $v_{lsr} \sim 16.5$ km s^{-1} and the velocity dispersion of most star groups are $\sim 15-20$ km s^{-1}, the average encounter velocity of the Sun with respect to nearby stars will be $V \sim (v_{lsr}^2 + \sigma_*^2)^{1/2} \sim 30$ km s^{-1}. The stellar flux is thus $\Phi_* = n_* V \simeq 7$ stars Myr^{-1} within one parsec (pc) (1 pc is the distance at which the radius of the Earth's orbit is seen under an one-second angle; 1 pc = 206265 AU). This value is consistent

Table 5.1: Stellar population in the Sun's neighborhood[(1)]

Class	$\rho_{*,i}$ ($M_\odot$ pc^{-3})	$n_{*,i}$ (pc^{-3})	$\sigma_{*,i}$ (km s^{-1})
giants	0.0006	0.0005	17.0
MS[(2)] $M_V < 2.5$	0.0031	0.0013	7.5
MS $2.5 < M_V < 3.0$	0.0015	0.0010	10.5
MS $3.0 < M_V < 4.0$	0.0020	0.0015	14.0
MS $4.0 < M_V < 5.0$	0.0024	0.0021	19.5
MS $5.0 < M_V < 8.0$	0.0074	0.0090	20.0
MS $M_V > 8.0$	0.014	0.05	20.0
white dwarfs	0.005	0.008	20.0
brown dwarfs	0.008	0.12	20.0

(1) Source: Holmberg and Flynn (2000)
(2) MS: main sequence stars with absolute visual magnitudes M_V in the range indicated.

with the lower limit of 3.5 stars Myr^{-1} found by García-Sánchez et al. (1999) from the *Hipparcos* satellite proper-motion and parallax data for nearby stars combined with ground-based measurements of their radial velocities. Brown dwarfs (BDs) are possibly more numerous than all the other types of stars together, although their small average mass (~ 0.065 $M_\odot$) makes their dynamical influence on Oort cloud comets very small as compared to stars. The only exception may be BDs that penetrate very deeply in the Oort cloud (the local density of BDs yields more than one hundred passages at distances smaller than 10^4 AU during the solar system age) knocking off encountering comets. But BD passages may affect only a small volume of the Oort cloud along the star's path, so we can neglect them in our calculations of stellar perturbations.

Since stellar perturbations occur at random, the cumulative change will sum quadratically. The cumulative change $\Delta v_{*,P}$ during an orbital revolution P can be computed by approximating a discrete sum of a large number of impulses to a continuous succession of impulses that can be integrated, namely

$$\Delta v_{*,P}^2 = P\left[\int_{D_m}^{D_L} \Delta v_1^2 s(D_\odot)dD_\odot + \int_{D_L}^{D_M} \Delta v_2^2 s(D_\odot)dD_\odot\right] \tag{5.4}$$

where Δv_1 and Δv_2 are for the approximations given by eqs.(5.3a) and (5.3b) respectively. $D_m = (2\Phi_* P)^{-1/2}$ is the minimum distance of closest approach of a star to the Sun expected during P; D_M is the maximum distance at which a passing star may have some dynamical influence on Oort cloud comets, it can be taken as infinity without too much error; and D_L is a somewhat arbitrary boundary separating the regimes in which approximations (5.3a) and (5.3b) apply. Of course, there exists a transition zone in which neither approximation (5.3a) nor (5.3b) are good, but we can avoid its consideration to make the problem analytically manageable, obtaining nevertheless results of the correct order of magnitude. Reasonable values for D_L are between r and $2r$, giving differences of no more than 30% (Fernández 1980a). If we integrate eq. (5.4) and consider an average $< \cos^2\beta >= 1/3$, we finally obtain

$$\Delta v_{*,P}^2 = 2K^2 P\Phi_*\left[\ln\left(D_L/D_m\right) + \frac{1}{6}\left(\frac{\bar{r}}{D_L}\right)^2\right] \tag{5.5}$$

where $K = 2G\overline{M}/V$ and $\overline{M}$ is the average stellar mass: $\overline{M} = 0.5$ $M_\odot$, $\bar{r}$ is the time-average heliocentric distance which is given by $\bar{r} = a(1 + e^2/2) \simeq 1.5a$ for a near-parabolic orbit.

5.3. Galactic tidal forces

It has long been recognized that tides from the Galaxy may have a significant influence on the shape and extent of the Oort cloud, though only in the last two decades their effect on the orbital motion of Oort cloud comets has been considered in some detail. The boundary of dynamical stability imposed by galactic tides can be simply evaluated by considering that all the mass of the Galaxy, $M_G = 1.3 \times 10^{11}$ $M_\odot$, is concentrated at its center (e.g. Chebotarev 1966) with the Sun moving around it at a distance of $r_G = 8.5$ kpc (Elmegreen 1998). The Sun will be surrounded by a region within which orbits will be dynamically stable against galactic tides. The outer boundary of this region will be defined by the condition that the relative velocity of a test particle with respect to the Sun becomes zero. The radius of the zero-velocity surface, also called the Hill's surface, is

$$r_t = \left(\frac{M_\odot}{3M_G}\right)^{1/3} r_G \simeq 240,000 \text{ AU}. \tag{5.6}$$

Yet the above potential is an oversimplification. Antonov and Latyshev (1972) adopted a more realistic model of the Galaxy in which the potential was expressed as

$$V_G = -\frac{1}{2}(\alpha x^2 + \gamma z^2), \tag{5.7}$$

where the x-axis is the radial direction from the galactic center, and the z-axis is perpendicular to the galactic plane, $\alpha = 4A(A - B)$ where A and B are the Oort constants describing the galactic rotation, and $\gamma = -2\pi G\rho_{disk}$ where ρ_{disk} is the density of the Galactic disk. In essence, the first term of eq. (5.7) describes the potential of the galactic bulge, whereas the second term describes the potential of the galactic disk. By applying this potential Antonov and Latyshev obtained a zero-velocity surface resembling a triaxial ellipsoid of semiaxes

$$x = 293 \times 10^3 \text{ AU},$$

$$y = 196 \times 10^3 \text{ AU},$$

$$z = 152 \times 10^3 \text{ AU}.$$

The z-axis turns out to be the shortest one (i.e. the shortest dynamical stability is along $\vec{z}$) which reflects the dominant influence of the galactic disk potential over that of the galactic bulge. Numerical experiments by Smoluchowski and Torbett (1984) confirmed the previous conclusion.

Thus, as a first approximation we can neglect the potential of the galactic bulge and consider only that of the galactic disk. The latter can be approximately modelled as a homogeneous disk of density ρ_{disk} in the mid-plane of the Galaxy, so its potential can be simply expressed as (Heisler and Tremaine 1986, Morris and Muller 1986, Torbett 1986)

$$U = U_o + 2\pi G\rho_{disk} z^2 \tag{5.8}$$

where U_o is a constant and z is the distance to the galactic mid-plane.

The estimate of the density of matter in the local galactic disk has been subject to several revisions in the last few decades. From the comparison of different gravitational potential models of the Galaxy with velocity dispersions of tracer stars, Bahcall (1984) derived a value of $\rho_{disk} = 0.185$ M$_\odot$ pc^{-3}. Later on this value was downwards revised by Kuijken and Gilmore (1989) to $\rho_{disk} = 0.10$ M$_\odot$ pc^{-3}. The *Hipparcos* satellite allowed the determination of very accurate stellar distances and proper motions within 125 pc for almost all stars brighter than apparent magnitude $m_v = 8$. From the *Hipparcos* data Crézé et al. (1998) selected a sample of A-type tracer stars from which they derived a local dynamical density of $\rho_{disk} = 0.076 \pm 0.015$ M$_\odot$ pc^{-3}, leaving no room for any disk-shaped component of dark matter (as we showed in the previous section, stars make up for about 0.036 M$_\odot$ pc^{-3} of the bulk mass density of the galactic disk, to which about 0.04 M$_\odot$ pc^{-3} of HI and H_2 interstellar clouds must be added). More recently, Holmberg and Flynn (2000) used a volume-complete sample of A and F stars, whose parallaxes and proper motions were measured from the *Hipparcos* satellite, to solve for the gravitational potential vertically in the local galactic disk. By comparing the *Hipparcos* measured space density with predictions from various disk models, they derive a value of 0.102 ± 0.010 M$_\odot$ pc^{-3} for the local dynamical mass density. From the previous results, we shall adopt in the following a value $\rho_{disk} = 0.10$ M$_\odot$ pc^{-3}, though we should bear in mind that this is the value measured *at present*. In the past, the mass density in the Sun's vicinity probably oscillated up and down around the current value, as the Sun experienced galactic radial and vertical excursions, which probably caused periodic modulations in the comet flux (Matese et al. 1995). The vertical oscillation of the Sun around the galactic mid-plane is expressed by means of the equation of motion $d^2z/dt^2 = -4\pi G \rho_{disk} z$, which has a half-period of $\tau_z = \sqrt{\pi/G\rho_{disk}} \sim 42$ Myr, reaching a maximum height of about 70 pc.

From the above potential, the tidal force of the galactic disk acting on a comet at a galactic latitude ϕ is

$$\vec{F}_{disk} = [(dU/dz)_c - (dU/dz)_\odot]\hat{z} = 4\pi G \rho_{disk} r \sin\phi \hat{z}, \qquad (5.9)$$

where r is the Sun-comet distance, $r \sin\phi = z_c - z_\odot$ is the difference between the distances of the comet and the Sun to the galactic mid-plane, and $\hat{z}$ is the unit vector perpendicular to the galactic plane.

We can now compute the change in the comet's perihelion distance q from the Lagrange's planetary equations under the Gaussian form

$$\frac{da}{dt} = \frac{2}{n\sqrt{1-e^2}}\left(Se\sin f + p\frac{T}{r}\right), \tag{5.10a}$$

$$\frac{de}{dt} = \frac{\sqrt{1-e^2}}{na}\left[S\sin f + T(\cos E + \cos f)\right] \tag{5.10b}$$

where n is the mean motion, e is the eccentricity, $p = a(1-e^2)$, f is the true anomaly, and E is the eccentric anomaly. S and T are the radial and transverse components of the perturbing force that are obtained from eq. (5.9), namely

$$S = F_{disk}\sin\phi, \tag{5.11a}$$

$$T = F_{disk}\cos\phi\cos\alpha. \tag{5.11b}$$

where α is the angle between the orbital plane and the plane perpendicular to the galactic disk containing the radius Sun-comet (Fig. 5.4). The radius Sun-comet is assumed to keep the same direction as the apsidal line through an orbital revolution, which is more or less fulfilled for near-parabolic comets.

Since $q = a(1-e)$ we have

$$\frac{dq}{dt} = (1-e)\frac{da}{dt} - a\frac{de}{dt} \tag{5.12}$$

Substituting eqs.(5.10a) and (5.10b) into eq. (5.12) and rearranging terms that depend on S and T, we finally obtain

$$\frac{dq}{dt} = \frac{S\sin f}{n}\left[2e\sqrt{\frac{1-e}{1+e}} - \sqrt{1-e^2}\right] + T\frac{\sqrt{1-e^2}}{n}\left[\frac{2(1-e)a}{r} - \cos E - \cos f\right]. \tag{5.13}$$

For near-parabolic orbits we have: $e \sim 1$, $1-e^2 \sim 2q/a$, and in general $f \sim \pi$. Bearing in mind that $\cos E \sim 1 - r/a$, and if we adopt as before a time-averaged heliocentric distance for the comet $\bar{r} \sim 1.5a$, we have $\cos E \sim -1/2$. Substituting these approximations together with $n = \mu^{1/2}a^{3/2}$ in eq. (5.13),where $\mu = GM_{\odot}$, we finally obtain

$$\frac{dq}{dt} \sim \frac{\sqrt{2qa}}{\mu^{1/2}}\left(\frac{2q}{r} + \frac{3}{2}\right)T \sim \frac{3}{2}\frac{a\sqrt{2q}}{\mu^{1/2}}T. \tag{5.14}$$

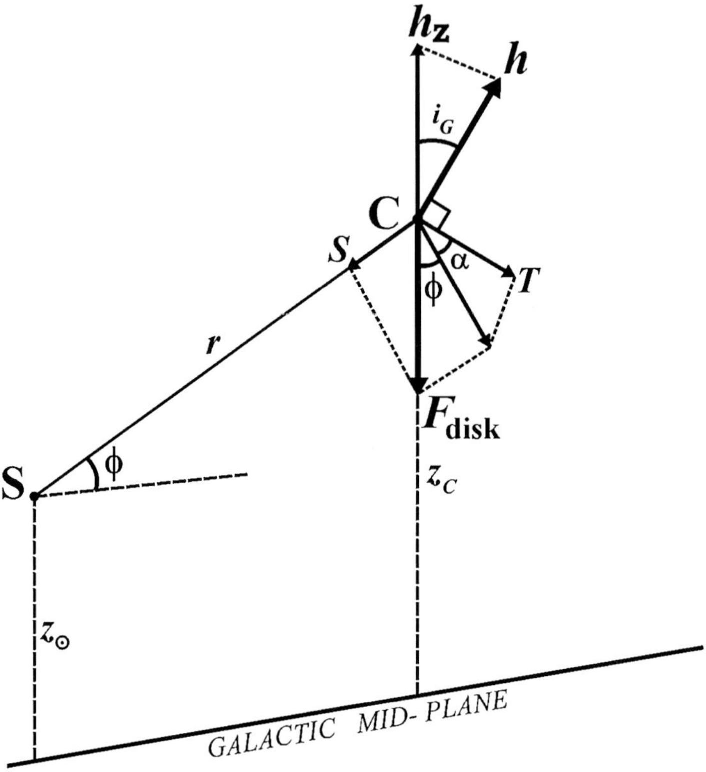

Figure 5.4. The tidal force of the galactic disk $\vec{F}_{disk}$ acting on comet C and its components radial S and transverse T. $\vec{h}$ is the orbital angular momentum vector and h_z its component along the vertical direction.

We note that eq. (5.14) is independent of the radial component of the perturbing force S.

By substituting eq. (5.9) into eq. (5.11b) we obtain for the transverse component T

$$T = 4\pi G\rho_{disk} r \sin\phi\cos\phi\cos\alpha. \tag{5.15}$$

Substituting this expression in eq. (5.14) and integrating it between zero and the orbital period $P = 2\pi\mu^{-1/2}a^{3/2}$, we obtain the change in the perihelion distance per orbital revolution

$$q_f^{1/2} = q_i^{1/2} + 4.5\sqrt{2}\pi^2 M_\odot^{-1}\rho_{disk} a^{7/2}\cos\alpha\sin 2\phi, \tag{5.16}$$

where q_i, q_f are the initial and final perihelion distances respectively. In the integration we assume that the other orbital elements of the comet, as well as α and ϕ, remain constant during P.

We can write $q_f = q_i + (\Delta q)_{disk,P}$, and if $(\Delta q)_{disk,P} << q_i$, eq. (5.16) becomes

$$(\Delta q)_{disk,P} = 12.7\pi^2 M_\odot^{-1} \rho_{disk} q^{1/2} a^{7/2} \cos\alpha \sin 2\phi, \tag{5.17}$$

where $q \sim q_i$. Equations (5.16) and (5.17) show that the change of the perihelion distance is maximum for a galactic latitude $\phi = 45°$, and negligible near the Galactic poles and the Galactic equator. Byl (1983) already noted that galactic effects on comets were strongest at mid-Galactic latitudes.

We note that the tidal force of the galactic disk cannot decrease the perihelion distance below a certain limiting value q_z. This is because $\vec{F}_{disk}$ acts always along $\hat{z}$, so the change in the comet's angular momentum $\Delta\vec{h} = \vec{r} \times \vec{F}_{disk}$ will lie on the galactic plane and will not affect the z-component, h_z, of the comet's orbital angular momentum (Fig. 5.4). Bearing in mind that $h_z = h \times \cos i_G$, where i_G is the inclination of the comet orbit with respect to the galactic mid-plane, that $\cos i_G = \cos\phi \times \sin\alpha$, and that $h^2 \simeq 2\mu q$ for a near-parabolic orbit, we have

$$h_z^2 = h^2 \cos^2 i_G \simeq 2\mu q \cos^2\phi \sin^2\alpha,$$

and if we define now q_z such that $h_z = 2\mu q_z$, we get

$$q_z \simeq q \cos^2\phi \sin^2\alpha, \tag{5.18}$$

From eq. (5.18) we can see that F_{disk} can decrease the perihelion distance to zero only if $\alpha = 0$ (provided that $\phi \neq \pm\pi/2$), i.e. if the orbital plane of the comet is perpendicular to the galactic plane. On the other hand, if $\alpha = 90°$ (or 270°), F_{disk} will not have any effect on q.

We can now compare the tidal force of the galactic disk with that exerted by the galactic bulge F_{bulge}. To this purpose, we can assume that all the mass of the galaxy, $M_G = 1.3 \times 10^{11}$ M$_\odot$, is concentrated in the galactic center. For a comet at a galatic latitude ϕ, galactic longitude λ, and at a distance r from the Sun, the tidal force of the galactic bulge can be approximately given by

$$\vec{F}_{bulge} \simeq \frac{2GM_G}{r_G^3} r \cos\phi \cos\lambda \hat{r}_G, \tag{5.19}$$

where $r_G = 8.5$ kpc, and $\hat{r}_G$ is the unit vector in the direction to the galactic center. Finally, if we consider an average galatic longitude $< \lambda >= 2/\pi$, we obtain

$$\frac{F_{disk}}{F_{bulge}} \simeq \frac{\pi^2 \rho_{disk} \tan\phi}{M_G/r_G^3} \simeq 4.7 \tan\phi, \tag{5.20}$$

Equation (5.20) shows that the tidal force of the galactic disk is indeed on average about five times greater than the tidal force of the galactic bulge. However, for special configurations (e.g. $\phi \simeq 0$, or $\alpha \simeq 90°$ or $270°$), the latter becomes dominant. Since $\vec{F}_{bulge}$ acts in a different direction that $\vec{F}_{disk}$, the former can help to decrease q below the limiting q_z found in eq. (5.18), so it can play a very important complementary role in bringing comets to the Sun's neighborhood.

5.3.1. Direct numerical integration

The simplified picture presented before is enough to illustrate the main dynamical features of Oort cloud comets under the action of the galactic potential. A more accurate treatment would require the direct integration of the equations of motion (e.g. Torbett 1986, Fouchard 2004). For the sake of completeness we will describe it in this section. Let us first define a fixed reference system $(\hat{x}, \hat{y}, \hat{z})$ in which $\hat{x}$ points toward the galactic center, $\hat{z}$ is normal to the galactic plane and points to the North galactic pole, and $\hat{y}$ completes the right-handed system. Let us now define a rotating frame $(\hat{x'}, \hat{y'}, \hat{z'} \equiv \hat{z})$ in such a way that it coincides with the fixed system at $t = 0$, and the x'-axis rotates with the angular velocity of the Sun around the galactic center, $\Omega_\odot$. The Sun is assumed to move in the galactic plane with uniform angular velocity. Therefore, at a certain time t we have

$$x' = x\cos(\Omega_\odot t) + y\sin(\Omega_\odot t),$$

$$y' = -x\sin(\Omega_\odot t) + y\cos(\Omega_\odot t).$$

The force $\vec{F}$ acting on the comet, expressed in the rotating frame, is (e.g. Heisler and Tremaine 1986)

$$\vec{F} = -\frac{\mu}{r^3}\vec{r} + (A - B)(3A + B)x'\hat{x'}$$
$$- (A-B)^2 y'\hat{y'} - [4\pi G\rho_{disk} - 2(B^2 - A^2)]z\hat{z}, \tag{5.21}$$

where A and B are the Oort constants introduced in eq. (5.7) for which we have the values: $A = 14.5$ km s^{-1} kpc^{-1}, and $B = -12$ km s^{-1} kpc^{-1} (e.g. Elmegreen 1998). The angular velocity $\Omega_\odot$ can be expressed in terms of A and B as $\Omega_\odot = B - A = -26.5$ km s^{-1} kpc^{-1}, or bearing in mind that 1 kpc $= 3.1 \times 10^{16}$ km, we get $\Omega_\odot \simeq -8.6 \times 10^{-16}$ s^{-1}.

The equations of motion are

$$\frac{d^2x}{dt^2} = -\frac{\mu}{r^3}x + (A-B)(3A+B)x'\cos(\Omega_\odot t) + (A-B)^2 y'\sin(\Omega_\odot t)$$

$$= -\frac{\mu}{r^3}x + \mathcal{C}_1 x'\cos(\Omega_\odot t) + \mathcal{C}_2 y'\sin(\Omega_\odot t), \tag{5.22a}$$

$$\frac{d^2y}{dt^2} = -\frac{\mu}{r^3}y + (A-B)(3A+B)x'\sin(\Omega_\odot t) - (A-B)^2 y'\cos(\Omega_\odot t)$$

$$= -\frac{\mu}{r^3}y + \mathcal{C}_1 x'\sin(\Omega_\odot t) - \mathcal{C}_2 y'\cos(\Omega_\odot t), \tag{5.22b}$$

$$\frac{d^2z}{dt^2} = -\frac{\mu}{r^3}z - [4\pi G\rho_{disk} - 2(B^2 - A^2)]z, \tag{5.22c}$$

where $\mathcal{C}_1 = (A-B)(3A+B)$ and $\mathcal{C}_2 = (A-B)^2$.

As showed above, we can neglect the action of the radial component and only consider the z-component of the galactic disk by setting $\mathcal{C}_1 = \mathcal{C}_2 = 0$ and $(B^2 - A^2) \approx 0$ in the above equations.

5.3.2. HAMILTONIAN FORMULATION

The use of the Hamiltonian equations averaged over one cometary orbit offer advantages with respect to direct numerical integrations in terms of computer time (Fouchard 2004). As described with the Lagrangian formalism, from the Hamiltonian model we can also get a quick diagnostic of the dyanmics of comets subject to the galactic potential. The Hamiltonian derived from the galactic potential is

$$\mathcal{H} = -\frac{\mu}{2a} - (A-B)(3A+B)\frac{x'^2}{2}$$

$$+ (A-B)^2\frac{y'^2}{2} + [4\pi G\rho_{disk} - 2(B^2 - A^2)]\frac{z^2}{2}. \tag{5.23}$$

Again, if we neglect the radial component, the Hamiltonian can be simplified to

$$\mathcal{H} = -\frac{\mu}{2a} + 2\pi G\rho_{disk} z^2. \tag{5.24}$$

Let us now define the orbital elements of the comet in the galactic frame $(\hat{x'}, \hat{y'}, \hat{z})$: inclination i_G, argument of the perihelion ω_G, and longitude of the ascending node Ω_G. We get the following relation

$$z = r \sin i_G \sin(\omega_G + f) = \frac{a(1-e^2)\sin i_G \sin(\omega_G + f)}{1 + e\cos f}, \tag{5.25}$$

where f is again the true anomaly of the comet. If we substitute eq. (5.25) in eq. (5.24) we obtain

$$\mathcal{H} = -\frac{\mu}{2a} + 2\pi G \rho_{disk} \frac{a^2(1-e^2)^2 \sin^2 i_G \sin^2(\omega_G + f)}{(1 + e\cos f)^2}. \tag{5.26}$$

We can neglect short-period perturbations, that depend on the fast variable f, by averaging the terms containing f (e.g. Heisler and Tremaine 1986)

$$\left\langle \frac{\sin^2(\omega_G + f)}{(1 + e\cos f)^2} \right\rangle = \frac{1 - e^2 + 5e^2 \sin^2 \omega_G}{2(1-e^2)^2}, \tag{5.27}$$

so the average Hamiltonian is

$$\mathcal{H}_{av} = -\frac{\mu}{2a} + \pi G \rho_{disk} a^2 \sin^2 i_G (1 - e^2 + 5e^2 \sin^2 \omega_G). \tag{5.28}$$

We can use Delaunay's set of canonical variables

$$\begin{aligned} &l, && L = \sqrt{\mu a}, \\ &\omega_G, && h = \sqrt{\mu a(1-e^2)} = L\sqrt{1-e^2}, \\ &\Omega_G, && h_z = h \cos i_G, \end{aligned}$$

where l is the mean anomaly of the comet.

The averaged Hamiltonian expressed in this set of variables is

$$\mathcal{H}_{av} = -\frac{\mu^2}{2L^2} + \frac{\pi G \rho_{disk}}{\mu^2} \frac{L^2}{h^2} (h^2 - h_z^2)[h^2 + 5(L^2 - h^2)\sin^2 \omega_G], \tag{5.29}$$

and the Hamiltonian equations of motion are

$$\frac{dL}{dt} = -\frac{\partial \mathcal{H}_{av}}{\partial l} = 0, \tag{5.30a}$$

$$\frac{dh}{dt} = -\frac{\partial \mathcal{H}_{av}}{\partial \omega_G}, \tag{5.30b}$$

$$\frac{dh_z}{dt} = -\frac{\partial \mathcal{H}_{av}}{\partial \Omega_G} = 0, \tag{5.30c}$$

$$\frac{dl}{dt} = \frac{\partial \mathcal{H}_{av}}{\partial L}, \tag{5.30d}$$

$$\frac{d\omega_G}{dt} = \frac{\partial \mathcal{H}_{av}}{\partial h}, \tag{5.30e}$$

$$\frac{d\Omega_G}{dt} = \frac{\partial \mathcal{H}_{av}}{\partial h_z}. \tag{5.30f}$$

We note that $dL/dt = dh_z/dt = 0$ because $\mathcal{H}_{av}$ does not depend on l and Ω_G. Therefore, L is constant, namely, the average change of the comet's semimajor axis over an orbital period is zero, and also the z-component of the orbital angular momentum, h_z, is constant, in agreement with was shown above.

Equation (5.30b) gives the variation of the comet's orbital angular momentum, i.e.

$$\frac{dh}{dt} = -\frac{5\pi G\rho_{disk}}{\mu^2}\frac{L^2}{h^2}(h^2-h_z^2)(L^2-h^2)\sin 2\omega_G. \tag{5.31}$$

Since $h^2 \simeq 2\mu q$ for a near-parabolic comet, we can obtain an equation for the variation of q similar to the one derived in eq. (5.14).

$$\frac{dq}{dt} = -5\sqrt{2}\pi q^{1/2}\mu^{-1/2}G\rho_{disk}a^2\sin^2 i_G \sin 2\omega_G. \tag{5.32}$$

Because of the chosen variables, eq. (5.32) does not depend explicitly on the galactic latitude as eq. (5.14). Yet, we can introduce the galactic latitude ϕ and longitude λ in eq. (5.32) by using the following relations: $\sin\omega_G = \sin\phi/\sin i_G$ and $\cos\omega_G = \cos\lambda\cos\phi$. In doing so, we can easily show that eq. (5.32) is also proportional to $\sin 2\phi$.

5.4. Penetrating encounters with giant molecular clouds

The spiral arms of the Galaxy are rich in gas and dust, a large part of this material is under the form of neutral atomic hydrogen (known as HI regions). The average density of HI regions is one atom cm^{-3} and their typical temperatures are $T \sim 100$ K. These clouds can contain up to 10^7 $M_\odot$. In still colder and denser regions atoms are shielded from the interstellar UV radiation that dissociates molecules, so atoms can combine to form molecules. Molecular clouds are formed in these regions with typical densities 50 H_2 molecules cm^{-3} (Blitz 1993) and temperatures 10-20 K (van Dishoeck et al. 1993), though they exhibit a clumpy structure in which most of their mass concentrates in dense cores with densitites $10^3 - 10^5$ H_2 cm^{-3} that are active regions of star formation (this point will be analyzed in Chapter 10). What is of our immediate interest for this section is that the mass in molecular clouds is concentrated enough to have very appreciable dynamical effects on Oort cloud comets during encounters. This is not the case of interstellar atomic clouds where the mass is too dilute to exert a strong perturbing effect, so these are of no concern for our inventory of external perturbers of the Oort cloud.

The largest molecular structures are called Giant Molecular Clouds (GMCs) with typical masses of the order of $1 - 2 \times 10^5$ $M_\odot$ and mean diameters ~ 45 pc. The mean separation among GMCs at the Sun's distance to the Galactic center is ~ 500 pc (Blitz 1993). Therefore, the frequency of penetrating encounters of the solar system with GMCs is quite low: between 1-10 over the solar system age (Bailey 1983). The importance of interstellar molecular clouds as major perturbers of the Oort cloud was first addressed by Biermann (1978) and Napier and Clube (1979). From numerical simulations Napier and Staniucha (1982) concluded that a primordial comet cloud would have been lost due to GMC perturbations, though their model considered rather low encounter velocities (5 to 10 km s^{-1}), thus favoring their disruptive influence.

A penetrating encounter of the Sun with a GMC, assumed to be spherical and of uniform density, radius R_{GMC} and mass M_{GMC}, will impart and impulsive change in the velocity of the comet at a distance r to the Sun of (Biermann 1978)

$$\Delta v_{GMC} = \frac{2GM_{GMC}}{v_{GMC}} \frac{r}{b^2} \left[1 - \left(1 - \frac{b^2}{R_{GMC}^2} \right)^{3/2} \right] \sin \zeta, \tag{5.33}$$

where $v_{GMC} \sim 20$ km s^{-1} is the typical encounter velocity with molecular clouds, $b(\leq R_{GMC})$ is the impact parameter, and ζ the angle between $\vec{r}$ and $\vec{v}_{GMC}$.

Hut and Tremaine (1985) argued that the energy change obtained from eq. (5.33) could be overestimated when the encounter time is longer than the orbital period P of the comet, namely when $b/v_{GMC}P \gtrsim 1$. This applies to penetrating encounters with GMCs of radius $R_{GMC} \sim 20$ pc (we further assume that $b \sim R_{GMC}$). Yet, Hut and Tremaine suggest that the clumpy structure of the material within a GMC may counteract somewhat the previous effect, so eq. (5.33) can still give results of the correct order of magnitude. Actually, during a penetrating encounter with a GMC the strongest dynamical effect may be expected to occur during close approaches to one of the massive clumps (Stern 1990).

5.5. Randomization of orbital inclinations

The orbital planes of comets reaching the Oort cloud will be randomized by external perturbers, even if they were originally strongly concentrated toward the ecliptic plane. We will estimate the time scale for fully randomization of a population of comets with original small inclinations. Of course, this time scale depends on the comet's semimajor axis a; the greater a, the greater the effect caused by external perturbers and, accordingly, the smaller the time scale for full randomization. If we decompose the perturbing force in the radial component R, transverse S, and normal W, it is clear that R and S will not affect the inclination since both lie in the orbital plane, so the change di/dt will depend only on W. The change in the inclination di during the interval dt due to the action of W, acting on a comet at a heliocentric distance r, can be obtained from the planetary equation under the Gaussian form

$$\frac{di}{dt} = \frac{Wr\cos\theta}{na^2\sqrt{1-e^2}} \tag{5.34}$$

where $\theta = \varpi - \Omega + f$, ϖ is the longitude of perihelion, Ω is the longitude of the ascending node, and f is the true anomaly, n is the mean motion.

Let us now consider a change Δi during Δt. The change in the normal component of the velocity vector will be given by $\Delta v_N = W \times \Delta t$. Furthermore, if the velocity's change Δv is randomly oriented, then the normal component is in statistical terms $\Delta v_N \approx \Delta v/\sqrt{3}$. Let us adopt as before an average value $\bar{r} = 1.5a$, and assume θ to have a random value in the interval $(0, 2\pi)$ so that $< \cos\theta >= 2/\pi$. We also have $n = \mu^{-1/2}a^{-3/2}$ and $1 - e^2 \approx 2q/a$ for a near-parabolic comet. By introducing these values in eq. (5.34) we get

$$\Delta i \sim \frac{3}{\sqrt{6}\pi}\frac{a\Delta v}{\mu^{1/2}q^{1/2}} \tag{5.35}$$

Let us assume that Oort cloud comets are only perturbed by passing stars, then Δv is given by eq. (5.5) and substituting this expression in eq. (5.35) we obtain the graphs of Fig. 5.5. These graphs show the change Δi in the comet's orbit for the considered time (we took the solar system age $\Delta t = 4.6 \times 10^9$ yr) as a function of the semimajor axis and for different perihelion distances. We note that in eq. (5.35) we are implicitly assuming that the orbital elements a and q remain constant during the integration, which of course is not true, though our

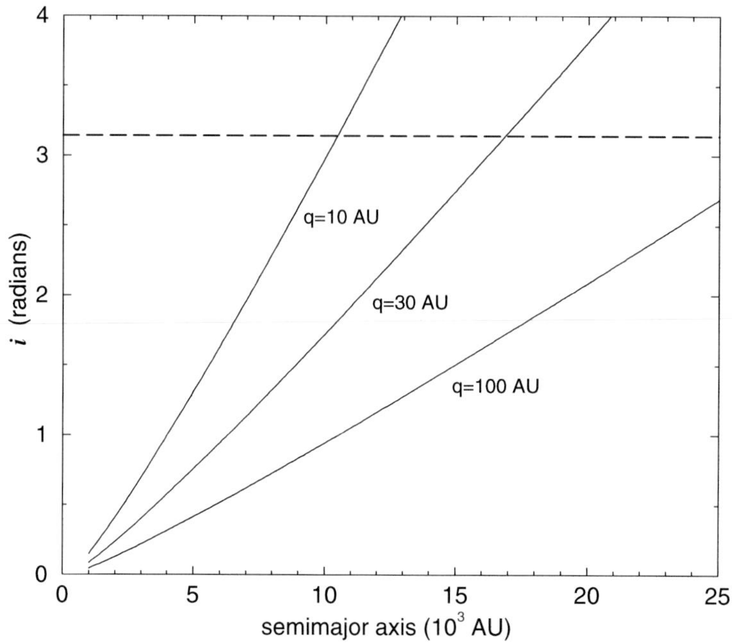

Figure 5.5. Average change in the orbital inclination of comets caused by stellar perturbations over the age of the solar system, and for three different perihelion distances. The dashed horizontal line for $\Delta i = \pi$ indicates the threshold for full randomization.

simplification should be precise enough to obtain results of the correct order of magnitude and, indeed, similar results were obtained by the author from a different approach (Fernández 1985a).

In addition to nearby stars, the inclinations are also perturbed by giant molecular clouds and galactic tidal forces. For the former we have a large uncertainty since the number of penetrating encounters with GMCs during the solar system age is itself uncertain. If we adopt a canonical number of five such encounters (e.g. Bailey 1983), we obtain a perturbation for Δi comparable with that produced by passing stars.

If we combine the effects of passing stars and GMCs, we find that Oort cloud comets with semimajor axes $a \gtrsim 10^4$ AU and perihelia $q = 30$ AU should have their orbital planes randomized after a residence time comparable with the solar system age. In this estimate we used as the criterion for full randomization that $\Delta i = \pi$ (Fig. 5.5). Oort cloud comets with somewhat greater q (say, $q \sim 100$ AU) may have avoided complete randomization, at least for semimajor axes up to $a \sim 2 \times 10^4$ AU. On the other hand, our study indicates that more tightly bound comets with $a \lesssim 10^4$ AU, scattered from the planetary region, should still keep "memory" of their primordial flattened, disk-shaped structure. Similarly, Oort cloud comets with $a \gtrsim 10^4$ AU, but with residence times in the cloud significantly shorter than the solar system age, should also keep a certain concentration toward the invariable plane of the solar system (or we can approximately take the ecliptic plane that form a small angle of about one and a half degrees with the invariable plane).

We can compare our simple analytical approach, as given by eq. (5.35), with the results obtained by Duncan et al. (1987) from numerical simulations in which the evolution of test comets under the action of the four giant planets, the tidal force of the galactic disk and random passing stars, was followed for the age of the solar system. The mean of the cosine of the inclination of the surviving comets approaches zero for semimajor axes $a \gtrsim 3 \times 10^3$ AU at 4.5×10^9 yr (Fig. 5.6), which indicates that the orbital planes of such comets have been randomized.

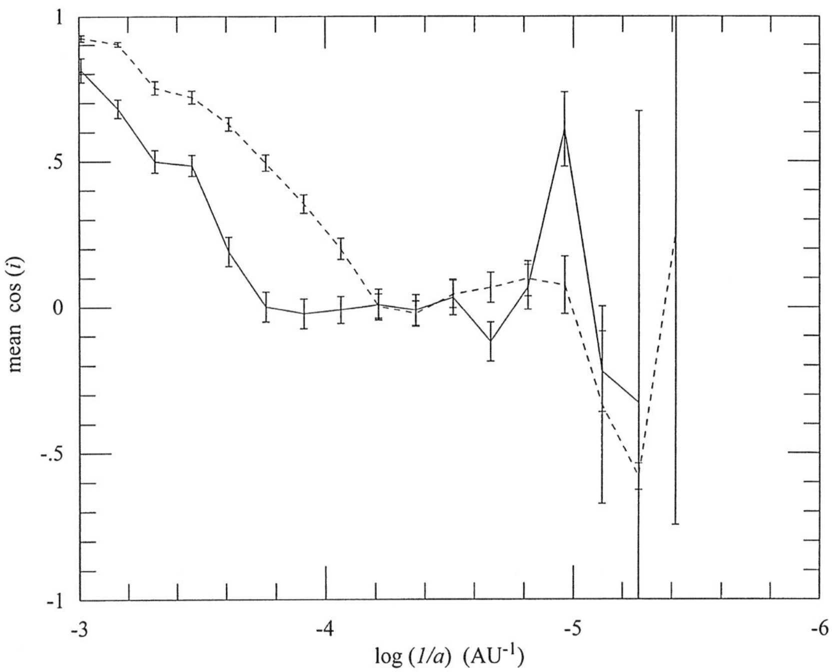

Figure 5.6. The mean of the cosine of the inclination of fictitious comets as a function of the orbital energy (expressed as the reciprocal of the semimajor axis). The dashed curve is for 1.0×10^9 yr, while the solid curve is for 4.5×10^9 yr (Duncan et al. 1987).

This is easily shown bearing in mind that the mean $\cos i$ is given by

$$\overline{\cos i} = \int_0^\pi \cos i f_i(i) di. \tag{5.36}$$

If there is a predominance of low-i orbits, $\cos i \to 1$, which is what is observed in Fig. 5.6 for comets with larger binding energies (smaller $1/a$). Full randomization implies that $f_i(i)di = 1/2 \sin i di$, which substituted in eq. (5.36) leads to $\overline{\cos i} = 0$.

Duncan et al's lower limit for the semimajor axes of fully randomized comets is somewhat smaller that the one shown in Fig. 5.5. Besides the different approaches to the problem, another reason that may help to explain the discrepancy with the results derived from eq. (5.35) is that Duncan et al. adopted a value $\rho_{disk} = 0.185$ $M_\odot$ pc^{-3} for the density of the galactic disk, which is almost twice the one most accepted at present (cf. Section 5.3).

Levison et al. (2005) have shown that the orbit's inclination can also change due to the precession of the galactic longitude of the ascending node, Ω_G, forced by tides of the galactic disk. For a body of semimajor axis a, eccentricity e, inclination and argument of perihelion with respect to the galactic plane i_G and ω_G respectively, the rate of precession of Ω_G can be derived from the Lagrange's planetary equation under the Gaussian form

$$\dot{\Omega}_G = \frac{W_G r \sin u}{na^2\sqrt{1-e^2}\sin i_G}, \tag{5.37}$$

where $u = \omega_G + f$, f being the true anomaly of the comet, and W_G is the normal component of the perturbing force which is given by

$$W_G = F_{disk} \times \cos i_G = 4\pi G \rho_{disk} r \sin\phi \cos i_G,$$

where F_{disk} is given by eq. (5.9).

By introducing this expression in eq. (5.37), bearing in mind that $n = \sqrt{GM_\odot} a^{-3/2}$, $\sin\phi = \sin i_G \sin\omega_G$, and that for a body in a near parabolic orbit we have $e \approx 1$, the time-average heliocentric distance $\bar{r} \approx 1.5a$, $\sqrt{1-e^2} \approx \sqrt{2q/a}$, and that f is generally close to π, so $\sin u \approx -\sin\omega_G$, we finally get

$$\dot{\Omega}_G \simeq -\frac{5\sqrt{2}\pi G\rho_{disk}}{\sqrt{GM_\odot q}} a^2 \cos i_G \sin^2\omega_G, \tag{5.38}$$

By defining $\tau_\Omega \equiv |\pi/\dot{\Omega}_G|$ as the time that takes Ω_G to precess an angle π, Levison et al. (2004) find

$$\tau_\Omega \simeq 4.3\times10^8 \text{ yr} \left(\frac{0.1M_\odot \text{ pc}^{-3}}{\rho_{disk}}\right) \left(\frac{q}{30 \text{ AU}}\right)^{1/2} \left(\frac{10^4 \text{ AU}}{a}\right)^2, \tag{5.39}$$

which shows that for $a \gtrsim 3000$ AU, the angle Ω_G of bodies with perihelia at the edge of the planetary region will precess 180° over time scales shorter than the solar system lifetime (4.6 Gyr).

The ecliptic inclination i will change as Ω_G precesses, according to

$$\cos i = \cos\epsilon_G \cos i_G + \sin\epsilon_G \sin i_G \cos(\Omega_G - \Omega_o), \tag{5.40}$$

where ϵ_G is the galactic oblicuity, namely the angle between the ecliptic and the galactic plane, and Ω_o is the longitude of the ascending node at the intersection of the ecliptic with the galactic plane. We can see in eq. (5.40) that large changes in the ecliptic inclination will occur only if the ecliptic is highly inclined with respect to the galactic plane. This is actually the case since $\epsilon_G \approx 120°$. Should the galactic oblicuity be low ($\epsilon_G \approx 0$ or π), the effect on i would be negligible (we can see that if $\epsilon_G \approx 0$ or π, the second term of the right-hand side of eq. (5.40) vanishes).

If the ecliptic inclination i is initially very low, as would be the case if comets formed in the protoplanetary disk, then $i \approx 0$, the comet's galactic inclination $i_G \approx \epsilon_G$ and $\Omega_G \approx \Omega_o$. The ecliptic inclination of comets scattered to the Oort cloud will raise to $i \approx 120°$ when $\Omega_G = \Omega_o + \pi$. As shown above, Ω_G will precess 180° on time scales shorter than the solar system age if $a \gtrsim 3000$ AU.

Therefore, the consideration of the forced precession of Ω_G by tides of the galactic disk lowers somewhat the inner boundary for full randomization of Oort cloud orbits, as compared to that found before that only considered stellar and GMC perturbations. Yet, we note that the precession of Ω_G takes the comet's orbit from $i \approx 0$ to a ceiling $\approx 120°$, so other effects are needed to drive orbits to $i \approx 180°$. The change in i_G caused by the tidal force of the galactic disk itself, and the other external perturbers can contribute to the full randomization of Oort cloud comet orbits with semimajor axes greater than a few thousands AU.

5.6. Distribution of eccentricities of a thermalized population

Comets in the Oort cloud subject to the action of external perturbers will experience an energy exchange leading to its thermalization. Even if comets initially had near-parabolic orbits with perihelia inside the planetary region (where they presumably formed), the action of external perturbers will gradually randomized their orbits. As a consequence, their perihelia will drift outside the planetary region and, at the same time, the eccentricities will diffuse from the edge $e \sim 1$ through all the range $(0, 1)$. Figure 5.7 illustrates the evolution of one of the fictitious comets of Duncan et al. (1987) in the parametric plane $(1/a, q)$. The comet starts in a tightly bound orbit with perihelion inside the planetary region, so at the beginning it is only subject to planetary perturbations. As discussed in the previous chapter, at this early stage the evolution can be described as a random-walk in the energy space. However, when the comet reaches Oort cloud distances, external perturbers remove its perihelion from the planetary region, so its later

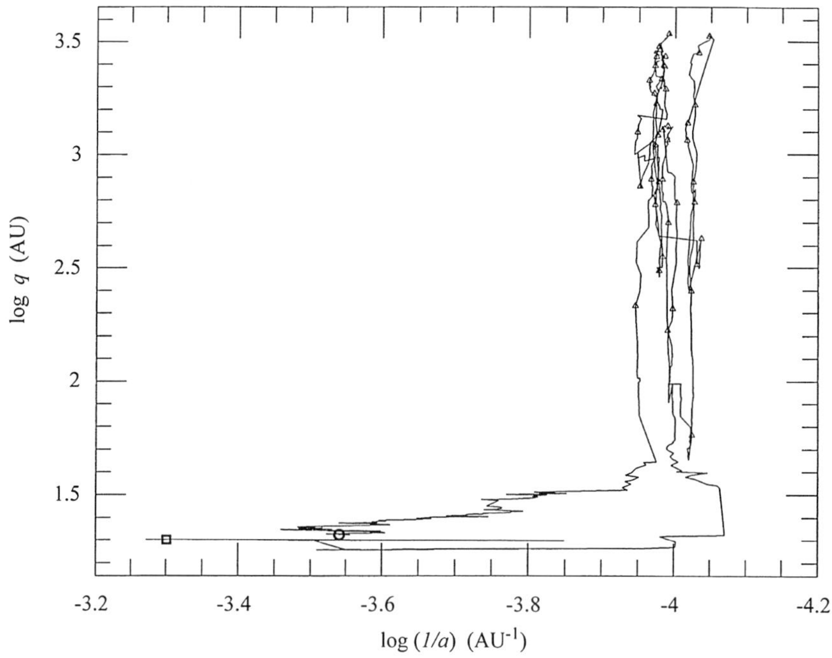

Figure 5.7. The evolution of a fictitious comet that starts with $q = 20$ AU and $a = 2000$ AU (square) and ends after 4.5×10^9 yr re-injected in the planetary region (circle). A triangle is plotted every 10^8 yr (Duncan et al. 1987).

evolution is mainly in q and i while keeping its orbital energy essentially constant. After long excursions of its perihelion distance that rises up to $\sim 3.2 \times 10^3$ AU, the comet is drawn back to the planetary region as a fresh "new" comet. As the perihelion distance of the comet increases, its eccentricity decreases from an initial $e \sim 1$ to values as low as $e \sim 0.65$.

We can envisage the Oort cloud population as a swarm of gaseous particles and apply the kinetic theory of gases to study the distribution of energy states. We will follow the treatment developed by Jeans (1919) for the case of binary stars subject to the perturbations of external stars. In our case, instead of two stars gravitationally bound, we consider the case of the Sun and a test comet of unit mass at a distance r to the Sun. Let us consider the components of the comet's velocity $\vec{v}$, such that v_x is along the radius vector $\vec{r}$, v_y is the transverse component in the orbital plane, and v_z is the normal. If under the continuous perturbations of passing stars Oort cloud comets reach a state of equipartition of energy and are at the same time subject to the Sun's gravitational potential $-\mu/r$, the distribution of v_x, v_y, v_z, and r can then be described by the Maxwell-Boltzmann distribution, namely

$$A \exp\left[-\frac{3}{2\bar{v^2}}\left(v_x^2 + v_y^2 + v_z^2 - \frac{2\mu}{r}\right)\right] dv_x dv_y dv_z 4\pi r^2 dr \tag{5.41}$$

where $\bar{v^2}$ is the mean-square velocity of the population of Oort cloud comets, assumed to be all of unit mass, and A is a constant. Furthermore, we have: $v_x = \dot{r}$ and $v_y^2 + v_z^2 = v_T^2$, where v_T is the tangential velocity. Let ϕ be the angle between the direction of v_T and the axis $\vec{y}$, then we have $v_y = v_T \cos\phi$ and $v_z = v_T \sin\phi$. In a thermalized state, all the directions of the velocity vector are equally probable, and for the same reason all the angles ϕ, so we can substitute the new variables v_T and ϕ by the transformation equation

$$dv_y dv_z = \frac{\partial(v_y, v_z)}{\partial(v_T, \phi)} dv_T d\phi = v_T dv_T d\phi. \tag{5.42}$$

Substituting eq. (5.42) into eq. (5.41) and integrating ϕ in the range $0 < \phi < 2\pi$ we obtain

$$8\pi^2 A \exp\left[-\frac{3}{2\bar{v^2}}\left(\dot{r}^2 + v_T^2 - \frac{2\mu}{r}\right)\right] d\dot{r} v_T dv_T r^2 dr. \tag{5.43}$$

Let us now introduce the new variables

$$E = \frac{1}{2}(\dot{r}^2 + v_T^2) - \frac{\mu}{r}, \qquad K = \frac{3}{2\bar{v^2}},$$

where E is the total orbital energy per unit mass. Substituting these expressions in eq. (5.43) we obtain

$$8\pi^2 A \exp(-2KE) d\dot{r} dv_T dv_T r^2 dr \tag{5.44}$$

Let us change now to the new variables

$$k = \frac{1}{rv_T}, \qquad \sigma = v_T - \frac{\mu}{rv_T},$$

and next transform the variables $\dot{r}$, v_T, and r to E, k, and σ. The Jacobian modulus of transformation is

$$\frac{\partial(E,k,\sigma)}{\partial(\dot{r},v_T,r)} = \frac{\dot{r}}{r^2 v_T},$$

from where we obtain

$$d\dot{r} dv_T dr = \frac{r^2 v_T}{\dot{r}} dE dk d\sigma. \tag{5.45}$$

We can easily derive the following relations among the different variables

$$v_T = \sigma + \mu k, \qquad r = \frac{1}{k(\sigma + \mu k)},$$

from which we obtain

$$\dot{r} = \left(2E - \sigma^2 + \mu^2 k^2\right)^{1/2}. \tag{5.46}$$

Substituting these variables in eq. (5.44) we obtain

$$\frac{16\pi^2 A}{k^4(\sigma + \mu k)^2(2E - \sigma^2 + \mu^2 k^2)^{1/2}} \exp(-2KE) dE dk d\sigma. \tag{5.47}$$

Since we want to apply the distribution law only to bound systems, the comet energy must fulfil the condition: $E < 0$ ($E = 0$ corresponds to a parabolic orbit and $E > 0$ to hyperbolic orbits). Furthermore, from eq. (5.46) we derive the following boundary condition: $2E - \sigma^2 + \mu^2 k^2 \geq 0$ and hence

$$-(2E + \mu^2 k^2)^{1/2} \leq \sigma \leq +(2E + \mu^2 k^2)^{1/2}. \tag{5.48}$$

Since v_T is always positive, we have the additional constraint: $v_T = \sigma + \mu k \geq 0$

Let us now introduce the variables

$$p^2 = \mu^2 k^2 + 2E, \qquad \sigma = p\cos\theta,$$

and substituting these expressions in eq. (5.46) we obtain

$$\frac{16\pi^2 A}{k^4(\sigma + \mu k)^2 p\sin\theta} \exp(-2KE) p\sin\theta d\theta dE dk. \tag{5.49}$$

We must first analyze the limits of θ. From the condition given by eq. (5.48) we obtain $-p \leq \sigma \leq +p$, whereby $-1 \leq \cos\theta \leq +1$, so θ can vary within the range $(0, \pi)$. Effecting the integration with respect to θ, i.e.

$$16\pi^2 A \exp(-2KE) dE k^{-4} dk \int_0^{\pi} \frac{d\theta}{(p\cos\theta + \mu k)^2}$$

we get the following expression

$$16\pi^3 A \frac{1}{\mu^2 k^2 \left(1 - \frac{p^2}{\mu^2 k^2}\right)^{3/2}} \exp(-2KE) dE k^{-4} dk. \tag{5.50}$$

As a final step, let us introduce the known orbital parameters: the angular momentum per unit mass $h = rv_T = 1/k$, and the eccentricity e related to h by means of $h^2 = \mu a(1 - e^2)$ and hence $e^2 = 1 - h^2/\mu a$. Bearing in mind that the orbital energy $E = -\mu/2a$, we get

$$e = \left(1 + \frac{2h^2 E}{\mu^2}\right)^{1/2} = \frac{p}{\mu k}.$$

Substituting k by e in eq. (5.50) we get the following distribution of the energies E and eccentricities e of the thermalized population of comets

$$16\pi^3 A\mu^3 \exp(-2KE)(-2E)^{-5/2} dE e de. \tag{5.51}$$

The form of this law shows that in the final steady state there is no correlation between the values of E and e. Whatever the values of E, the law of distribution of eccentricities follows the simple expression

$$f_e(e) de = 2e de, \tag{5.52}$$

which means that all values of e^2 are equally probable, so its mean $< e^2 >= 0\,5$. The example of Fig. 5.8, again drawn from Duncan et al.

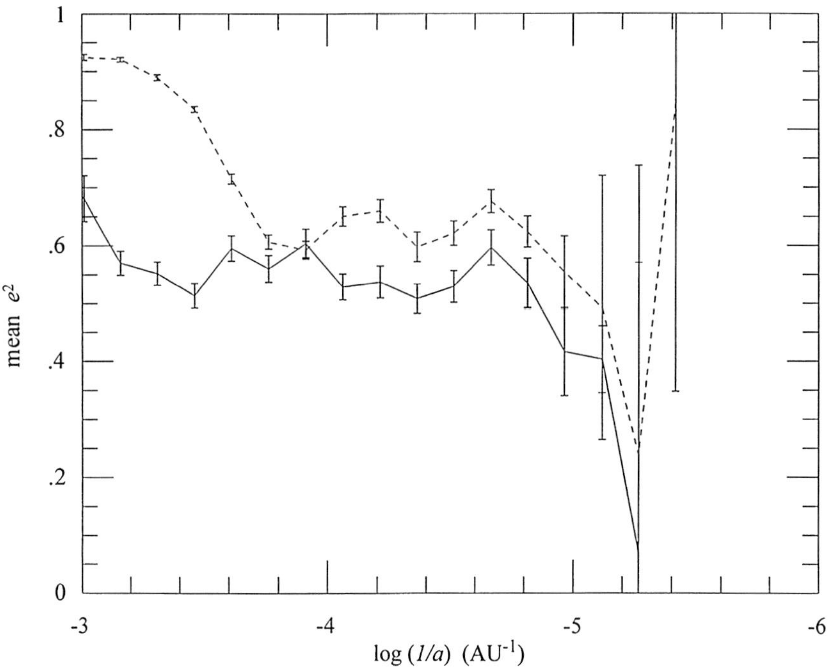

Figure 5.8. The mean-square eccentricity of fictitious comets as a function of the orbital energy (expressed as the reciprocal of the semimajor axis). The dashed curve is for 1.0×10^9 yr, while the solid curve is for 4.5×10^9 yr (Duncan et al. 1987).

(1987), shows very nicely how comets that reach the Oort cloud are thermalized with increasing time, process that is described by the mean $< e^2 >$ that tends to 0.5 for $a \gtrsim 3 \times 10^3$ AU at 4.5×10^9 yr.

From eq. (5.52) we can obtain the fraction of Oort cloud comets with eccentricities between e and unity as

$$F_e = 1 - e^2. \tag{5.53}$$

Bearing in mind that $e = 1 - q/a$, from eq. (5.53) we can obtain the fraction of comets with perihelion distances smaller than q

$$F_q = \frac{2q}{a}\left(1 - \frac{q}{2a}\right). \tag{5.54}$$

If $q << a$ we approximately have: $F_q \sim 2q/a$.

5.7. The outer boundary of the Oort cloud

As we showed above, tides from the galactic bulge set a very broad boundary of $\sim 200,000$ AU for the sphere of influence of the Sun. Tides from the galactic disk have a large influence on the evolution of the comet's angular momentum (and thus on q), but very little on the comet's energy. Therefore, passing stars and penetrating encounters with GMCs will set the outer boundary of the Oort cloud, R_{oort}. The requirement is that the energy gained by an Oort cloud comet during the solar system age reaches the escape energy at the distance R_{oort}, namely

$$\Delta v_{rms} = v_{esc} = \left(\frac{2GM_{\odot}}{R_{oort}}\right)^{1/2}. \tag{5.55}$$

Δv_{rms} applied to stellar pertubations can be obtained from eq. (5.5) integrated during $T = 4.6 \times 10^9$ yr, or by means of eq. (5.33) for GMCs, considering the r.m.s. of the change for a number $N(\sim 5)$ encounters.

For stellar perturbations we find an $a_{esc} \sim R_{oort}/1.5 \simeq 1.3 \times 10^5$ AU, so comets with initial semimajor axes $a > a_{esc}$ would have by now been lost to interstellar space. This is in fairly good agreement with results from other authors. For instance, Weissman (1980) obtained a r.m.s. velocity perturbation $\Delta v_{rms} = 1.7 \times 10^{-3} T^{1/2}$ m s^{-1}, where T is the time in years. If we introduce $T = 4.6 \times 10^9$ yr, we obtain $\Delta v_{rms} \simeq 115$ m s^{-1}, which corresponds to the escape velocity at $\sim 1.35 \times 10^5$ AU. Yet, the limit imposed by penetrating encounters with GMCs is much more stringent: it constraints the stability boundary of the Oort cloud to about one fourth the boundary imposed by stellar perturbations, i.e. to about 3×10^4 AU (Fernández and Ip 1991). This value can be compared with the observed maximum separations between members of wide binary stars in the Galaxy, which is coincidentally of the order of a few times 10^4 AU (e.g. Latham et al. 1991). Hut and Tremaine (1985) found that the combined effect of stars and GMCs set the semimajor axis for stability at $\sim 10^4$ AU over the solar system age. Therefore, comets in the outer region of the Oort cloud should have stayed there for time scales significantly shorter than the solar system age. Figure 5.9 is a blow-up of the $(1/a_{orig})$-distribution shown in Fig. 5.1, but restricted to the sample of "new" comets with $(1/a)_{orig} < 100 \times 10^{-6}$ AU^{-1} and those with original hyperbolic orbits with $(1/a)_{orig} > -100 \times 10^{-6}$ AU^{-1}. We find a concentration in the range $2 \times 10^{-5} < 1/a_{orig} < 5 \times 10^{-5}$ AU^{-1}, which corresponds to semimajor axes $2 - 5 \times 10^4$ AU.

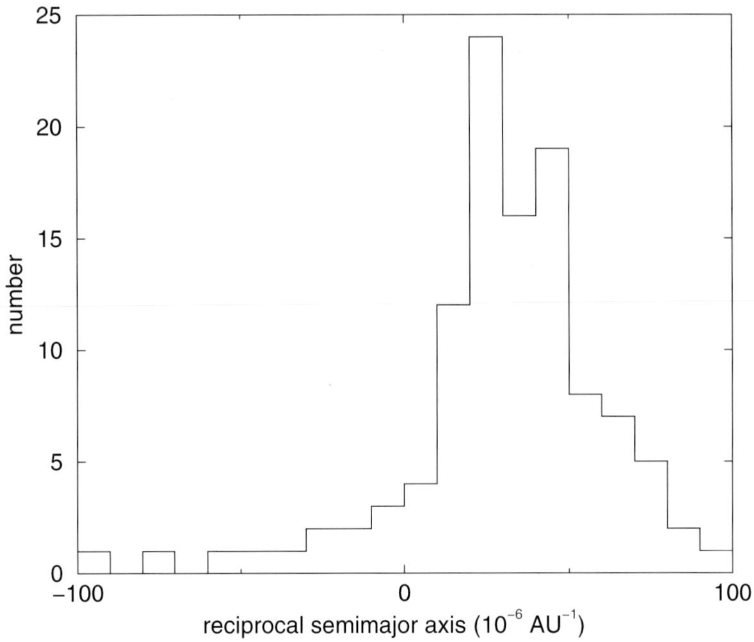

Figure 5.9. Distribution of $1/a_{orig}$ of "new" comets. Only comets of quality classes 1A and 1B were considered from the Marsden and Williams's (2003) catalogue.

This range is in good agreement with the values quoted before for wide binary stars and the stability radius.

In summary, the fact that we define a stability boundary does not mean that the Oort cloud is empty beyond it, since the outer portions will be continuously replenished with comets from the "stable" inner portions of the Oort cloud and from other sources in the planetary region still active, such as the trans-neptunian belt (see Section 8.11). These comets will slowly gain energy under the action of external perturbers and diffuse outwards in a process similar to the thermal escape of gaseous molecules from the outer layers of a planetary atmosphere, which is kept in a quasi steady state by the continuous supply of molecules from the lower atmosphere. In this regard we note that nearly 40% of the original semimajor axes of new comets have $a > 3 \times 10^4$ AU (see Fig. 5.9), indicating that they come from the outer Oort cloud. Nevertheless we should expect a significant drop in the number of comets in the outer portions of the Oort cloud as they have ever decreasing dynamical lifetimes.

THE FLUX OF NEW COMETS: QUIESCENT AND EXCITED STAGES

As we showed in the previous chapter, external perturbers will slowly change the perihelion distances of Oort cloud comets in such a way that some perihelia will drift into the planetary region. If planetary perturbations were neglected, the diffusion of perihelion distances of small-q Oort cloud comets by external perturbers would lead to a uniform q-distribution. This can be seen bearing in mind that the fraction of thermalized comets with perihelia smaller than q is $F_q \sim 2q/a$ for $q << a$ (cf. eq. (5.42)), so the number of comets with perihelion distances in the range $(q, q + dq)$ is

$$f_q(q)dq = \left(\frac{dF_q}{dq}\right) dq = \frac{2}{a}dq, \tag{6.1}$$

which is independent of q.

However, the action of planetary perturbations will change the uniform q-distribution because the Jovian planets - mainly Jupiter and Saturn - will eject the Oort cloud comets coming to their vicinity before their perihelia can drift to the inner planetary region. This effect will cause a severe depletion of Oort cloud comets whose perihelia fall in the region interior to Jupiter's orbit. We will analyze in this chapter how the Jovian planets, and very especially Jupiter and Saturn, operate as a barrier against the injection of Oort cloud comets in the inner planetary region, and the temporal variation in their flux following changes in the field of external perturbers, which has the most dramatic expression in the production of comet showers.

6.1. The drift of the perihelion distance under external perturbers

Comets stored in the Oort cloud have their perihelia outside the planetary region safe from the action of planetary perturbations. However, this does not mean that they do not evolve dynamically, since they are continuously subject to galactic tidal forces and perturbations from nearby stars. The action of external perturbers can decrease some perihelia to values around or below the radius of Neptune's orbit. In such a case, planetary perturbations can become significant again, with the consequence of the random walk of the comet's orbital energy that can either raise the comet's semimajor axis to the outer portions of the Oort cloud, eject the comet or remove it from the Oort cloud altogether to a more strongly bound elliptic orbit. From eq. (5.16), the change in q per orbital revolution of an Oort cloud comet of semimajor axis a due to the tidal force of the galactic disk is

$$\frac{(q_f - q_i)}{q_i} = 2Aq_i^{-1/2} + A^2 q_i^{-1}, \qquad (6.2)$$

where q_i and q_f are the initial and final values of q during an orbital revolution, and

$$A = 4.5\sqrt{2}\pi^2 M_\odot^{-1} \rho_{disk} a^{7/2} \cos\alpha \sin 2\phi.$$

As noted, the effect depends on the galactic latitude ϕ of the comet's aphelion direction.

Equation (6.2) can be applied as such only if the orbital plane of the comet is perpendicular to the galactic plane (i.e. if $\alpha = 0$). As discussed in Section 5.3, when $\alpha \neq 0$, the force F_{disk} will only change the component of the orbital angular momentum h_o lying in the galactic plane. Therefore, if we write $\vec{h} = \vec{h}_o + \vec{h}_z$, and introduce q_o, q_z such that $h_o = 2\mu q_o$ and $h_z = 2\mu q_z$, we get $q = q_o + q_z$, where $q_o = q\cos^2\alpha$ and $q_z = q\sin^2\alpha$, so the change in the perihelion distance given by eq. (6.2) will only affect q_o, leaving q_z unchanged. Obviously if $\alpha = 0$, $q_z = 0$.

In a similar manner, we can compute the cumulative change in the transverse component of the comet's velocity during P, as due to stellar perturbations. It is given by

$$(\Delta v_{*,P})_T^2 \simeq \Delta v_{*,P}^2 < \cos^2\psi >= \frac{2}{3}\Delta v_{*,P}^2 \qquad (6.3)$$

where $\Delta v_{*,P}^2$ is given by eq. (5.5), and ψ is the angle between $\Delta\vec{v}_{*,P}$ and its transverse component. The change in the transverse component Δv_T is related to the change in the perihelion distance Δq through the equation

$$\frac{\Delta q}{q} = 2\frac{\Delta v_T}{v_T} + \frac{\Delta v_T^2}{v_T^2} \qquad (6.4)$$

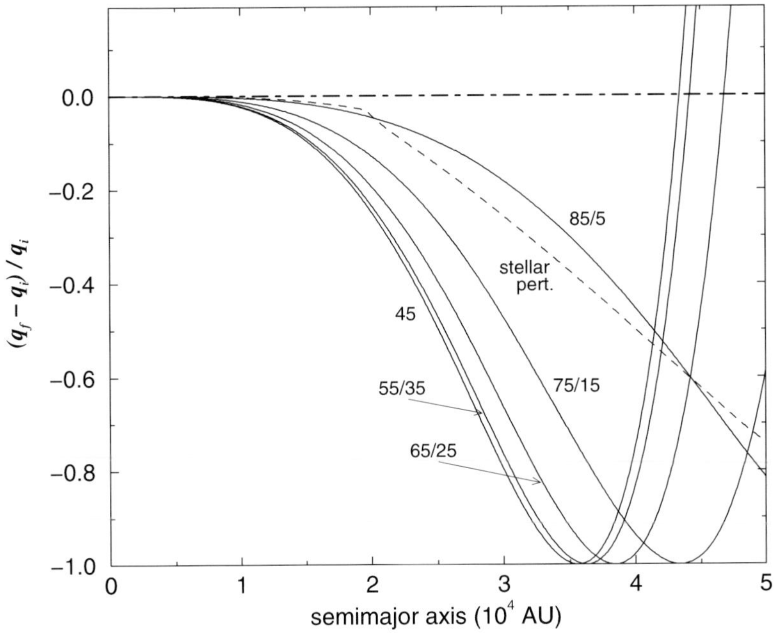

Figure 6.1. Relative change in the perihelion distance of Oort cloud comets per orbital revolution, as due to the tidal force of the galactic disk and stellar perturbations, as a function of their semimajor axis. The initial perihelion distance of the comets is $q_i = 15$ AU. We consider the tidal force of the galactic disk for five galactic latitudes of the comets' aphelion points: 45°, 35°/55°, 25°/65°, 15°/75°, 5°/85°.

In Fig. 6.1 we show the change in the perihelion distance $(q_f - q_i)/q_i$, as produced by stellar perturbations and by the tidal force of the galactic disk, as a function of the semimajor axis and for different galactic latitudes of the aphelion points. The plots confirm the dominant influence of the galactic disk on the injection of Oort cloud comets in the planetary region, except for those whose aphelion points lie near the galactic equator or the galactic poles ($\phi \lesssim 5°$, or $\phi \gtrsim 85°$ North or South). It is also shown that the change in the perihelion distance by external perturbers (at least under the current galactic environment) is negligible for $a \lesssim 10^4$ AU. For the computations of eq. (6.2) we adopted an average $< \cos\alpha >= 2/\pi$. Strictly speaking, the decrease in q plotted in Fig. 6.1 only affects the component q_o, so it cannot decrease below $q_z \simeq q < \sin^2\alpha >= 0.5q_i$. Yet, stellar perturbations and the tidal force of the galactic bulge can further decrease q to values close to zero. We can then say confidently that comets whose semimajor axes fulfill the condition $(q_f - q_i)/q_i \sim -1$ (cf. Fig. 6.1) are able to pass from the trans-saturnian region straight to the Earth's neighborhood after one revolution.

6.2. The Jupiter-Saturn barrier

In the outer planetary region planetary perturbations are very weak, so new comets can "bounce" back and forth between this region and the Oort cloud several times before being definitely removed, either by planetary perturbations or by external perturbers. Yet, most comets crossing or approaching Jupiter's orbit will be removed from the Oort cloud after a single passage, owing to the powerful gravitational field of Jupiter and, to a lesser degree, Saturn. We can see this in terms of the typical energy change per orbital revolution ϵ_t (see Fig. 4.1): when ϵ_t exceeds the binding energy of the Oort cloud comet, it will very likely be removed from the Oort cloud after a single revolution. Jupiter and Saturn therefore appear as very efficient dynamical barriers of Oort cloud comets to prevent them, but a tiny fraction, from reaching the inner planetary region through a smooth drift in their perihelion distances. This is what is known as the *Jupiter-Saturn barrier*. For Oort cloud comets approaching Saturn (say $q \sim 15$ AU), a smooth drift in q means that $|q_f - q_i|/q_i << 1$ (change per orbital revolution), which is fulfilled for semimajor axes $a \lesssim 2.5 \times 10^4$ AU (Fig. 6.1). The practical effect is that very few Oort cloud comets with $a \lesssim 2.5 \times 10^4$ AU can manage to cross the Jupiter-Saturn barrier as compared to those passing by the outer planetary region. Numerical experiments that considered samples of test comets under the action of planetary, stellar perturbations, or Galactic tidal forces confirm the previous conclusion (see, e.g., Fernández 1982; Weissman 1985; Yabushita and Tsujii 1991). The importance of the "Jupiter barrier" was already highlighted by Wetherill (1994) who conjectured about what could happen in planetary systems where massive Jupiters failed to form. From Monte Carlo simulations of the evolution of test bodies in such systems, Wetherill found that the flux of comets in the region of the terrestrial planets would be ~ 1000 times greater, thus providing a collisional environment extremely dangerous for the survival of life. Therefore, the Jupiter-Saturn barrier has play a fundamental role since it provided a safe heaven in the Sun's neighborhood where life could fluorish.

To reach the Earth's neighborhood (thereby becoming potentially observable), comets drifting in q must therefore overcome the Jupiter-Saturn barrier. Let us consider an Oort cloud comet whose perihelion has drifted right up to the Jupiter-Saturn barrier ($q \sim 15$ AU). It should be noted that ϵ_t, and thus the boundary of the barrier, also depends somewhat on the comet's inclination, but this is nevertheless not very important for the matter discussed here. To reach a perihelion distance small enough that the comet may become "observable" in the next orbital revolution of period P, say $q < q_{obs}$ (where q_{obs} is about a few AU), the comet must overshoot the Jupiter-Saturn barrier. As shown in the previous section, the requirement for such an overshooting is that $(q_i - q_f)/q_i \sim 1$ after P. An inspection of Fig. 6.1 shows that this is fulfilled for $a \sim 3.5 \times 10^4$ AU, considering both the tidal force of the galactic disk and the smaller contributions from stellar perturbations and the tidal force of the galactic bulge. We note that this holds for mid-galactic latitudes ($\phi \sim 20 - 70$ degrees) where the tidal force of the galactic disk has the strongest effect. It becomes apparent that the transition from the "smooth-drift" mode of perihelion distances to the "overshooting" mode is very fast at around $a \simeq 2.5 - 3.5 \times 10^4$ AU, owing to the strong dependence of the galactic tidal force on a.

As discussed in Section 5.6, Oort cloud comets have been thermalized over the solar system age by the action of external perturbers, say for semimajor axes greater than about 10^4 AU. But not all the directions of the velocity vectors of thermalized comets are possible, for Oort cloud comets crossing the Jupiter-Saturn barrier will be quickly removed by planetary perturbations. Since the velocity vectors of these comets fall very close to the solar direction, there will be an empty region in the velocity phase space known as the *loss cone* (Hills 1981, Torbett 1986). For a thermalized comet population, the fraction of comets of semimajor axis a having perihelion distances $q < q_L$ is $F_L \simeq 2q_L/a$, provided that $q_L << a$ (cf. eq. (5.42)). We have $q_L \simeq 15$ AU for the outer edge of the Jupiter-Saturn barrier. The loss cone will have an angular radius of $2F_L^{1/2}$ radians and the solar direction as the axis. The loss cones of comets with semimajor axes a whose perihelion distances can experience changes $\Delta q/q_L \sim 1$ per orbital revolution $P = a^{3/2}$ will be filled at any time, since comets within the loss cone that are lost by planetary perturbations will be replaced by those refilling the loss cone after P. As said, this will occur for semimajor axes $a > a_{fill} \sim 3.5 \times 10^4$ AU.

Some Oort cloud comets with $a < a_{fill}$ will anyway find their way to the inner planetary region through a multiple-step process, in which q drifts inwards and the energy kicks by planetary perturbations are weak enough for the comets to return to more or less the same region of the Oort cloud from which q continues drifting inwards. There is thus a very small -but not null- probability that Oort cloud comets with $a < a_{fill}$ reach the observable region through several weak steps within the Jupiter-Saturn barrier. It is clear that as a approaches a_{fill} the Jupiter-Saturn barrier becomes more "permeable" since Oort cloud comets require less steps to reach perihelion distances $q < q_{obs}$. The passage through the Jupiter-Saturn barrier is a process that resembles the penetration of the Coulomb barrier in reactions of nuclear fusion (though the physical principles are quite different here).

6.3. The "inner" and "outer" Oort cloud. The possible existence of an inner core

The change $\Delta q/q$ for Oort cloud comets with semimajor axes $a \lesssim 10^4$ AU becomes negligible (see Fig. 6.1), so these comets -if they exist at all- will remain in the deep freeze, unperturbed for long periods of time, at least until a very close stellar passage or a penetrating encounter with a giant molecular cloud will awake them from their dynamical dormancy. Therefore, we will define the inner boundary of the classical Oort cloud at $a_{min} \sim 10^4$ AU. It is thus clear that it is not a boundary in the sense that the circumsolar space is empty within it, but in the sense that even if comets are stored there, they are not able to reach the observable region under the current galactic environment. We will come back to this point below.

Our derivation of a_{min} and a_{fill} allows us to define three regions in the Oort cloud: (i) an *inner core*, still in the hypothetical realm which, if it exists, is safe from the action of external perturbers during quiescent stages; (ii) the *inner Oort cloud* for comets with semimajor axes in the range $a_{min} < a < a_{fill}$, namely those whose diffusion in q inward stops at the Jupiter-Saturn barrier; and (iii) the *outer Oort cloud* for comets with semimajor axes $a > a_{fill}$, namely those that can overshoot the Jupiter-Saturn barrier from the outer planetary region straight to the Earth's neighborhood in a single step. Once Oort cloud comets reach the outer planetary region, the small energy kicks by Uranus and Neptune can transfer some of them from the inner to the outer Oort cloud and viceversa. Therefore, it is clear that the limit between the inner and outer Oort cloud population is quite permeable. For instance, an inner Oort cloud comet approaching the Jupiter-Saturn barrier can have first its semimajor axis raised to the outer Oort cloud by planetary perturbations from where it can overshoot the Jupiter-Saturn barrier straight to the observable region in the next orbital revolution. Actually, this is the most common way to transfer the perihelia of inner Oort cloud comets to the observable region, as compared to the diffusion process through the Jupiter-Saturn barrier at nearly constant semimajor axis of very low probability.

There is not a defined upper limit for the outer Oort cloud, though comets with larger a will be on increasingly unstable orbits. As shown in Section 5.7, the stability boundary of the Oort cloud can be set at $a \sim 3 \times 10^4$ AU (i.e., it roughly coincides with the boundary inner/outer Oort cloud). Yet the outer cloud population is quite substantial since comets diffusing from the inner cloud make up for the comet losses. We can set a practical limit for the extension of the outer Oort cloud at $a_{max} \sim 10^5$ AU, beyond that stellar perturbations and tides from the galactic disk and the galactic bulge itself (cf. eq. (5.6)) will remove comets very quickly. We show in Fig. 6.2 a sketch depicting the different regions of the Oort cloud and the different transfer modes of comets

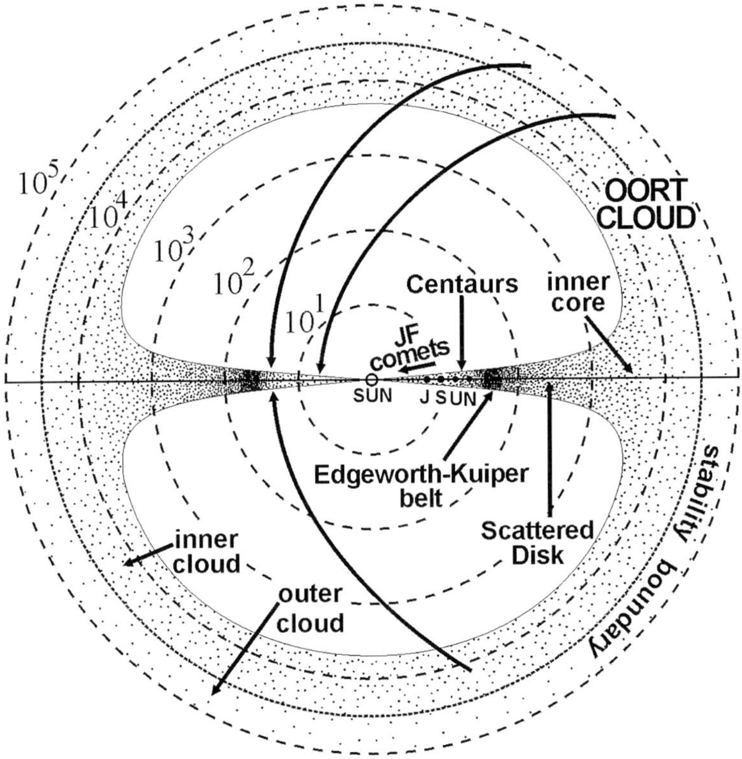

Figure 6.2. Sketch showing the Oort cloud and its different regions and the trans-neptunian belt. The curved arrows indicate tha main dynamical routes from the Oort cloud to the planetary region. JF comets and Centaurs come from the trans-neptunian belt via diffusion by planetary perturbations.

to the planetary region. Besides the Oort cloud, we include the trans-neptunian (or Edgeworth-Kuiper belt) and the scattered disk which constitute other important comet reservoirs that will be analyzed in Chapter 8.

There is still the question of what happens with comets enclosed in an inner core, i.e. with $a < a_{min}$ and perihelia beyond the planetary region. The direct access to this putative population is very difficult since under steady-state conditions their perihelia cannot drift into the planetary region. Yet, if the inner core population contained Ceres- or Pluto-sized bodies, we would be able to detect directly some of them if they happened to be near perihelion at present. Actually object 2003 VB_{12} (Sedna), discovered by Brown et al. (2004), is the first inner core candidate. Its orbital elements are: $q = 76 \pm 4$ AU, $a = 480 \pm 40$ AU, and $i = 11.^{\circ}9$.

A large number of comets (and of mass) might indeed be hidden in such an inner core, but we cannot tell anything about it because there is not a direct link between such a population and the new comets we observe. It is nevertheless quite possible that, from time to time, strong perturbations of the Oort cloud (for instance during a very close stellar encounter or a penetrating encounter with a GMC) will fill the loss cones of comets in the inner core, thus producing a sudden increase in the influx rate of Oort cloud comets that we will call a *comet shower*. This point was first discussed by Hills (1981). Going further, we may ask what is the spatial structure of comets in the inner core. It is believed that comets formed together with the planets in the protoplanetary disk that surrounded the early Sun, so they should have been originally distributed in a flat, disk-shaped structure, like the trans-neptunian belt, the fossil record of such a protoplanetary disk. External perturbers might have not been strong enough to fully randomize comet orbits in the inner core throughout the solar system lifetime (cf. Fig. 5.5), so such comets might have kept a flattened structure toward the ecliptic plane (see Fig. 6.2). Therefore, there might be a continuous distribution of bodies from the trans-neptunian, or Edgeworth-Kuiper belt (see Chapter 8) to the inner core, forming an increasingly broader disk, as external perturbations become stronger, to merge finally with the spherically-symmetric Oort cloud.

6.4. The fluxes of Oort cloud comets in the inner and outer planetary region

As discussed before, comets from the outer Oort cloud are able to reach the Earth's neighborhood by overshooting the Jupiter-Saturn barrier. We will neglect for the moment the contribution of new comets from the inner Oort cloud that reach the Earth's neighborhood via smooth diffusion through the Jupiter-Saturn barrier. The steady-state flux of Oort cloud comets per unit of perihelion distance that reach the Earth's neighborhood will then be given by

$$\dot{n}_{Earth} = \frac{1}{q_L} \int_{a_{fill}}^{+\infty} F_L \frac{1}{P} \Gamma(a) da \qquad (6.5)$$

where $F_L = 2q_L/a$, $P = a^{3/2}$, and $\Gamma(a)da$ is the number of Oort cloud comets with semimajor axes in the range $(a, a + da)$. Numerical simulations show that comets in the Oort cloud have a distribution of semimajor axes of the kind

$$\Gamma(a)da \propto a^{-\gamma} da \qquad (6.6)$$

where the exponent γ is unknown, though we can constrain possible values to the range 2-4 (Bailey 1983, Fernández and Ip 1987). Different values of γ reflect different degrees of central condensation of the Oort cloud. From numerical simulations that included planetary perturbations, stellar encounters and tidal forces of the galactic disk, Duncan et al. (1987) obtained a density profile of Oort cloud comets $\propto r^{-3.5}$. Bearing in mind that $r \sim 1.5a$ and that the distribution of semimajor axes is related to the density profile by $\Gamma(a)da \propto r^2 n(r) dr$, Duncan et al.'s empirical law would roughly correspond to a distribution $\propto a^{-1.5}$, i.e. somewhat smoother that our quoted range. Yet, it is very likely that penetrating encounters with molecular clouds have contributed to the depletion of the outer layers of the Oort cloud, thus leading to a steeper decrease in the number of comets with a, i.e. a larger value of the exponent γ (in absolute value). Admittedly, we know very little about the structure of the Oort cloud, so the power-law expressed by eq. (6.6) is only a working hypothesis. It is nevertheless very useful to explore models of the Oort cloud with different degrees of central condensation.

As mentioned above, comets from the inner Oort cloud must diffuse through the Jupiter-Saturn barrier to reach the Earth's neighborhood, which is a low-probability process. Nevertheless, since there are many more comets in the inner Oort cloud than in the outer Oort cloud, the fraction of new comets that reach the Earth's neighborhood coming from the inner Oort cloud may be significant. The Marsden and Williams's (2003) Catalogue of Cometary Orbits brings 46 new comets with $q < 3$ AU of quality class 1A and 1B, from which 24 come from the inner Oort cloud and 22 from the outer Oort cloud. The ratio inner/outer Oort cloud comets might range between ~ 3 and 35, depending on γ. If $\sim 10\%$ of the inner Oort cloud comets can leak through the Jupiter-Saturn barrier to the observable region, we would get a ratio for the fluxes inner/outer Oort cloud comets $\sim 0.3 - 3.5$, which is in fairly good agreement with the observed inner/outer ratio ~ 1. Thus, the flux of new comets in the Earth's neighborhood given by eq. (6.5) has to be increased by a factor f, i.e. $\dot{N}_{Earth} = f \times \dot{n}_{Earth}$, where $f \sim 2$.

The influx rate of new comets in the Uranus-Neptune region (say, with $15 < q < 30$ AU) per unit of q will come from the contribution of both the inner and outer Oort cloud, namely

$$\dot{N}_{U-N} = \frac{1}{q_2 - q_1} \int_{a_{min}}^{+\infty} F_{U-N} \frac{1}{P} \Gamma(a) da \tag{6.7}$$

where $q_1 = 15$ AU, $q_2 = 30$ AU, and $F_{U-N} = 2(q_2 - q_1)/a$.

By solving the integrals of eqs.(6.5) and (6.7) and dividing between them we obtain

$$\frac{\dot{N}_{U-N}}{\dot{N}_{Earth}} = \frac{1}{f} \left(\frac{a_{min}}{a_{fill}} \right)^{-(\gamma+3/2)} \tag{6.8}$$

By introducing in eq. (6.8) the numerical values of a_{min}, a_{fill} and f discussed above, we obtain for the flux of Oort cloud comets in the outer planetary region a factor 30-380 times greater than that in the inner planetary region, depending on the degree of central condensation of the Oort cloud (expressed here by the range of exponents $\gamma = 2-4$). For comparison, we show in Fig. 6.3 the flux-distributions of fictitious Oort cloud comets (in number of comets per unit of perihelion distance) obtained from numerical simulations by Fernández (1982) and Yabushita and Tsujii (1991). Their models included the perturbations of the Jovian planets when comets enter into the planetary region. As regards to the external perturbers, the former author considered stellar perturbations while the latter authors considered the tidal force of the

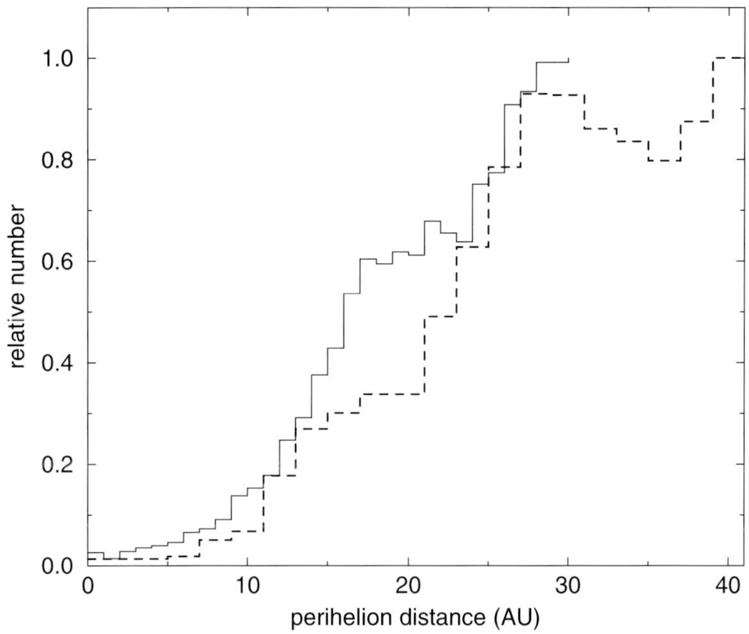

Figure 6.3. Distribution of perihelion distances of samples of fictitious Oort cloud comets obtained from numerical simulations by Fernández (1982) (solid histogram) and Yabushita and Tsujii (1991) (dashed histogram).

galactic disk. Their numerical results are in reasonable agreement with our simple numerical estimate from eq. (6.8). The numerical simulations also show the sharp drop in the flux of Oort cloud comets from the outer to the inner planetary region, though the step function at $q = 15$ AU, as we implicitly assumed for our simple model of eq. (6.8), is only a very rough, though convenient, approximation to describe the drop. In actuality, the flux slowly decreases with decreasing q in the outer planetary region, and then it drops very fast for q within the range $\sim 10 - 20$ AU, which can be defined as the width of the Jupiter-Saturn barrier centered around $q \sim 15$ AU. For $q \lesssim 10$ AU the decrease proceeds again at a much lower pace all the way down to the Sun.

We remark that the passage rate of Oort cloud comets in the Earth's neighborhood and, concomitantly, the boundaries a_{min} and a_{fill} of the Oort cloud are a function of the field of galactic perturbers. The Sun's galactocentric distance has varied through its lifetime between $\sim 8 - 9$ kpc and its vertical distance to the galactic midplane by about ± 100 pc. Thus, it is very likely that the Sun has passed through regions of different galactic densities ρ_{disk} which modulated the strength of the

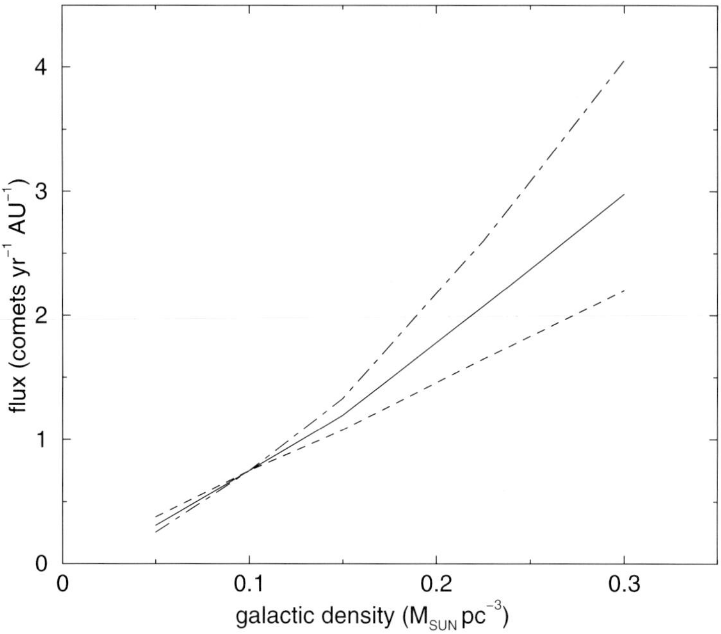

Figure 6.4. Computed fluxes of Oort cloud comets in the Earth's neighborhood as a function of the local density of the galactic disk. The computations were performed for an Oort cloud model with different degrees of central condensation expressed by the exponent γ. The fluxes have been normalized to a standard flux of 0.75 comets per yr and per AU (cf. Section 2.7) for a local galactic density of 0.1 $M_\odot$ pc^{-3}.

tidal force of the galactic disk and, consequently, the flux of Oort cloud comets (Matese et al. 1995). In Fig. 6.4 the fluxes of Oort cloud comets in the Earth's neighborhood, derived from eq. (6.5), are plotted as a function of the galactic density and for different central concentrations of the Oort cloud. It is shown that for a plausible range of galactic densities, the comet flux can vary by as much as a factor of five. It is thus clear that even under steady conditions there may be a strong modulation in the comet flux as the Sun passes through regions of different mass densities following its radial and vertical excursions in the Galaxy.

6.5. Comet showers

Occasional strong perturbations of the Oort cloud, caused either by a very close stellar passage (say, at distances $\lesssim 10^4$ AU or smaller), or by a penetrating encounter with a GMC, will produce the sudden refilling of the usually empty loss cones of comets in the inner Oort cloud, and even in the core. A comet shower will ensue that will last until the quiescent pre-encounter situation is restored, a time scale that is of the order of the orbital period of the comets discharged from the refilled loss cones.

Let us consider in the first place a star of mass M and relative velocity V that penetrates very deeply within the Oort cloud, reaching a distance of minimum approach to the Sun $D_{\odot}$. We can estimate the volume of the Oort cloud within which the loss cones are refilled. The central axis of such a volume coincides with the star's path, and the distances of closest approach of comets within the volume to the star will be in general $D << D_{\odot}$.

From eq. (5.2) we have that the impulse received by the comet from the star is

$$\vec{\Delta v} = \vec{\Delta v_c} - \vec{\Delta v_{\odot}} \sim \vec{\Delta v_c} = \frac{2GM}{VD}\frac{\vec{D}}{D} \tag{6.9}$$

The transverse component of the impulse Δv_T (perpendicular to $\vec{r}$ and contained in the orbital plane of the comet) is

$$\Delta v_T = \Delta v \sin\eta \cos\nu = \frac{2GM}{VD} \sin\eta \cos\nu \tag{6.10}$$

where η is the angle between $\vec{D}$ and $\vec{r}$, and ν the angle that the orbital plane of the comet forms with the orbital plane containing $\vec{r}$ and $\vec{D}$ (Fig. 6.5). The comet is assumed to be at a heliocentric distance $r \sim (D_{\odot}^2 + u^2)^{1/2}$ (valid as long as $D << D_{\odot}$) where $u \equiv \overline{K_{\odot}K_c}$.

Let us define $D_F = D_F(u)$ as the distance to the passing star such that comets will have their loss cones refilled if $D < D_F$. By setting the condition $\Delta q/q = 1$ in eq. (6.4), required for the loss cone to be refilled, and introducing there the expression of Δv_T given by eq. (6.10) with the numerical values $M = 0.5$ $M_{\odot}$, $q = 15$ AU, and $V = 30$ km s^{-1}, we obtain the boundary

$$D_F \simeq \frac{D_{\odot}}{4.57}\left[1 + \left(\frac{u}{D_{\odot}}\right)^2\right]^{1/2}, \tag{6.11}$$

where we have considered for the angles ν and η the average values: $< \cos\nu >= 2/\pi$ and $< \sin\eta >= \pi/4$. From this condition we have that all comets within the elemental disk perpendicular to $\vec{V}$ centered

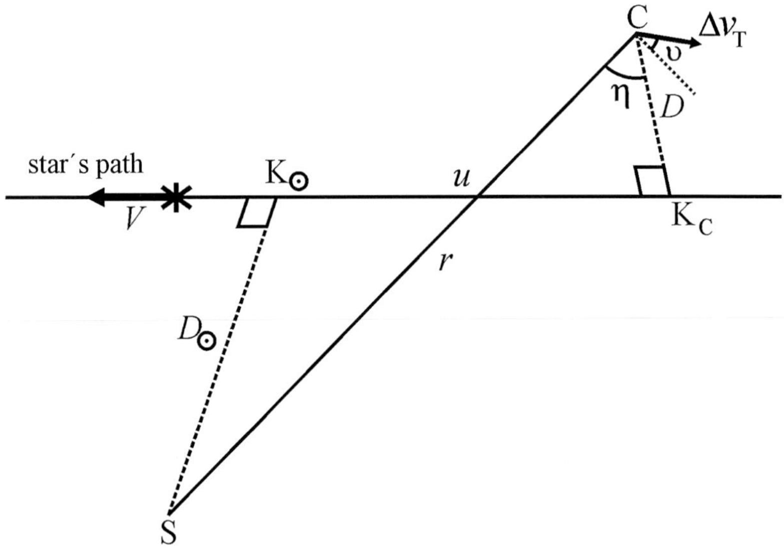

Figure 6.5. Geometry of a close star's passage. S: Sun, C: comet.

on K_c of radius $D_F = D_F(u)$, thickness du, and at a distance u to $K_\odot$ will have their loss cones refilled (Fig. 6.6). Furthermore, from eq. (6.6) and bearing in mind that $r \sim 1.5a$, we find that the number density of comets at a heliocentric distance r, will follow a law: $\Gamma'(r) \propto r^{-(\gamma+2)}$, so the number of comets within the elemental ring $2\pi DdDdu$ is: $2\pi\Gamma'(r)DdDdu$. The influx rate of new comets (per unit of perihelion distance), triggered by the star's close passage, will then be given by

$$\dot{n}_{close} = \frac{2}{q_L}\int_0^{u_{max}} du \int_0^{D_L} F_L 2\pi\Gamma'(r)D\frac{1}{P}dD, \tag{6.12}$$

where $F_L = 2q_L/a$ and $u_{max} \sim 1.5a_{fill}$ (namely for $u > u_{max}$ the star moves in the outer Oort cloud where comets have their loss cones always refilled anyway). Furthermore, we have $P = a^{3/2} \approx (D_\odot^2 + u^2)^{3/4}/1.5^{3/2}$. We note that D_F depends in a complex manner on the angles η and ν, so the perturbed volume will not be axisymmetric around the star's path, but this was simplified by averaging η and ν in eq. (6.11).

A star passing at a closest distance to the Sun $D_\odot$ will perturb Oort cloud comets mostly of semimajor axis $a \sim D_\odot/1.5$, owing to the increase of the number density of comets for smaller heliocentric distances. As a consequence, the ensuing enhancement in the flux of new comets will have an intense phase of a duration on the order of $P = a^{3/2} \sim (D_\odot/1.5)^{3/2}$. Stars will pass at a closest distance to the

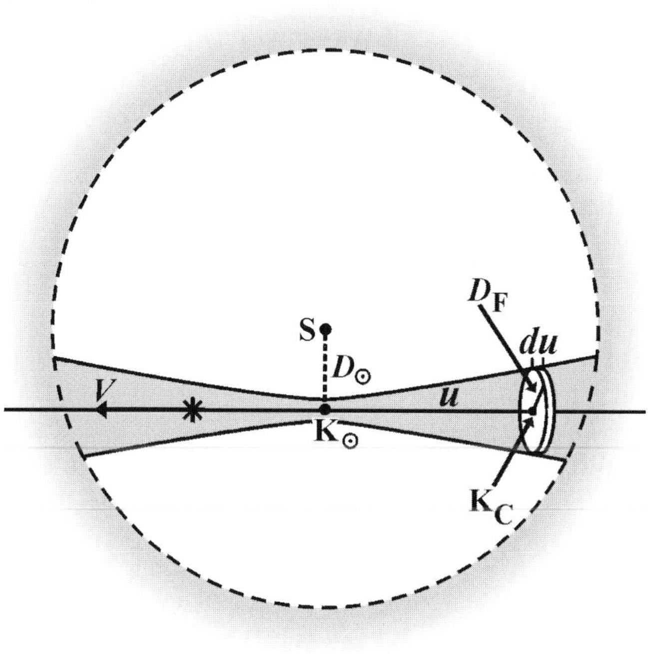

Figure 6.6. Shape of the volume perturbed by the passing star that causes the refilling of the loss cones of comets within the volume. Comets located beyond the dashed circle of radius $1.5a_{fill}$ have their loss cones permanently filled, so the star has no effect in their flux.

Sun $D_\odot$ at average intervals

$$\Delta t = \frac{1}{2n_* D_\odot^2} \simeq \frac{3 \times 10^9}{D_\odot^2} \text{ Myr}, \tag{6.13}$$

where $D_\odot$ is expressed in AU. For instance, from this equation we find that a close encounter at $D_\odot = 10^4$ AU will occur at mean intervals $\Delta t \sim 30$ Myr.

Strong perturbations affecting the inner core of the Oort cloud may also be expected from penetrating encounters with GMCs (see the previous chapter). Let $(\Delta v_{GMC})_T = \Delta v_{GMC} \times \cos\zeta \times \cos\theta$ be the change in the transverse velocity of a comet of semimajor axis a. Δv_{GMC} is given by eq. (5.21) where ζ is the angle between $\vec{r}$ and $\vec{v}_{GMC}$, and θ is the angle between the orbital plane of the comet and the plane containing $\vec{r}$ and $\vec{v}_{GMC}$. The loss cones of Oort cloud comets with semimajor axes $a > a_G$ will be suddenly refilled, where again the value of the minimum semimajor axis a_G is set by the condition $\Delta q/q = 1$. The influx rate of

Oort cloud comets (per unit of perihelion distance) due to a penetrating encounter with a GMC will then be given by

$$\dot{n}_{GMC} = \frac{1}{q_L} \int_{a_G}^{a_{fill}} F_L \frac{1}{P} \Gamma(a) da \qquad (6.14)$$

which is valid provided that $a_G < a_{fill}$. We note that a_G is a function of the angles ζ and θ, so the comet shower triggered by the encounter with the GMC will not have radial symmetry.

Comet fluxes in the Earth's neighborhood under quiescent conditions are compared with the increases produced during strong perturbations of the Oort cloud. The results (normalized to a quiescent influx rate = 1) are shown in Table 6.1. We can see that increases of two-three orders of magnitude are posible, depending on the degree of central condensation of the Oort cloud represented in the table by different values of the index γ.

Table 6.1: Flux of Oort cloud comets in the inner planetary region (normalized to a quiescent influx rate = 1)

Trigger mechanism \ γ	2	3	4
Close star's passage ($D_\odot = 10^4$ AU)	9.3	43	200
GMC	68	260	989

Monte Carlo models carried out by Heisler et al. (1987) and Heisler (1990), that incorporate both stellar and galactic tide perturbations, show very nicely the production of comet showers at average intervals of several 10^7 yr (Fig. 6.7). As Heisler's numerical results show, for $a = 10^4$ AU 98.6% of the comets with $q < 2$ AU will enter during showers. This is in agreement with was shown above: it is very difficult for these comets to overshoot the Jupiter-Saturn barrier during quiescent stages, and only a tiny fraction will manage to leak through the barrier inwards leading to a very low passage rate of new comets. As a strong perturbation occurs, the loss cone of comets with $a = 10^4$ AU is suddenly refilled and subsequently discharged, producing the shower. For $a = 2 \times 10^4$ AU this percentage decreases to 34% as the Jupiter-Saturn barrier becomes more permeable, thus allowing a more substantial steady flux of new comets via diffusion in q through the barrier. For $a = 3 \times 10^4$ AU showers are clearly missing, indicating

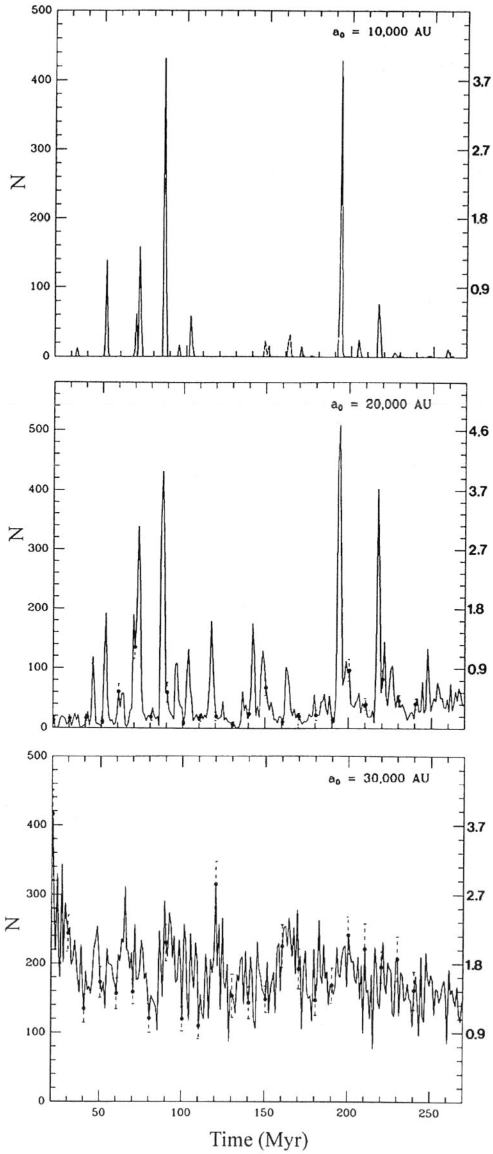

Figure 6.7. Monte Carlo simulations of comet showers that record the time-variation in the influx rate of new comets with $q < 2$ AU. The initial semimajor axis of the computed comets is indicated at the upper right corner of each panel. The left hand axis gives the flux per Myr from a cloud with 10,930,000 comets. The right hand axis gives the fluxes in comets yr^{-1} for an Oort cloud population of 10^{11} comets (Heisler 1990).

that the respective loss cones are filled, irrespective of the occurrence of an occasional very strong perturbation.

6.6. Do comet showers reflect in the impact cratering record?

The Earth has been subject to a steady bombardment of interplanetary bodies since the time of its formation. At the beginnings the projectiles were the residual planetesimals left after the formation of the terrestrial planets and the giant Jupiter. When this local source was depleted, the projectiles came from more dynamically stable niches of the solar system, such as the asteroid belt, Jupiter's Trojans, the trans-neptunian belt and the Oort cloud. All the sources inside or close to the planetary region are only subject to planetary perturbations that slowly remove bodies from their stable niches, usually via mean-motion and/or secular resonances, thus becoming potential colliders with any of the planets or the Sun. This flux keeps more or less steady through time, and only the release of a giant object from one of the stable niches might rise the probability of a megaimpact. The Oort cloud is a quite different source in the sense that its dynamics essentially depends on external, stochastic perturbers, and not on planetary perturbations. As shown above, very strong external perturbations on the Oort cloud can trigger comet showers with the consequent great enhancement in the number of projectiles.

We can learn about the impact history of the Earth and the other planets, satellites and minor bodies, through the impact craters on their surfaces. In particular, a sudden increase in the number of craters of a certain age might be the fossil record of a comet shower that occurred at that time. There are at present ~ 140 hypervelocity impact craters on Earth dating from pre-Cambrian times (~ 2000 Myr ago) to nearly the present (Grieve and Shoemaker 1994). This crater sample supplies a valuable record of the collisional environment of the Earth in the past. A couple of questions are relevant: (i) how do comets and asteroids contribute to the cratering record, and (ii) could past comet showers reflect in some clusterings of crater ages? As regards question (i), we can just say that comets are responsible for only a small fraction of all impact craters, perhaps about 10%. A discussion of this topic is left for Section 11.4.

The overall cratering rate on the Earth is given by the empirical relation (Bailey 1990)

$$\dot{N}_{obs}(\geq D_c) = (3 \pm 2) \times 10^{-6} D_{20}^{-\nu} \ \mathrm{yr}^{-1}, \tag{6.15}$$

where $\nu \simeq 2.0 \pm 0.2$ and $D_{20} = D_c/20$ km. If we take $D_c = 10$ km, we obtain a total production of about 6000 craters > 10 km over a period of 500 Myr. Grieve and Shoemaker (1994) estimated an average cratering rate of $5.6 \pm 2.8 \times 10^{-15}$ km^{-2} yr^{-1} for $D \gtrsim 20$ km, which multiplied by the Earth's surface (5.11×10^8 km^2) gives an overall cratering rate in agreement with eq. (6.15). Thus, the sample of well-dated impact craters formed during this period is scarcely $\approx 0.5\%$ of this value.

During the eighties there was a flurry about possible periodicities in biological mass extinctions. The debate started when Alvarez et al. (1980) suggested that the anomalous abundance of iridium in the Cretaceous/Tertiary (K/T) boundary layer could be explained by the collision of a comet or asteroid with lethal consequences, since it led to the mass extinction that wiped out the dinosaurs and, in general, all animals weighing above 25 kg. The basis for this hypothesis was that iridium is an element very rare in the Earth's crust but much more abundant in heavenly bodies like comets and asteroids. Alvarez et al.'s hypothesis paved the way to several theories that associated the K/T mass extinction - and perhaps others - to collisions of comets and asteroids with the Earth. A few years later, Raup and Sepkoski (1984) argued that these mass extinctions had a period of 26 Myr. Alvarez and Muller (1984) found that such a putative periodicity in mass extinctions was correlated with a possible periodicity in the cratering rate, presumably caused by comet showers, which they believed was the only potential impact source to be periodic.

Several mechanisms for triggering periodic comet showers were proposed, in particular: (1) a solar companion star *Nemesis*, on an eccentric, 26-Myr-period orbit, that perturbs the Oort cloud every time it passes perihelion (Davis et al. 1984, Whitmire and Jackson 1984); (2) transits of the solar system through the galactic plane, as it oscillates up and down along the vertical direction, during which it crosses dense regions that strongly perturbed the Oort cloud; and (3) an as-yet undetected trans-neptunian planet X on a highly-inclined orbit whose precession of perihelion drives it to the trans-neptunian belt at periodic intervals of 28 Myr (Whitmire and Matese 1985). In the latter case the perturber triggers a shower with comets from the trans-neptunian belt. All these mechanisms presented serious dynamical objections that were reviewed by other authors (e.g., Tremaine 1986). Furthermore, the claim that biological mass extinctions and the cratering rate are both periodic and in phase was challenged by different statistical analyses by Heisler and Tremaine (1989) and Fogg (1989) among others. The interest on this problem faded afterwards, and we can say that it has only a historical interest today.

Even relaxing the constraint that comet showers must be periodic, the question is still how intense a comet shower should be to reflect in crater statistics. Let $\dot{N}_{ss}$ be the steady-state cratering rate on the Earth from Earth-crossing objects. Let $\dot{N}_{sh}$ be the time-dependent cratering rate due to comets from a shower. If we assume that close stellar passages at, say $D_{\odot} \sim 10^4$ AU, are the main cause of comet showers, then these may occur at average intervals of $\Gamma \sim 30$ Myr (see Table 6.1). The duration of a comet shower -at least during the intense phase - is of the order of the orbital period of shower comets, typically of about $T_{sh} \sim 1$ Myr. The number of craters produced during a comet shower should be at least comparable to the background craters produced during Γ, otherwise we will tend to pick mostly background craters which will tend to blur possible clusterings in crater ages caused by impacting shower comets, i.e.

$$\dot{N}_{sh} \times T_{sh} \sim \dot{N}_{ss} \times \Gamma. \tag{6.16}$$

Now, the contribution of comets to the steady-state cratering rate is about 10%, from which only about 10% may correspond to LP comets coming from the Oort cloud, the rest being JF comets coming from the trans-neptunian belt (this point will be further discussed in Section 11.4). Therefore, the cratering rate from LP comets $\dot{N}_{LP} \sim 0.01\dot{N}_{ss}$, so by introducing this value in eq. (6.16) we get

$$\dot{N}_{sh} \sim 3,000\dot{N}_{LP} \tag{6.17}$$

Therefore, the intensity of the comet shower should be at least about 3,000 times greater than the steady-state comet flux to show up in crater statistics. This turns out to be at least one order of magnitude higher than the showers triggered by a star passage at $D_{\odot} = 10^4$ AU. This result agrees with that from Bailey et al. (1987) who found that close star passages could trigger comet showers at mean intervals of 30 Myr (not periodically), but only if the Oort cloud had an extremely massive inner core $\gtrsim 10^4$ $M_{\oplus}$. Weissman (1990) has analyzed a large sample of well-dated craters with diameters ≥ 10 km and does not find any evidence of clusterings of crater ages attributable to comet showers. As mentioned, this sample is still too small to reveal weak crater-age clusterings attributable to past very intense comet showers occurred over time spans of 500 Myr or so.

6.7. The current passage rate of Oort cloud comets: does it represent a quiescent stage or an excited one?

A steady-state supply of Oort cloud comets, as due to the dominant action of tides of the galactic disk, will produce a pattern of aphelion points on the celestial sphere concentrated at mid-galactic latitudes. As shown before (cf. eq. (6.2)), the change in the perihelion distance of Oort cloud comets by the tidal force of the galactic disk is proportional to $\sin 2\phi$, i.e. it is maximum for a galactic latitude $\phi = 45°$. By contrast, comet showers will not show such a galactic dependence, but the concentration of aphelion points will be related instead to the path of the passing star or GMC.

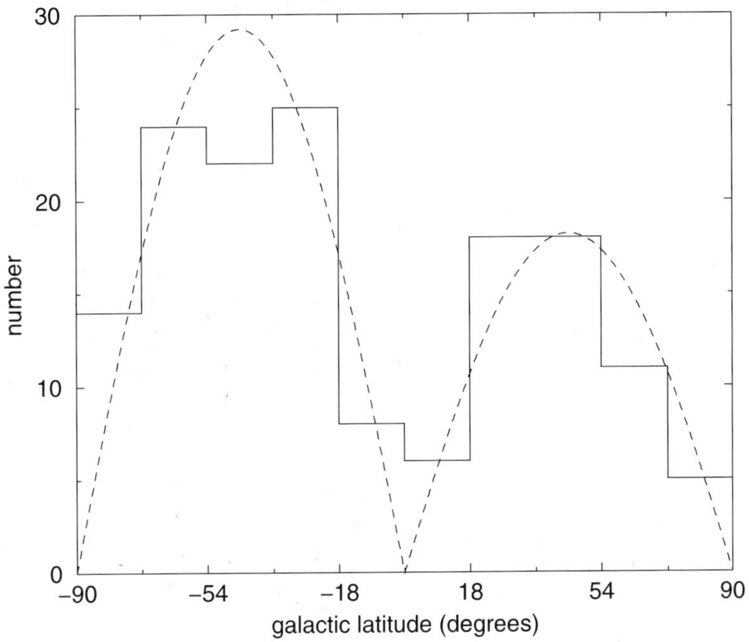

Figure 6.8. Distribution of galactic latitudes of 151 new and young comets with $a_{orig} > 500$ AU. Since these comets have at most a few passages by the inner planetary region, it is considered that they still preserve their original aphelion directions. Sinusoidal curves have been fitted to the histograms.

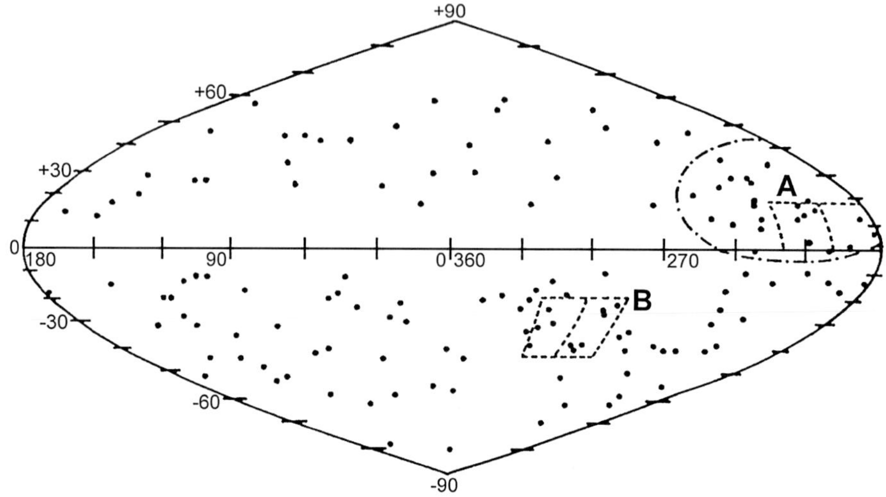

Figure 6.9. Distribution of apelion points on the celestial sphere of 142 new and young comets with $a_{orig} > 500$ AU. A and B denote two zones showing an excess of aphelion points (Fernández and Ip 1991).

The observed aphelion distribution of new and dynamically young comets follows a pattern reflecting the influence of the galactic disk potential, showing minima near the galactic poles and galactic equator and maxima at mid-galactic latitudes (Delsemme 1987) (Fig. 6.8). This regular pattern argues against a recent comet shower, that would have disrupted it, and suggests that the frequency of comet passages may be at present close to its quiescent bottom level (Fernández and Ip 1991). Comets injected during a shower might greatly exceed the background supply of Oort cloud comets, brought mainly by galactic tides, in such a way that the galactic dependence in the distribution of aphelion points could be severely weakened or erased altogether. There are indeed some weak aphelion clusterings (Fig. 6.9) that might reflect close stellar passages in the recent past (e.g. Biermann et al. 1983, Lüst 1984), though such clusterings seem to encompass only a minor fraction of the overall aphelion sample.

THE JUPITER FAMILY

Jupiter family comets are those with the shortest orbital periods. The boundary with the comets of longer periods is usually set at $P = 20$ yr, which corresponds to a semimajor axis $a = 7.37$ AU. This boundary is somewhat arbitrary and has changed with time. Thus, Russell (1920) considered as comets of the Jupiter's "family" those with periods between 5 and 9 yr, while Everhart (1972) set the boundary at $P = 13$ yr. These comets have been formerly called "short-period" (SP) comets, but now this denomination has a slightly different meaning, since some of the SP comets may not belong to the Jupiter family as we will see in more detail below. Levison and Duncan (1997) have assembled Jupiter family comets and the more distant Centaurs into a broader category of "ecliptic" comets in reference to their low inclinations. In order to keep a certain consistency in our definitions, we will stick in the following to the name "Jupiter family" (JF) comets as applied to those with periods $P < 20$ yr, but with an additional condition for their Tisserand constant as we will explain below. The appropriateness of the boundary at $P = 20$ yr will be discussed in Section 7.9.

Because of their short orbital periods, most of the discovered JF comets have been observed in more than one apparition, and in some cases through all the orbit up to their aphelia. This orbital coverage has allowed us to gather a very valuable wealth of physical data on this population. Furthermore, several JF comets have been or will be the targets of space missions which will help to greatly increase our knowledge about their physical nature.

7.1. Statistics of comet discoveries and orbital properties

The first JF comet recognized as such was D/1770 L1 (Lexell) whose period was found to be of only 5.6 years, though comet La Hire, observed in 1678, was later identified with a previous passage of the JF comet 6P/d'Arrest (Carusi et al. 1991). There are 227 comets with $P < 20$ yr discovered through the end of 2002 (Marsden and Williams 2003) from which we have left aside 29P/Schwassmann-Wachmann 1, because its orbit lies entirely beyond Jupiter, so strictly speaking it should be classified as a Centaur (see Section 8.6). The shortest orbital period so far recorded corresponds to 2P/Encke with 3.3 yr. The discovery rate of JF comets has been steadily increasing: while no JF comet discovered prior 1892 had $q > 2$ AU, deep sky surveys have led to the discovery of a growing number of distant JF comets, so that at present about half of the observed JF population have $q > 2$ AU (Fig. 7.1). This trend suggests that a large fraction of the population of distant JF comets ($q \gtrsim 2$ AU) has still to be discovered. Systematic surveys for near-Earth objects (NEOs), such as LINEAR, NEAT, LONEOS and Spacewatch, are responsible for the discovery of most of JF comets since 1998. These surveys also account for the dramatic increase in the discovery rate in the last few years, not only of JF comets but also of LP comets.

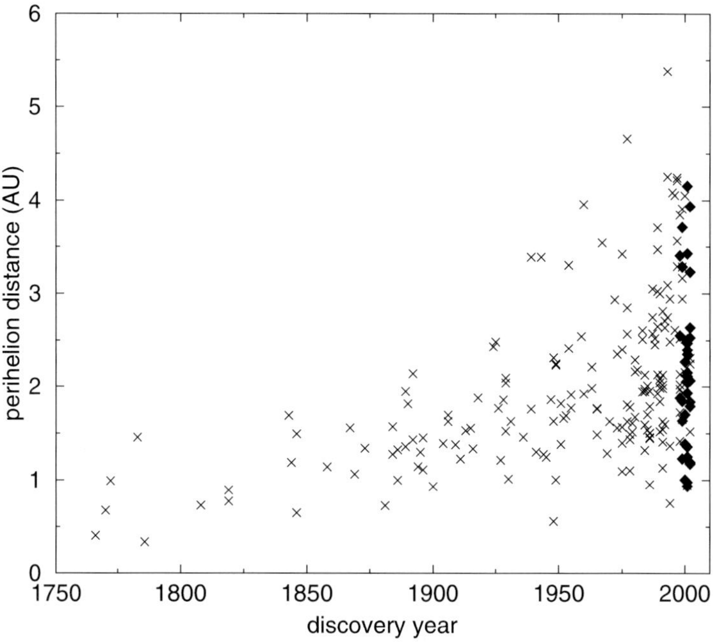

Figure 7.1. The discovery rate of Jupiter family comets defined as those with orbital periods $P < 20$ yr and Tisserand parameter $T > 2$. Filled diamonds represent JF comets discovered by the search projects LINEAR, NEAT and LONEOS. The sample has been taken from Marsden and Williams's (2003) catalogue.

Since Jupiter controls the dynamical evolution of JF comets, their dynamics can be analyzed under the framework of the three-body problem: Sun-Jupiter-comet. To this purpose, a convenient parameter is the Tisserand invariant T, introduced in eq. (4.44), which is derived from the Jacobi's integral of the restricted three-body problem. Since

Jupiter's orbit is not circular and other planets besides Jupiter perturb the comet, its Tisserand parameter will slowly change with time. Nevertheless, the change in T will be much smaller than those of the orbital elements, so T will still be a useful parameter to trace back the provenance of the comet.

The Tisserand parameter of most comets with orbital periods $P < 20$ yr and $q < 2$ AU falls within the range $2 < T < 3$, while most JF comets with $2 < q < 5$ AU have Tisserand parameters even more constrained to the narrow range $2.5 < T < 3$. By contrast, most HT and LP comets have $T < 2$ with very little overlapping with JF comets. It is easy to show that most LP or HT comets should have $T < 2$, provided that they come from the capture of near-parabolic comets by Jupiter. From eq. (4.44) we can see that the Tisserand parameter for a near-parabolic comet ($a \to \infty$) is $T \simeq 2\sqrt{2} \cos i (q/a_J)^{1/2}$. Since most of the observed LP/HT comets have $q < 2$ AU, they will fulfill the condition $q/a_J \lesssim 1/2$, whereby $T \lesssim 2$. Should JF comets come from the capture of near-parabolic comets, they would also have $T < 2$ since T has to be (more or less) conserved through the capture process, which again argues against a common origin for JF comets and HT/LP comets. Consequently, a Tisserand parameter $T \simeq 2$ defines the boundary between LP comets coming from the Oort cloud and JF comets coming from the trans-neptunian belt (Carusi et al. 1987) as will be analyzed in Section 7.9. In fact, there are four comets with periods $P < 20$ yr whose Tisserand parameters are $T < 2$; they are: 8P/Tuttle, 96P/Machholz 1, 126P/IRAS and P/1994 X1 (McNaught-Russell). It is quite possible that they are interlopers within the Jupiter family, whose origin is in the Oort cloud.

It is thus appropriate to define a JF comet as one that fulfills both conditions: orbital period $P < 20$ yr and Tisserand parameter $T > 2$, since the combination of both criteria should enclose a more homogeneous population from the same source region. In theory $T = 3$ defines an upper limit for the Tisserand parameter, since for $T > 3$ encounters with Jupiter are not possible (see eq. (7.25) below). There are two anomalous cases in the list of discovered JF comets: 107P/Wilson-Harrington and 133P/Elst-Pizarro, whose Tisserand parameters are: $T = 3.086$ and $T = 3.184$ respectively. Yet, their true nature has been very controversial and they may well be asteroids (a more thorough discussion on these objects will be presented in Section 9.3). Therefore, we will not consider them in the ensuing discussion on JF comets. Actually, because Jupiter's orbit is slightly elliptic, T values slightly above three are still allowed for *bona fide* JF comets. We can mention the case of 2P/Encke ($T = 3.018$) which has possibly been subject to strong nongravitational forces that decreased its semimajor axis (and thus increased its T) (Fernández et al. 2002). There are a few other cases with T above three, but all have $T < 3.1$. We can then say with confidence that JF comets should have Tisserand parameters in the interval $2 < T \lesssim 3.1$.

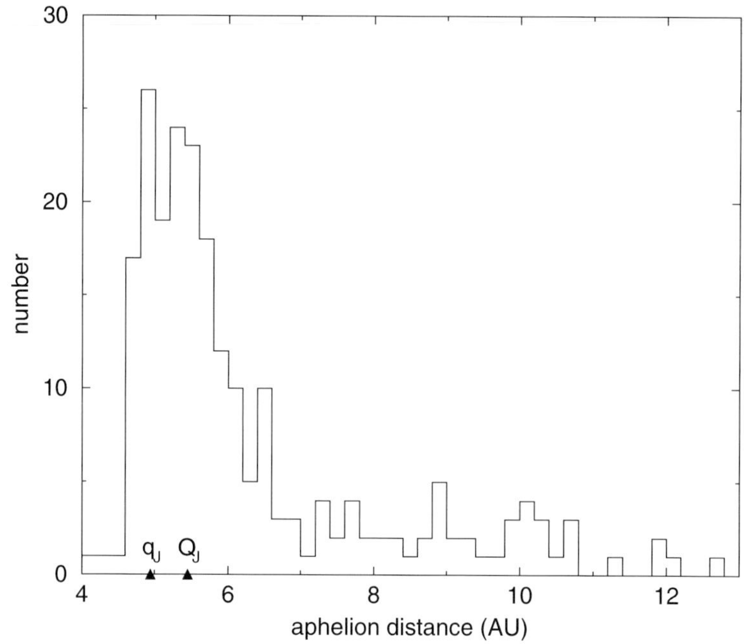

Figure 7.2. Distribution of aphelion distances of the sample of Jupiter family comets taken from Marsden and Williams's (2003) catalogue. Jupiter's perihelion and aphelion distances (q_J, Q_J) are also indicated.

The dominant influence of Jupiter on the dynamical evolution of JF comets can be seen by the clustering of their aphelion distances around the radius of Jupiter's orbit (Fig. 7.2), and by the clustering of their arguments of perihelion around $\omega = 0°$ and $\omega = 180°$ (Fig. 7.3). The combination of these two orbital features permits the occurrence of frequent close encounters between Jupiter and JF comets, since most of their aphelion points lie close to the line of nodes, and thus to the plane of Jupiter's orbit (strictly speaking, the line of nodes is the intersection of the comet's orbital plane with the ecliptic plane and not with Jupiter's orbital plane, but these two planes form an angle of only $\sim 1.3°$). With such a favorable geometry, a close encounter will occur

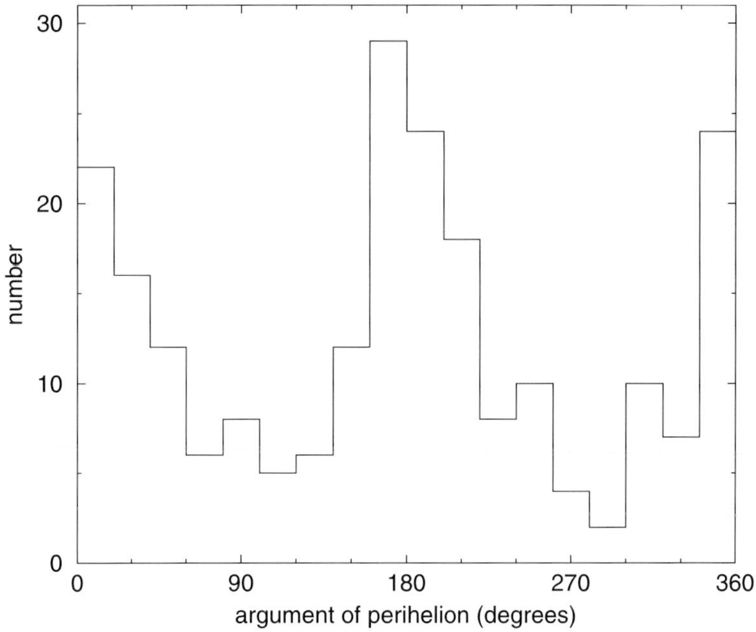

Figure 7.3. Distribution of arguments of perihelion of the sample of Jupiter family comets taken from Marsden and Williams's (2003) catalogue.

every time the comet is near aphelion and Jupiter happens to be in the part of its orbit close to the comet's aphelion.

Some JF comets are found to librate temporarily around mean motion resonances as for instance 9P/Tempel 1 (see Section 4.4), though they will tend to decouple after some time. A typical comet behavior thus consists of brief passages through resonant states and quick jumps or "hops" between different resonances (Belbruno and Marsden 1997). Furthermore, low-velocity, long-lasting encounters of comets with Jupiter may lead to their capture as temporary satellites. These are particularly the cases of 39P/Oterma, 74P/Smirnova-Chernykh, 82P/Gehrels 3 and 111P/Helin-Roman-Crockett, all of which have the distinctive properties of being near the 3:2 resonance with Jupiter and Tisserand parameters $T \sim 3$, indicating the possibility of low-velocity encounters with Jupiter (cf. eq. (7.25)).

7.2. Determination of comet sizes: methods

Estimates of comet sizes have formerly relied upon their absolute total magnitudes H_T (e.g. Hughes 1988, Bailey 1990). Such estimates are quite uncertain bearing in mind that H_T is a poorly defined quantity that depends on the assumed slope n of the heliocentric comet's lightcurve ($n = 4$ is usually assumed which leads to the well-known total magnitude H_{10}, see Section 2.4). The ideal would be to measure directly the magnitude of the comet nucleus and to derive from here a comet size by assuming a certain geometric albedo. Eventually, combined observations in the visible and thermal infrared could allow the determination of both, size and albedo, as has been done for many asteroids. We shall next review the most important methods that have so far been applied, with greater or lesser success, to measure the comet radius.

7.2.1. Close-up imaging from flyby missions

This is the best procedure to determine unambiguously the size, albedo, and shape of a comet nucleus. Unfortunately, it depends on expensive space missions that can reach only a few targets. So far, only the comets shown in Table 7.1 have been imaged from close distance (see more details in Section 3.3 and Figs. 3.1 and 3.3). The effective radius of the Table defines the radius of a sphere whose volume is equal to that of the comet (usually modelled as a triaxial ellipsoid of semiaxes a, b, c).

Table 7.1: Comets with known size, shape and albedo determined from flyby missions

comet	R_N effective (km)	$2a \times 2b \times 2c$ (km)	p_v	source
1P/Halley	5.2	$16.0 \times 8.4 \times 8.2$	0.04	(1)
19P/Borrelly	2.2	$8.0 \times 3.2 \times 3.2$	0.03 ± 0.005	(2)
81P/Wild 2	2.1	$2.1 \times 4.0 \times 3.3$	0.03 ± 0.015	(3)

(1) Keller et al. (1987)
(2) Soderblom et al. (2002)
(3) Brownlee et al. (2004)

7.2.2. Photometry of comets of "stellar" appearance

Coma contamination has been a severe shortcoming in all attempts to determine reliable magnitudes of the comet nucleus. However, with the advent of CCD cameras and the use of medium to large-size telescopes, able to follow JF comets all along the orbit, the goal of estimating good nuclear magnitudes started to appear as feasible. Comets can now be observed far from the Sun where they show little activity or no activity at all, in other words, they look "stellar". University of Arizona's astronomer Elizabeth Roemer was among the pioneer comet observers that tried to collect a good database of nuclear magnitudes. She used photographic plates taken with the 40-inch astrometric reflector at the Flagstaff station which produced nearly "stellar" images of comets generally observed at large heliocentric distances. Her work covered approximately two decades around the sixties and seventies. During the eighties some researchers started to work with the novel CCD cameras aimed at getting a first physical picture of a comet nucleus, including its size, albedo, fraction of active surface area and rotation period. Periodic comets were the candidates for these type of studies because their apparitions could be predicted beforehand. Among the pioneer research we can mention the CCD photometry of 1P/Halley by Jewitt and Danielson (1984) at distances greater than 8 AU. These authors found an upper limit of ~ 7.4 km for the effective radius of Halley's nucleus if a geometric albedo $p_v = 0.04$ was assumed. The close match between this estimate and the later *in situ* measurement of the comet size showed the usefulness and potential of the new CCD technique for ground-based determinations of comet sizes at large heliocentric distances. Following the Halley campaign, new CCD photometric observations of a few JF comets were carried out at near aphelion, as was the case of 2P/Encke (Jewitt and Meech 1987), 49P/Arend-Rigaux (Brooke and Knacke 1986), and 28P/Neujmin 1 (Campins et al. 1987). The latter two comets combined photometry with spectroscopy in the infrared. The bulk data on nuclear magnitudes comes today from photometry of comets that appear stellar on CCD images (see an overview of observational works and results in Lamy et al. 2004).

7.2.3. Coma substraction

For active comets that come close to the Earth it can still be possible to obtain high-resolution images that permit to discriminate the light from the solid nucleus from that from the coma. James Scotti, one of the most prolific observers of comet magnitudes, has been a pioneer in the use of this technique. His comet campaign has actually been a byproduct of his work within the Spacewatch project, led by Tom Gehrels, aimed primarily at the discovery and follow-up of NEOs. This project has been carried out most of the time with a 91-cm, f/5 Newtonian located at Kitt Peak, Arizona. In order to solve the problem of coma contamination in the determination of nuclear magnitudes, Scotti introduced a crude method of coma substraction. In his approach, the average coma background per pixel is estimated by hand and then substracted from the brightest ten or so pixels, where the nucleus is assumed to be located, or about two arcseconds radius from the photometric center. In all this, an optically thin coma is assumed. Despite the crudeness of the method, a large percentage of Scotti's nuclear magnitudes are in remarkably good agreement with later more sophisticated measurements, showing the usefulness and pertinence of coma substraction methods.

More recently, Philippe Lamy and co-workers have introduced a more refined method of coma substraction applied to comet images of very high spatial resolution, taken with the Hubble Space Telescope when the target comets are close to the Earth. Their approach is to maximize the contrast between the brightness of the "stellar" source (the comet nucleus) and the coma that expands and dilutes in the field of view as the comet gets closer to the Earth. Astronomical images are blurred by the atmospheric turbulence (the *seeing*) and the instrumentation (telescope and detector). These effects are encompassed in the *Point Spread Function* (PSF) of the system. The image in the focal plane is obtained as the convolution of the astronomical image with this PSF. For instance, a stellar image (point source) will be described by the Dirac delta function $\delta(\rho)$, where ρ is the projected distance to the image center. If a stellar image is convolved with the PSF, a bell-shaped profile is obtained in the focal plane. Lamy and co-workers model the brightness contribution of the coma and the unresolved nucleus by a function of the form

$$B(\rho) = \frac{c_1}{\rho} + c_2\delta(\rho), \tag{7.1}$$

where c_1 and c_2 are parameters to be derived from the fitting between model and observations, and the coma is assumed to have a spatial density that varies with ρ^{-2}, so the surface brightness decreases as ρ^{-1}. This function was then convolved with the PSF to get a theoretical bright-

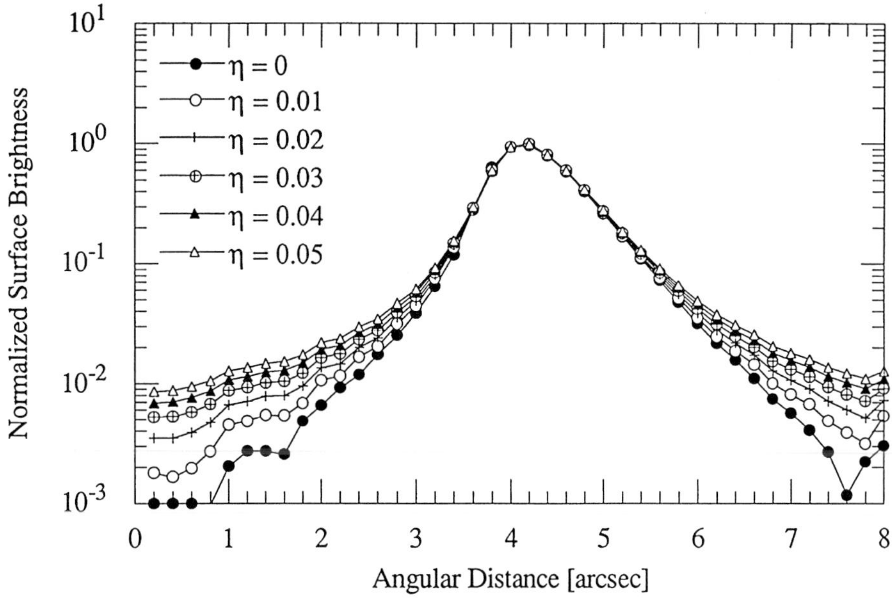

Figure 7.4. Theoretical profiles obtained from the convolution with the seeing of model images consisting of a point-source nucleus plus a coma of surface brightness $B(\rho) = K/\rho$, where K is a constant. The parameter η represents the ratio of the flux density scattered by the coma to the flux density scattered by the nucleus, so the profile for $\eta = 0$ will correspond to the "stellar" source (Luu and Jewitt 1992).

ness profile, which should be broader than the one obtained for a stellar source. We can see in Fig. 7.4 how a growing coma contribution reflects in a broader profile of the source. Values of the parameters c_1 and c_2 were obtained by fitting the theoretical profile to the observations. In this way, it was possible to discriminate the brightness contribution from the nucleus. By applying this technique, Lamy and co-workers have achieved remarkably good measurements of nuclear magnitudes (a summary of their results is presented in Lamy et al. 2005), although in some cases comets were observed at high phase angles, in which case the determination of the absolute nuclear magnitude depends on quite uncertain corrections for phase angle. For instance, if the correction for phase angle is 0.04 mag deg^{-1}, an observation at phase angle $\alpha = 90°$ would imply a correction of 3.6 mag, so changes in the phase law may introduce changes in the estimated absolute nuclear magnitude well above one mag. This problem is avoided when comets are observed far from the Sun, which usually implies small phase angles.

The coma substraction method implicitly rests on the assumption of an optically thin coma. Were the nucleus shrouded by an opaque coma, there would be no hope of optically detecting and measuring it. Let us now check if one expects to find optically thin or thick comas when comets come close to the Earth, that usually corresponds, unfortunately, to their most active period. We will follow the procedure described in Tancredi et al. (2000).

Let Q_d be the dust production rate in number of particles s^{-1}. For a spherically symmetrical dust coma with a uniform radial outflow velocity v_d, the number density of grains at a distance x from the nucleus is

$$n_x = \frac{Q_d}{4\pi x^2 v_d} \; . \tag{7.2}$$

and the extinction coefficient is

$$\tau_x = n_x \pi a^2 = \frac{Q_d a^2}{4 v_d x^2} \; , \tag{7.3}$$

where a is the radius of the optically important dust particles.

The equation of radiative transfer gives the change in the intensity I when the radiation passes from x to $x + dx$. It is given by

$$\frac{dI}{I} = -\tau_x dx = -\frac{Q_d a^2}{4 v_d x^2} dx. \tag{7.4}$$

Let I_o be the intensity at the nucleus surface $x = R_N$. Thus, by integrating eq. (7.4) from R_N to $x \to \infty$ we obtain the overall extinction by the dust coma, i.e.

$$\int_{I_o}^{I} \frac{dI}{I} = -\int_{R_N}^{x} \tau_x dx = -\frac{Q_d a^2}{4 v_d} \int_{R_N}^{x} \frac{dx}{x^2}$$

that leads to

$$\frac{I}{I_o} = \exp\left(-\frac{Q_d a^2}{4 v_d R_N}\right), \tag{7.5}$$

and considering the extinction in magnitudes instead of intensity we obtain

$$\Delta m = m - m_o = -2.5 \log\left(\frac{I}{I_o}\right) = \frac{2.5 \log e Q_d a^2}{4 v_d R_N} \simeq 0.27 \frac{Q_d a^2}{v_d R_N} \; . \tag{7.6}$$

Let now Q'_d be the dust production rate in mass. We then have $Q'_d = m_d Q_d$, where $m_d = \frac{4}{3}\pi a^3 \rho_d$ is the mass of each individual grain and $\rho_d \sim 1$ g cm^{-3} its mass density. If ψ is the dust to gas mass ratio of the material released by the comet, we have $Q'_d = \psi Q'_g$, where Q'_g is the gas production rate in mass which, under the assumption that the main gas component is water, can be computed as

$$Q'_g = 4\pi R_N^2 f Z m_w, \tag{7.7}$$

where f is the fraction of active surface area, Z is the gas production rate per unit area, and m_w the mass of the water molecule. Thus, we can compute Q_d as

$$Q_d = \frac{3\psi R_N^2 f Z m_w}{a^3 \rho_d}, \tag{7.8}$$

and substituting eq. (7.8) in eq. (7.6) we get

$$\Delta m \simeq 0.81 \frac{\psi f Z m_w R_N}{a \rho_d v_d}. \tag{7.9}$$

The dust outflow velocity can be expressed as $v_d \simeq 4.5\times10^4 (Z/Z_o)^{1/2}$ $[R_N(\mathrm{km})]^{1/2}$ cm s^{-1} (see a discussion in Fernández et al. (1999) and references therein), where Z_o is the gas production rate per unit area at $r = 1$ AU. For a comet of visual Bond albedo $A_v = 0.04$ we have $Z_o \simeq 3.2\times10^{17}$ mol cm^{-2} s^{-1} (cf. Fig. 9.1). By substituting the previous expression of v_d in eq. (7.9) we obtain

$$\Delta m \simeq 1.8 \frac{\psi f Z^{1/2} Z_o^{1/2} m_w}{a \rho_d} [R_N(\mathrm{km})]^{1/2}, \tag{7.10}$$

where Z_o, Z, m_w, a and ρ_d are expressed in c.g.s. units.

As an example, let us consider a typical JF comet of $R_N = 1$ km and perihelion distance $q \simeq 1.5$ AU. As we discussed in Section 3.9, the optically important grains in the coma have radii $a \sim 1$ μm. The gas production rate near perihelion is $\sim 1.2 \times 10^{17}$ mol cm^{-2} s^{-1}. Furthermore, if the comet has an active surface area of $f \simeq 0.1$ and a dust to gas mass ratio $\psi \simeq 0.5$ (cf. Section 3.10), we obtain

$$\Delta m \simeq 6 \times 10^{-4}.$$

The extinction is then $\sim 10^{-3}$ mag, i.e. too low to affect the estimated magnitude of the nucleus. More sophisticated models, in which the dust particles are released from the nucleus with a velocity a few times smaller than v_d and then accelerated by the sublimating gases to the terminal velocity v_d (e.g. Grün and Jessberger 1990), can yield somewhat greater extinctions, but still too low to be of great concern. Optically thick comae might only be expected in unusually large ($R_N \gtrsim 10$ km) and active ($f \sim 1$) comets, or during outbursts. In conclusion, in most reasonable scenarios we deal with optically thin comas in which the light from the solid nucleus can reach the observer with negligible extinction, which warrants the applicability of coma-substraction methods.

7.2.4. Thermal infrared

This technique relies heavily on modelling. Firstly, it is necessary to discriminate between the IR contribution of the nucleus from that of the coma by means of a procedure similar to that described in 7.2.3 for the visible. Afterwards we need to model the nucleus itself, what fraction of its surface is active and what fraction is covered by an insulating dust mantle. In essence what we measure is the thermal flux $F(\lambda)$ of the nucleus (after substracting the coma contribution) at a geocentric distance Δ. This flux is related to the Planck function $B(\lambda, T)$ by means of the equation

$$F(\lambda) = \epsilon \int\int B[\lambda, T(\theta, \phi, p_v)]\Omega dS, \tag{7.11}$$

where ϵ is the infrared emissivity, $T(\theta, \phi, p_v)$ is the surface temperature where θ and ϕ are the latitude and longitude measured from the subsolar point, $\Omega = (1/\Delta^2)\cos\theta\cos(\phi - \alpha)$ is the solid angle, α the phase angle, and $dS = R_N^2 \cos\theta d\theta d\phi$ an element of surface (see, e.g., Groussin et al. (2004a,b)).

The problem is further complicated since the surface temperature is different in the active and in the inactive zones (in the active zones part of the solar energy is spent in sublimating the surface ices, so the equilibrium temperature is lower than in the inactive zones). Therefore, the thermal flux can be decomposed in two terms, i.e.

$$F(\lambda)_{nucl} = (1 - f)F_{inac} + fF_{act}, \tag{7.12}$$

which introduces a new unknown f, i.e. the fraction of active surface area. The problem can then be solved only if we have another observational datum, as for instance the water production rate Q_{H2O} which is related to the fraction f (see eq. (9.7) in Chapter 9).

The several requirements as regards the infrared observations of comets, in particular that have to be close to Earth (in order to obtain a good resolution to discriminate the nucleus contribution from the coma), and to have at the same time measurements of gas production rate, make this method of very limited use.

7.2.5. Radar

In this method, a burst of microwaves is sent to the target comet and the power of the returned echo is measured. Therefore the intensity of the signal to be measured goes as Δ^{-4}, indicating that the method can be applied only to comets that approach the Earth to less than a few tenths AU. Radar has been used extensively in planetary astronomy for studies of the Moon, planets, satellites and asteroids. Radar offers the possibility to observe directly the nucleus of a comet, since the returning echo from the solid nucleus far exceeds that from the dust particles and plasma in the coma (Kamoun et al. 1982b).

There are notwithstanding severe problems to the applicability of this method to comets. To the requirement of closeness to Earth mentioned before, we must add the need of very accurate ephemeris, not only for aiming the radar correctly to the target, but for setting the receiver to the correct Doppler-shifted frequency for which we must know accurately the velocity of the comet. Furthermore, the returned echo will be broadened due to the nucleus rotation, so a knowledge of the rotation speed and orientation of the spin vector is also required, as well as the surface scattering properties.

The first - marginal - radar detection was of 2P/Encke in 1980 with the Arecibo Observatory radar system ($\lambda = 12.6$ cm) (Kamoun et al. 1982a). These authors estimated a radius of $1.5^{+2.3}_{-1.0}$ km. The uncertainty was very high because of all the problems mentioned above. In the last twenty years a few radar detections of comets have been achieved, although its usefulness for the nucleus size determination remains as very marginal.

7.3. The size distribution

The application of the above-mentioned methods, especially photometry of distant inactive comets and coma substraction, has led to a substantial volume of data on nuclear magnitudes, in particular of JF

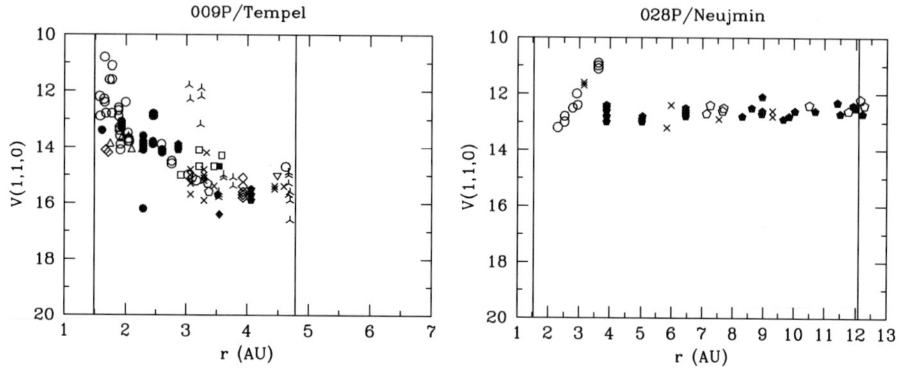

Figure 7.5. Plots of measured absolute nuclear magnitudes versus the heliocentric distances of the measurements for the two Jupiter family comets 9P/Tempel 1 and 28P/Neujmin 1. The different symbols are for different observers (open: before perihelion, filled: after perihelion) (details about observers are provided in Tancredi et al. 2000). The vertical lines indicate the minimum and maximum heliocentric distances of the comets attained during the period in which the observations were carried out.

comets. The first catalog of absolute nuclear magnitudes of JF comets, that included the contributions from several experienced observers of the last few decades, was compiled by Fernández et al. (1999) and Tancredi et al. (2000). The catalog has recently been updated by Tancredi et al. (2005). In several cases the data sets were found to be very self-consistent, thus allowing to extract meaningful values for the nuclear magnitudes. Figure 7.5 shows two examples of JF comets that have many estimates of their absolute nuclear magnitudes by different authors at different heliocentric distances. In the case of 9P/Tempel 1, we see that close to perihelion the absolute (visual) nuclear magnitudes $H_N = V(1,0,0)$ appear as brighter due to the strong coma contamination. As r increases, the values of H_N become fainter indicating that the measured magnitude approaches the true absolute magnitude of the nucleus (estimated at $H_N = 15.6$), as the comet activity decreases to undetectable levels. On the other hand, the measured nuclear magnitudes of 28P/Neujmin 1 are most of the time nearly constant, at $H_N \simeq 12.7$, indicating that it is very little active, even when it is near perihelion. The catalog of absolute nuclear magnitudes of JF comets compiled by Tancredi et al. (2005) is presented in Table A2.1 of Appendix 2.

From the estimated absolute nuclear magnitudes we can determine the cumulative luminosity function (CLF) and the size distribution of

JF comets in a straightforward manner, without resorting to dubious correlations between total magnitudes and sizes. Shoemaker and Wolfe (1982) used the more restricted sample of Roemer's data of absolute nuclear magnitudes to derive a linear relationship $\log N_c(H_N) = C + 0.4H_N$, up to the absolute blue magnitude 16, where $N_c(H_N)$ is the cumulative number of comets brighter than H_N and C is a constant. Unfortunately, Roemer's nuclear magnitudes were too bright because of unaccounted coma contamination, which makes Shoemaker and Wolfe's result more of a historic value rather than a meaningful one. We can now re-discuss the CLF of JF comets with a much more reliable data set of nuclear magnitudes. In order to obtain a good CLF we must have a sample as unbiased as possible, so it seems appropriate to restrict it to comets with perihelion distances $q < 2$ AU which are more prone to be detected and studied. In agreement with Shoemaker and Wolfe study, Fernández et al. (1999) found a CLF that follows a linear relation

$$\log [N_c(H_N)] = C_1 + \gamma H_N. \tag{7.13}$$

Bearing in mind that $H_N = k - 2.5 \log B_N$, and that $B_N \propto R_N^2$, where B_N is the nucleus brightness, from the CLF we can derive the cumulative size distribution (CSD)

$$\log N_R(R_N) = C_2 - s \log R_N, \tag{7.14}$$

where $N_R(R_N)$ is the number of comets with radii $> R_N$, and $s = 5\gamma$.

The cumulative luminosity functions of two samples of JF comets with perihelion distances smaller than given values are plotted in Fig. 7.6. The values of H_N were drawn from Table A2.1 as derived by Tancredi et al. (2005). From the linear fits to the CLFs, Tancredi et al. (2005) estimate a slope of $\gamma = 0.56 \pm 0.03$ for the JF population. The observed cumulative number progressively departs from the linear fit for comets fainter than $H_N \simeq 16.5$, which may be explained as due to the growing incompleteness of the sample of smaller JF comets. From this value of γ we can derive an exponent $s = 5\gamma = 2.80 \pm 0.15$ for the CSD. We note that the conversion from the CLF to the CSD is straightforward as long as we assume a constant geometric albedo p_v. There are very few determined p_v (see Table 7.1 and Section 7.6 below) and certainly it varies from comet to comet. Nevertheless, p_v seems to be very low for all the comets, so we can adopt an average $p_v = 0.04$ that, for the time being, seems to be accurate enough for our discussion here.

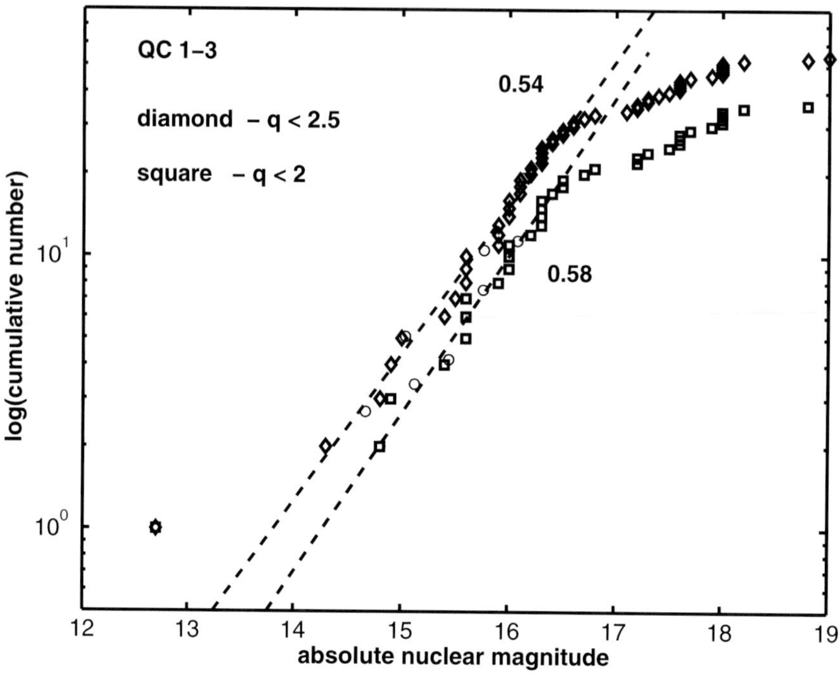

Figure 7.6. The cumulative luminosity function of JF comets with perihelion distances: $q < 2.5$ AU (diamonds), and $q < 2$ AU (squares). Only comets of Quality Classes 1-3 (as defined in Table A2.1) are considered. The value of the slope is indicated beside each linear fit (Tancredi et al. 2005).

There have also been other recent estimates of the slope s of the CSD of JF comets (a list of these estimates and their respective references is given in Table 7.2). We can see a stricking discrepancy between the value of s obtained by Tancredi et al. and those from the other authors, which are smaller, i.e. their CSDs are flatter. Yet, Tancredi et al. argue that if all the samples of JF comets used by different authors are restricted to $q < 2$ AU, if we make allowance for some discrepant magnitudes of a few large comets (and thus with a large weight in the CLF), and/or if other presumably non-JF comets are removed from their samples (e.g. Halley-type comets or asteorids suspected to be of cometary origin), then all the linear fits give slopes s close to 2.5, more in line with the value presented above.

We can then be satisfied to have been able to assemble, in a matter of a few years, enough good data of sizes of cometary nuclei to discuss their size distribution on a much more solid basis than was possible before. However, despite the advances in this matter, a note of caution is necessary. First, the linear fits are with very few points, so we need to have more complete comet samples down to smaller radii to obtain more reliable results. Second, the sample of comets with $q < 2$ AU represents a physically evolved population, so its original size distribution could have been changed due to sublimation, outbursts and splittings (in some cases leading to the production of daughter comets). In the future, it will be very interesting to compare this sample with others less evolved physically, as for instance, comets in the Jupiter's region or beyond.

Table 7.2: Estimated power-law exponents of some cumulative size distributions

Population	Exponent s	source
JF comets	2.80 ± 0.15	(1)
	1.66 ± 0.016	(2)
	1.6 ± 0.1	(3)
	1.59 ± 0.03	(4)
	1.91 ± 0.06 for $2 < R_N < 5$ km	(5)
	1.45 ± 0.05 for $1 < R_N < 10$ km	(5)
NEAs	1.8	(6)
Trojans	2.0 ± 0.3	(7)
Collisional model	2.5	(8)
Collisional model	2.5 - 3.0	(9)

(1) Tancredi et al. (2005)
(2) Lamy et al. (2005)
(3) Lowry et al. (2003)
(4) Weissman and Lowry (2004)
(5) Meech et al. (2004)
(6) Bottke et al. (2000)
(7) Jewitt et al. (2000)
(8) Dohnanyi (1969)
(9) Kenyon and Bromley (2004)

The power-law CSD of index 2.8 found by Tancredi et al. (2005) is somewhat higher than the theoretical distribution of index 15/6 expected for self-similar collision cascades as predicted by Dohnanyi (1969) for main-belt asteroids, but it is in reasonable agreement with Kenyon and Bromley's (2004) model for icy bodies in the trans-neptunian belt. Kenyon and Bromley actually obtained two power-laws; the range of exponents quoted in Table 7.2 is for small objects with radii $\lesssim 0.1 - 1$ km. For larger bodies (radii $\gtrsim 10 - 100$ km) they obtain a larger exponent 3.5. The size distribution of other collisionally evolved populations of comet-sized bodies, like near-Earth asteroids (NEAs) and Trojans, have also been derived from sky surveys. The slopes s are found to be smaller than the slope found here for the JF population: $s = 1.8$ for NEAs (Bottke et al. 2000), and $s = 2.0 \pm 0.3$ for Trojans (Jewitt et al. 2000). Of course, we do not expect to find the same CSD for the different populations of minor bodies in the solar system given their different physical nature and collisional environment. We do not include studies of the size distribution of trans-neptunian objects because these have been limited to 100-km class objects and, therefore, much larger than comets. The size distribution of trans-neptunian objects will be studied in the next chapter.

7.4. Nongravitational forces and masses

Since comets have no known satellites and no spacecraft have so far had a rendezvous with a comet to measure the gravitational pull from the comet on the spacecraft, the only effect that can give a direct estimate of the mass of the comet nucleus M_N is the nongravitational (NG) acceleration $\vec{J}$. This was already noted by Whipple (1950) in his model of NG forces acting on a comet nucleus. If Q is the gas production rate (in molecules s^{-1}), $\vec{u}$ is the effective outflow velocity, and m is the average molecular mass, the conservation of momentum leads to

$$M_N \vec{J} = -Qm\vec{u}. \qquad (7.15)$$

The effective outflow velocity is given by $u = \zeta(8kT/\pi m)^{1/2}$, namely the thermal speed multiplied by a factor ζ that takes into account the degree of collimation of the flux, collisions among the gas molecules, and recondensation on the surface (Wallis and Macpherson 1981, Rickman 1986). Rickman (1989) estimated $\zeta \simeq 0.5 - 0.6$. As discussed (see Section 4.3), for periodic comets the most noticeable NG effect is the delay or advance of the perihelion passage or, in other words, the variation in the orbital period ΔP given by

$$\Delta P = \frac{6\pi\sqrt{1-e^2}}{n^2}\left[\frac{e}{p}\int_0^P J_r \sin f\,dt + \int_0^P \frac{J_t}{r}dt\right], \tag{7.16}$$

where J_r and J_t are the radial and transversal components of the NG acceleration, and $p = a(1-e^2)$ is the parameter.

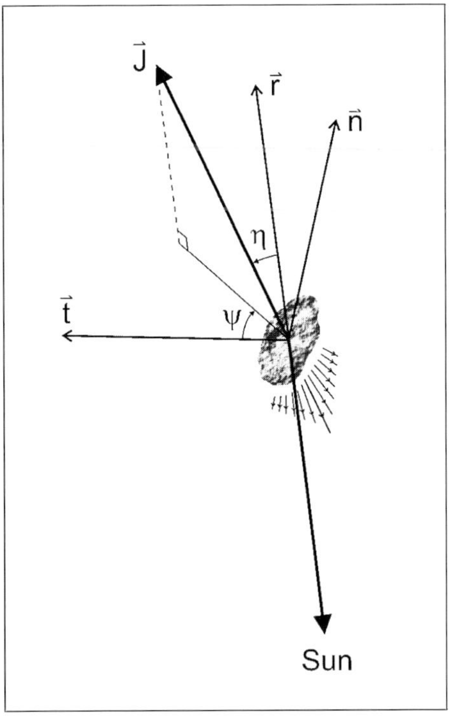

Figure 7.7. The direction of the jet force described by the angles ψ and η in a reference frame defined by the radial direction $\vec{r}$, the transverse component in the orbital plane $\vec{t}$ and the normal $\vec{n}$.

The direction of the jet force can be described by the angle η with respect to the antisolar direction, and an azimuthal angle ψ in the plane perpendicular to $\vec{r}$ (Fig. 7.7). We can then write

$$J_r = J\cos\eta, \tag{7.17a}$$

$$J_t = J\sin\eta\cos\psi. \tag{7.17b}$$

η can be identified with the thermal lag angle, so it will be a function of the nucleus spin's period and on the thermal inertia. In general,

η and ψ will vary with the orbital position, i.e. $\eta = \eta(t)$, $\psi = \psi(t)$. Unfortunately, we cannot know the values of these angles, unless we can study the comet nucleus *in situ*. Therefore, we depend on educated guesses and averages to sort this problem out. The angle η depends on the thermal inertia of the outermost layers of the nucleus, the rotation period and the heliocentric distance. For most reasonable combinations of the values of these parameters we obtain $\eta \lesssim 30°$ (Rickman 1986, Rickman et al. 1987), so we can take $< \cos\eta >\sim 1$. Let η_e be the effective lag angle such that $\sin\eta_e =< \sin\eta\cos\psi >$ (e.g. Festou et al. 1990). If we assume that ψ has a random value in the range $(0, 2\pi)$, we get $\sin\eta_e \simeq \pm 0.1$

By substituting J_r and J_t of eq.(7.16) by eqs.(7.17a) and (7.17b) respectively, and then substituting J by eq.(7.15), we finally obtain for the comet mass

$$M_N \simeq \frac{6\pi\sqrt{1-e^2}}{n^2\Delta P} m < u > \left[\frac{e}{p}\int_0^P Q\cos\eta\sin f dt + \int_0^P \frac{Q}{r}\sin\eta\cos\psi dt\right]$$

$$\simeq \frac{6\pi\sqrt{1-e^2}}{n^2\Delta P} m < u > \left[\frac{e}{p}\int_0^P Q\sin f dt + \sin\eta_e\int_0^P \frac{Q}{r}dt\right], \qquad (7.18)$$

where we have considered an average effective outflow velocity $< u >$ for all escaping molecules during an orbital revolution of period P. We have $< u >\simeq 0.25$ km s^{-1} (Wallis and Macpherson 1981).

The computation of M_N will require to know from observations the shape of the curve $Q = Q(t)$. For a few cases there are good measurements of gas production rates at different orbital positions, but in general we have to rely on lightcurves, as those shown for comets C/1995 O1 (Hale-Bopp) and 1P/Halley in Figs. 3.4 and 3.5 respectively. In the latter case we have to prescribe some formula that approximately converts total magnitudes to gas production rates. From least-square linear fits, Festou (1986) derives the following empirical law

$$\log Q_{OH}(\mathrm{s}^{-1}) = 32.0 - 0.4h_T, \qquad (7.19a)$$

where $h_T = m_T - 5\log\Delta$ is the heliocentric total magnitude, m_T is the apparent total magnitude, and Δ is the geocentric distance. Yet, from the study of another sample of periodic and nonperiodic comets, Jorda et al. (1992) obtained a linear relation with different numerical coefficients

$$\log Q_{H2O}(\mathrm{s}^{-1}) = 30.74 - 0.24h_T, \qquad (7.19b)$$

where the authors assumed $Q_{H2O} = 1.1Q_{OH}$. Festou obtained the Q_{OH} data from OH emission measurements at 3064 Å by the *International Ultraviolet Explorer*, whereas Jorda et al. obtained their data from the 18-cm radio line measured with the Nançay radiotelescope.

The different empirical laws, expressed by eqs. (7.19a) and (7.19b), illustrate the uncertainties involved in formulas relating $\log Q_{H2O}$ with h_T. For $h_T \simeq 8.1$ we obtain the same value of $\log Q_{H2O}$ from either eq. (7.19a) or eq. (7.19b), but increasingly diverging values of $\log Q_{H2O}$ are obtained for lower or higher h_T values. A major source of uncertainty is undoubtedly the dependence of the coma luminosity on both the gas production rate and the dust production rate. Some comets may be dust-rich in which case the dust coma contributes most of the visual brightness.

Given the uncertainties in several parameters involved in eq.(7.18), it is only possible to make a rough estimate of cometary masses. We stress that the only observed NG effect required for the application of this model is the variation of the orbital period ΔP, so it can only be applied to comets observed in more than one apparition, in particular to JF comets that are the bulk of multiple-apparition comets. If in addition to the comet mass we have the nucleus size determined by one of the methods described before, we can then compute the mass density of the nucleus, a fundamental physical parameter to learn about the geochemistry of the nucleus and the compactness of the material or, in other words, its porosity. It is interesting to note that the results obtained for comets 1P/Halley, 22P/Kopff and several others JF comets, indicate densities below 1 g cm^{-3}, or even below 0.5 g cm^{-3}, thus suggesting a very fluffy material (Wallis and Macpherson 1981, Rickman 1986, Rickman et al. 1987, Rickman 1989).

7.5. Lightcurves, shapes and rotation periods

There is a good photometric coverage for a few JF comets at large heliocentric distances where they do not show measurable activity. It has then been possible to obtain good rotational lightcurves for these comets, which may be associated to changes in the photometric cross-section by the rotation of a body of irregular shape, and/or variations in the surface albedo. Yet, the albedo did not show significant variations in a few comets for which it was possible to measure it at different rotational phases by combining data from the thermal infrared and the visible (Jewitt et al. 2003). This seems to discard the albedo as an important cause for rotational lightcurves. If we then attribute the lightcurve to the rotation of a body of elongated shape, that may be fitted to a prolate ellipsoid of semiaxes a and $b = c$ rotating about one of the minor axis, the ratio between the maximum and minimum brightness will be given by the ratio of the maximum to the minimum geometric cross-section, namely $\pi ab/\pi b^2 = a/b$. This will be related to the amplitude of the lightcurve (given as the difference of nuclear magnitudes corresponding to the maximum and minimum brightness Δm_N) by

$$\frac{a}{b} = 10^{-0.4\Delta m_N}. \tag{7.20}$$

Good lightcurves have been obtained for about a dozen JF comets for which the ratio a/b has been derived from the previous equation. Table 7.3 brings the list of comets for which b/a and P have been determined from their lightcurves. The table also includes the effective radius $R_{N,eff}$, as defined in 7.2.1. Figure 7.8 brings the example of the lightcurve of 48P/Johnson obtained by Jewitt and Sheppard (2004).

TABLE

Table 7.3: Sizes, shapes and spin periods of JF comets derived from their lightcurves[(*)]

Comet	$R_{N,eff}$ (km)	b/a	P (hr)
2P/Encke	2.4 ± 0.3	0.56 ± 0.03	15.8 ± 0.08
9P/Tempel 1	2.6 ± 0.2	0.58 ± 0.10	41.5 ± 0.5
10P/Tempel 2	5.8 ± 0.5	0.52 ± 0.05	8.95 ± 0.01
19P/Borrelly	2.8 ± 0.3	0.40 ± 0.07	25.0 ± 0.5
22P/Kopff	2.3 ± 0.3	0.66 ± 0.05	12.91 ± 0.05
28P/Neujmin 1	9.7 ± 0.8	0.63 ± 0.05	12.67 ± 0.05
31P/Schwassmann-Wachmann 2	3.1 ± 1.0	0.63 ± 0.05	5.58 ± 0.03
46P/Wirtanen	0.6 ± 0.02	0.83 ± 0.04	6.0 ± 0.3
48P/Johnson	2.6 ± 0.2	0.74 ± 0.03	29.0 ± 0.04
49P/Arend-Rigaux	5.0 ± 0.5	0.52 ± 0.05	13.56 ± 0.16
143P/Kowal-Mrkos	5.7 ± 0.6	0.66 ± 0.03	17.10 ± 0.01

(*) The references of these data, except for 48P/Johnson, are found in Jewitt et al. (2003) and in Jewitt and Sheppard (2004) for 48P/Johnson

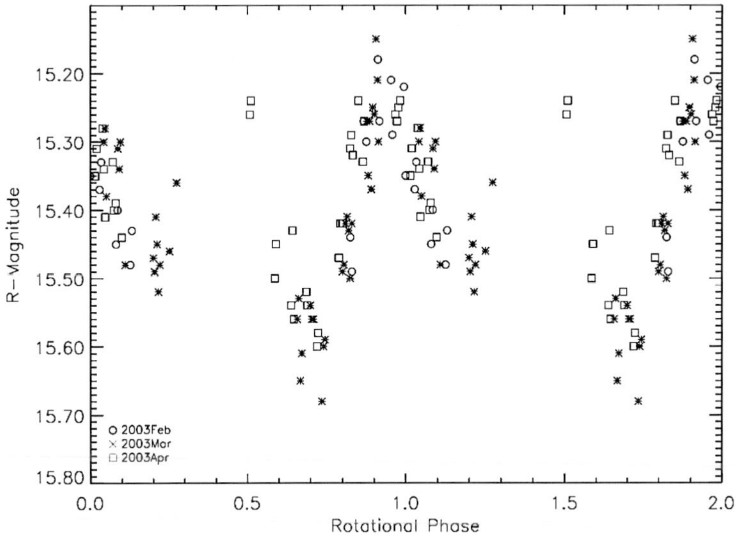

Figure 7.8. Lightcurve of 48P/Johnson phased to a common double-peaked period $P = 28.996$ hr (Jewitt and Sheppard 2004).

The results show that the nuclei of JF comets are on average more elongated than small main belt asteroids (SMBAs) and fragments produced by catastrophic disruption of targets in the laboratory. The mean axial ratio projected onto the plane of the sky is $\overline{b/a} = 0.61 \pm 0.04$ for JF comet nuclei, while it is $\overline{b/a} = 0.74$ for main belt asteroids of comparable size (Jewitt et al. 2003). The distribution of photometric amplitudes of JF comet nuclei, SMBAs and fragments produced in the laboratory are shown in Fig. 7.9.

Why could comet nuclei be so elongated as compared to other solar system bodies and collisional fragments? Jewitt et al. (2003) discard several possible observational biases (as, e.g., albedo variations), and note that the nuclei of comets Halley and Borrelly are indeed very elongated (see Table 7.1), so they conclude that very elongated shapes are indeed a real effect. Jewitt et al. provide a physical reason to explain why such an extreme shapes occur so frequently in terms of a sublimation that proceeds uniformly over the entire surface of the comet nucleus. Thus, if we assume that a comet nucleus has a (moderate) initial axial ratio $(a/b)_o$ and that after entering the planetary region it loses a layer of thickness Δz per orbital revolution due to sublimation, after N revolutions the nucleus axes will have shrunk to: $a = a_o - N\Delta z$ and $b = b_o - N\Delta z$, which holds provided that $N\Delta z < b_o$. The axial ratio then becomes

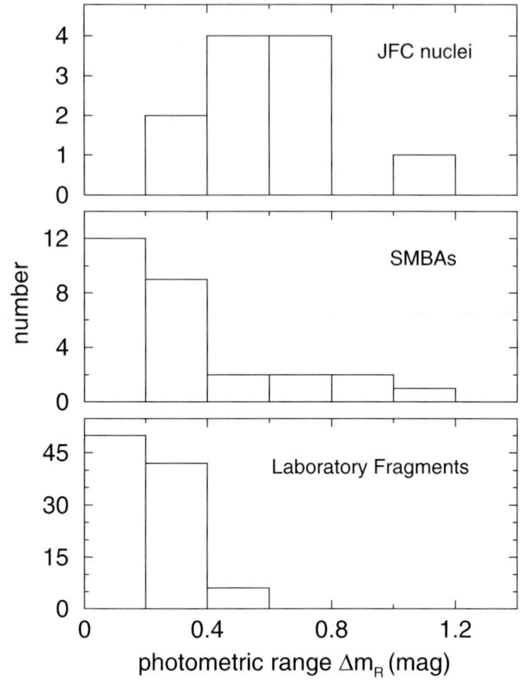

Figure 7.9. Distribution of photometric (red) amplitudes of JF comet nuclei, small main belt asteroids, and fragments of targets collisionally disrupted in the laboratory (figure adapted from Jewitt et al. 2003).

$$\frac{a}{b} = \left(\frac{a}{b}\right)_o \left(\frac{1 - N\Delta z/a_o}{1 - N\Delta z/b_o}\right), \tag{7.21}$$

since $b_o < a_o$ we have that $a/b > (a/b)_o$.

Jewitt et al.'s explanation seems to be plausible, provided that the erosion rate is controlled by sublimation and that it proceeds uniformly over the entire surface. This leaves the problem of the role played by splittings in the mass loss and ultimate demise of comet nuclei, since splittings may not favor the formation of more elongated bodies and/or highly elongated daughter comets. Another possibility is that some phase-dependent residual activity of the comet nucleus distorts and artificially enlarges the magnitude difference between peaks and troughs of the lightcurve. In this regard, we note that some of the effective radii of Table 7.3 are somewhat larger than those presented in Table A2.1 of Appendix 2 or, in the case of 19P/Borrelly, of the true effective radius as derived from the Deep Space 1 mission (Fig. 7.10).

We can evaluate what is the relationship among the axial ratio a/b, the rotation period P and the density ρ_N of the comet nucleus for

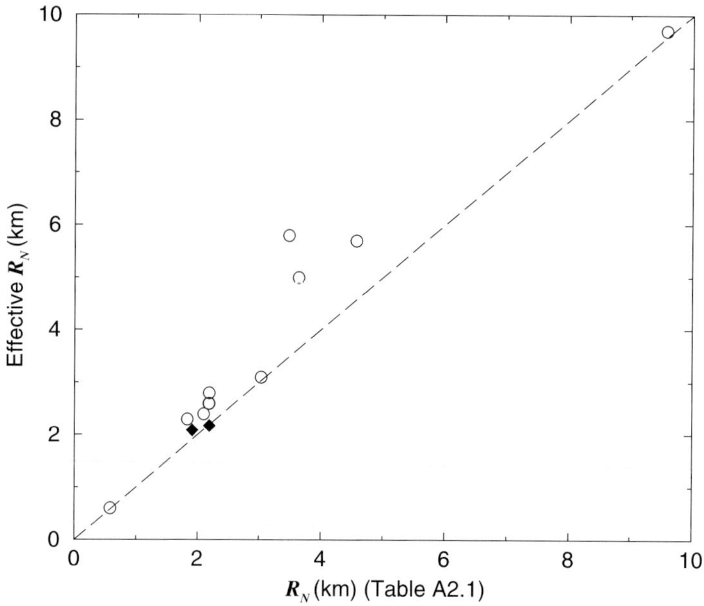

Figure 7.10. Estimated effective nucleus radii from Table 7.3 (open circles) and those of 19P/Borrelly and 81P/Wild 2 obtained from flyby missions (filled diamonds), versus the nucleus radii of the same comets from Table A2.1 of Appendix 2.

rotation stability. A criterion for the critical limit of the rotation period $P_c(= 2\pi/\omega_c)$ can be given by equating the acceleration of gravity at the surface with the centrifugal acceleration at the equator. For a prolate spheroid of semiaxes a, $b = c$, the acceleration of gravity at the tip of the long axis is decreased with respect to that of a sphere of radius a by a factor b/a, so the condition for gravitational instability becomes (Pravec and Harris 2000)

$$\frac{Gmb}{a^3} = \omega_c^2 a,$$

which leads to

$$P_c \approx \frac{3.3 \text{ h}}{\sqrt{\rho_N}} \sqrt{a/b}, \qquad (7.22)$$

where $m = 4/3\pi a^3 \rho_N$.

In Fig. 7.11 we plot the axial ratio a/b versus the orbital period for a sample of NEAs taken from Pravec and Harris (2000), and for a sample of JF comet nuclei taken from Jewitt et al. (2003). Most NEAs have

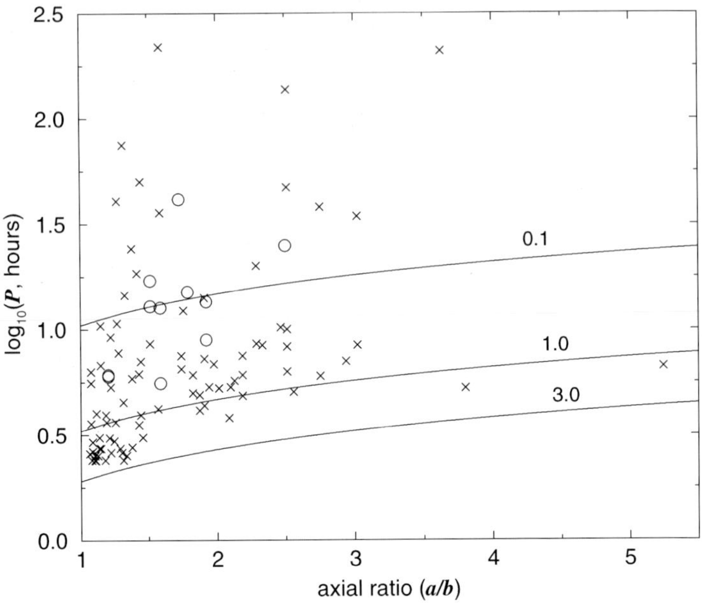

Figure 7.11. The axial ratio derived from amplitude lightcurves versus rotation period for near-Earth asteroids (x), and JF comet nuclei (open circles). The samples have been taken from Pravec and Harris (2000) and Jewitt et al. (2003) respectively. The solid curves are the limits for rotational instability of strengthless bodies for densities 0.1, 1.0 and 3.0 g cm^{-3}.

diameters $D < 10$ km and all of them have $D > 0.2$ km, so their range of sizes overlaps that of the JF comet nuclei. We also plot the curves for rotational instability for strengthless bodies of densities 0.1, 1.0 and 3.0 g cm^{-3}. We do not see here a clear trend for JF comet nuclei to be on average more elongated than NEAs, though many NEAs get close to the stability limit for $\rho = 3$ g cm^{-3}, while all of the JF comet nuclei fall in the region of rotational stability for densities < 1 g cm^{-3}, in agreement with the low nuclear densities derived from nongravitational forces (cf. previous section). On the other hand, Margot et al. (2002) estimate that about 16% of NEAs larger than 0.2 km may be binary systems. Binary systems and contact binaries, in which two bodies stuck together, may also produce lightcurves with very wide amplitudes. Such binary systems presumably form after the collisional disruption of a parent asteroid. Therefore, it is possible that contamination with binary systems biased the lightcurves of NEAs toward larger amplitudes.

7.6. Color, albedo and phase function

Colors have already been determined for several comet nuclei, mainly in the V, R bands. Most of them belong to the Jupiter family. The mean value is found to be $< V - R > \simeq 0.42$ (Meech et al. 2004, Lamy et al. 2005), which is somewhat redder than the solar color $(V - R)_{\odot} = 0.35$, but substantially less red than the average color index of trans-neptunian objects $< V - R >_{TNO} = 0.61$. We note however that there is a high dispersion in the observed colors of JF comets, going from $(V - R) = 0.02 \pm 0.22$ for 14P/Wolf to 0.62 ± 0.08 for 6P/d'Arrest (Lamy et al. 2005).

By combining data from the thermal infrared with that from the visible, it has also been possible to derive the geometric albedos of a few comets, most of them JF comets. If we neglect sublimation of ices and heat conduction into the nucleus interior, the solar radiation absorbed by the comet nucleus of Bond albedo A_v will be re-emitted as thermal radiation. Furthermore, if we assume that the nucleus surface reaches an uniform equilibrium temperature, the thermal energy will be isotropically re-radiated, so the observer at a distance Δ to the comet will measure an infrared flux ϵ_{IR} given by

$$\epsilon_{IR} \simeq \frac{1}{4}\frac{(1 - A_v)F_{\odot}R_N^2}{r_{UA}^2\Delta^2}, \tag{7.23}$$

where the different parameters have the same meaning as in eq.(3.3). In the other extreme case in which we assume that only the diurnal hemisphere re-radiates thermal energy, instead of a factor 1/4 we should insert a factor 1/2. In a more realistic situation we would have a factor in between 1/4-1/2. Equation (2.7) provides another relation among the nuclear visual magnitude, the geometric albedo p_v and the nucleus radius R_N. An additional equation: $A_v = p_v q$ relates the Bond and geometric albedos, where q is the phase integral. A solution of the problem then requires to define the phase integral.

By applying the above technique, comet nuclei are found to be very dark, with albedos in the range $\sim 0.02 - 0.06$ (Lamy et al. 2005). These extremely low albedos agree with those found in situ from spacecrafts for comets 1P/Halley, 19P/Borrelly and 81P/Wild 2.

As said before, for some problems we need to know the relation between the Bond and geometric albedos, which depends itself on the phase function $\phi(\alpha)$ (cf. Sections 2.4 and 3.4). This is not a trivial problem since the determination of $\phi(\alpha)$ requires to measure the reflected sunlight on the bare nucleus at diferent phase angles α. Unfortunately, comets can be observed at large α only when they are close to the Sun ($r \lesssim 1.5$ AU), and thus usually very active. Therefore, good photometric measurements at large phase angles can only be obtained from flyby missions, or from Earth-based high-resolution images in which a coma-substraction method is applied.

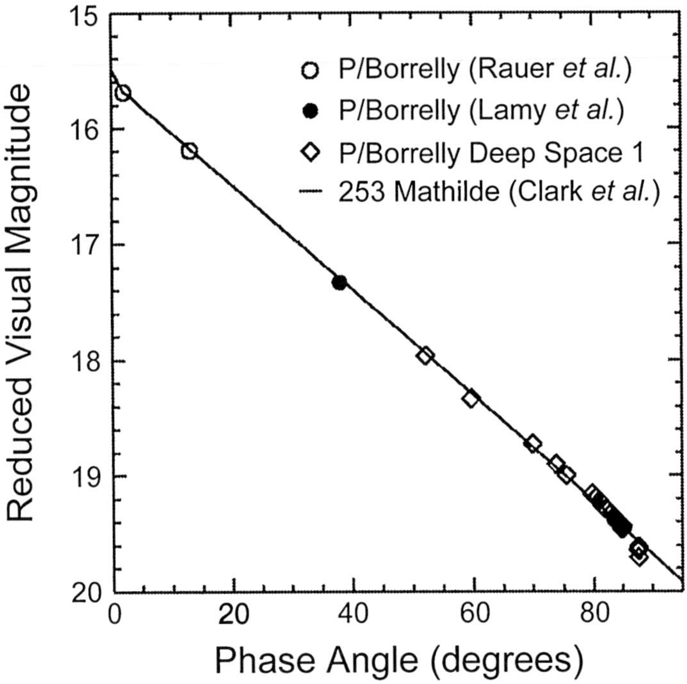

Figure 7.12. Phase curve for the nucleus of 19P/Borrelly. The plotted values combines Earth-based observations at small phase angles with observations with the Miniature Integrated Camera and spectrometer (MICA) aboard the *Deep Space 1* spacecraft at large phase angles. The straight line is the phase dependence observed for asteroid 253 Mathilde (Soderblom et al. 2002).

Figure 7.12 shows the phase curve for 19P/Borrelly that combines Earth-based observations at small phase angles with those from the *Deep Space 1* spacecraft at phase angles between 52° and 88° (Soderblom et al. 2002). The measured magnitudes at different phase angles fit very well a linear relation of slope $\beta = 0.04$ mag deg^{-1}. The dark carbonaceous asteroid 253 Mathilde, encountered by the NEAR spacecraft in 1997, shows a similar phase curve. The large low-active comet 28P/Neujmin 1 has also been observed within a range of phase angles ranging from near-opposition to $\alpha \sim 15°$. For this comet, Delahodde et al. (2001) found a shallower linear phase law with a phase coefficient $\beta = 0.025 \pm 0.006$ mag deg^{-1}. Yet, it is too premature to assign too much weight to this determination which should require nuclear magnitude measurements over a much wider range of α. From a collection of photometric and infrared observations of 2P/Encke from different sources in the wide range $2.5° < \alpha < 117°$, Fernández et al. (2000) derive a linear phase function $\beta = 0.06$ mag deg^{-1}. From these examples we conclude that a phase coefficient $\beta = 0.04$ mag deg^{-1} seems to be a typical value to adopt in general, unless a more accurate value is measured for a given comet.

If we assume that the linear phase function is applicable to all the range $0 < \alpha < 180°$, and remembering that $2.5 \log \phi(\alpha) = -\beta\alpha$ (cf. Section 2.4), we obtain a phase integral (cf. eq.(3.2))

$$q_\alpha = \int_0^\pi 10^{-\beta\alpha/2.5} \sin \alpha d\alpha \simeq 0.37, \qquad (7.24)$$

which allows us to relate the Bond and geometric albedos through the equation $A = pq_\alpha$.

7.7. Dynamics of close encounters

As mentioned, most aphelia of JF comets lie very close to Jupiter's orbit, so their dynamics is dominated by close encounters with Jupiter. Some JF comets may fall temporarily in resonances that keep them safe from suffering close encounters with Jupiter, but only for very short time spans. It is also possible that nongravitational forces decouple some JF comets from Jupiter's influence zone. The paradigmatic case is 2P/Encke whose aphelia $Q = 4.1$ AU is well inside Jupiter's orbit. Fernández et al. (2002a) computed the probability that a JF comet acquires an Encke-type orbit at any time during its dynamical evolution, assuming that a nongravitational force similar to that acting on Encke acts on the comet. The probability was found to be 2×10^{-3} per comet, so if the whole population of JF comets is of the order of several thousands, there might be at any time not more than a handful of JF comets in Encke-type orbits. But except for these few exceptional cases, for the rest repeated close encounters with Jupiter will be the norm, producing strong perturbations in their orbits. As a consequence, dynamical lifetimes of JF comets are relatively short, of $\sim 2 \times 10^5$ yr

(Fernández et al. 2002a), as compared with other populations, as for instance NEAs which is of $\sim 10^7$ yr (Gladman et al. 1997).

The encounter velocity U of a body with respect to Jupiter can be expressed in terms of the Tisserand invariant T as (Öpik 1951)

$$U = (3 - T)^{1/2}, \tag{7.25}$$

where U is expressed in terms of Jupiter's orbital velocity v_J. However, because of the ellipticity of Jupiter's orbit and the perturbations of the other planets, T and U will slowly change with time. Multiple encounters of a small body with Jupiter will cause a secular increase of U due to the Fermi acceleration mechanism (Arnold 1965). The variations are nevertheless small compared with the variations of the orbital elements of the perturbed body, so T and U can still be useful parameters for discussing the dynamics of JF comets and their source regions. Furthermore, the time scale for a significant increase of U is much longer than the lifetimes of JF comets. From these considerations we shall assume in the following that the magnitude of $\vec{U}$ is preserved after an encounter with Jupiter, changing only its direction in a random manner.

Let us characterize the encounter velocity vector $\vec{U}$ of a JF comet after an encounter with Jupiter by its magnitude U and the angles θ and ϕ defined in Fig. 7.13. The semimajor axis a will only depend on U and θ through the equation (Öpik 1951)

$$A^{-1} = 1 - U^2 - 2U\cos\theta \tag{7.26}$$

where $A = a/a_J$.

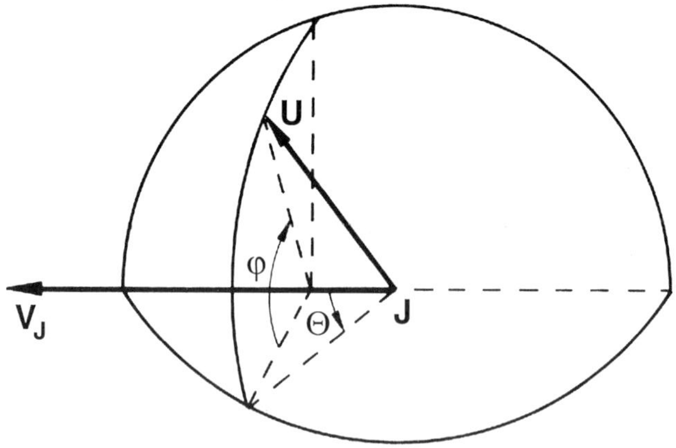

Figure 7.13. Geometry of a comet encounter with Jupiter.

We can derive the perihelion distance q of the cometary orbit after the encounter as a function of U and the angles θ and ϕ, obtaining (Fernández 1984)

$$q = a\left\{1 - \sqrt{1 - A^{-1}[1 + U^2(1 - \sin^2\theta\cos^2\phi) + 2U\cos\theta]}\right\}. \qquad (7.27)$$

For elliptic orbits we have to set the condition $A > 0$, which leads to

$$\theta > \theta_{ej} = \cos^{-1}\left[Min\left(1, \frac{1 - U^2}{2U}\right)\right]. \qquad (7.28)$$

The angle θ_{ej} defines the "ejection cone" whose axis is $\vec{v_J}$. If after an encounter $\vec{U}$ falls within the ejection cone, the comet will be ejected in a hyperbolic orbit. Thus, the volume within the ejection cone will remain empty. Of course, for $U < \sqrt{2} - 1$ we have $\theta_{ej} = 0$, so comet ejection is not possible. In principle the angles θ and ϕ can have values in the ranges: $[\theta_{ej}, \pi]$ and $[0, 2\pi]$ respectively.

For an ensemble of comets having exit vectors $\vec{U}$ with a distribution of magnitudes $f_U(U)$ and angular distributions $f_\theta(\theta)$ and $f_\phi(\phi)$, the fraction of comets with values of (U, θ, ϕ) in the ranges: $[U, U + \delta U]$, $[\theta, \theta + \delta\theta]$, and $[\phi, \phi + \delta\phi]$ is

$$F(U, \theta, \phi)\delta U\delta\theta\delta\phi = f_U(U)\delta U f_\theta(\theta)\delta\theta f_\phi(\phi)\delta\phi. \qquad (7.29)$$

These comets will have perihelion distances in the range: $[q, q + \delta q]$ which can be computed from eq.(7.27), where the distributions of the angles θ and ϕ are

$$f_\theta(\theta)d\theta = \frac{\sin\theta}{1 + \cos\theta_{ej}}d\theta, \qquad (7.30a)$$

$$f_\phi(\phi)d\phi = \frac{d\phi}{2\pi}, \qquad (7.30b)$$

which hold under the assumption that the vector $\vec{U}$ is randomly oriented, which is fulfilled for the observed sample of JF comets (Lowrey 1973). In repeated encounters with Jupiter, the Jovicentric velocity will reach a kind of equipartition in all directions (Öpik 1965), so the distribution of U can be described by a three-dimensional Maxwellian law

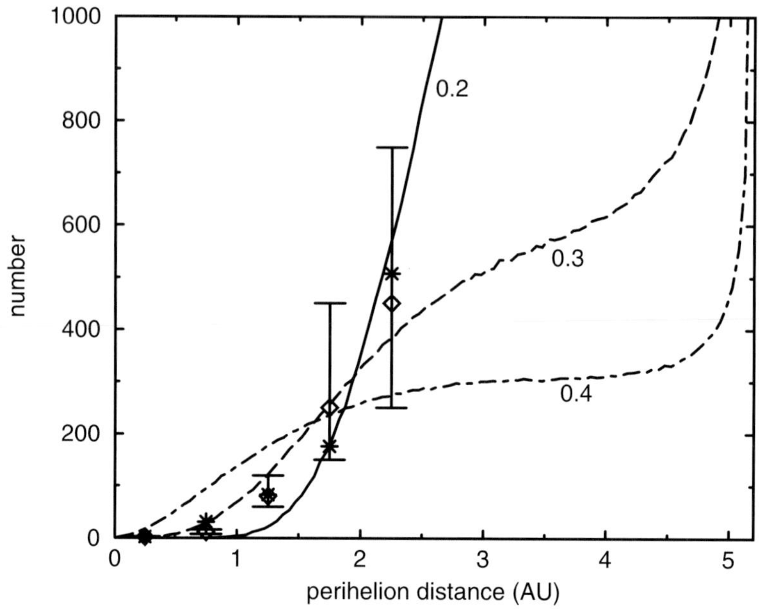

Figure 7.14. Estimated number of JF comets within q-intervals of 0.5 AU (diamonds). The curves are for theoretical q-distributions for mean velocities 0.2, 0.3 and 0.4 (in units of Jupiter's orbital velocity, assumed to be circular). The asterisks are values estimated by Levison and Duncan (1997). All the values are normalized to a comet population of 800 for $q < 2.5$ AU (Fernández et al. 1999).

$$f_U(U) = \frac{32}{\pi^2 \bar{U}^3} \exp\left(-\frac{4U^2}{\pi \bar{U}^2}\right) U^2 dU, \tag{7.30c}$$

where $\bar{U}$ is the mean encounter velocity.

By introducing the distributions (7.30a), (7.30b) and (7.30c) in eq.(7.29) we can compute $F(U, \theta, \phi)$ from which it is possible to derive the distribution of perihelion distances $f_q(q)$.

A set of solutions for the distribution for $f_q(q)dq$ of the JF population are presented in Fig. 7.14 for different values of $\bar{U}$. On dynamical grounds, JF comets under the gravitational control of Jupiter should exhibit a dropoff toward smaller q that depends on the average encounter velocity with the planet (Fernández 1984). For a typical encounter velocity $\sim 0.2 - 0.3$ (in units of Jupiter's circular velocity), the expected dropoff is fairly steep since for such low velocities it is very difficult for Jupiter to scatter a comet into a small-q orbit. On the other hand, if the encounter velocities were higher ($U \sim 0.4 - 0.5$), the

q-distribution would tend to be much flatter since scattering into small-q orbits becomes much more likely. Periodic comets captured from near-parabolic orbits should have encounter velocities $U > \sqrt{2} - 1$. By contrast, JF comets captured from heliocentric orbits of rather small eccentricities and inclinations should have smaller encounter velocities ($U \simeq 0.1 - 0.3$). For instance, 29P/Schwassmann-Wachmann 1 has an encounter velocity with Jupiter $U = 0.124$. Brunini (2004) has simulated the transfer process of fictitious bodies from the trans-neptunian belt to Jupiter's vicinity. The bodies started and evolved in low-inclination orbits, in a multi-stage process in which they are first under the gravitational control of Neptune, a fraction of them are then handed down to Uranus, then to Saturn, and finally to Jupiter. Most of the orbits that reach Jupiter are of the type of 29P/Schwassmann-Wachmann 1, i.e. confined between Jupiter and Saturn. Brunini obtains a computed U-distribution in agreement with eq.(7.30c) with a mean encounter velocity $\bar{U} = 0.28$.

7.8. The population size

It is still very difficult to make a reliable estimate of the JF population, mainly due to incompleteness of the sample of JF comets and uncertainties in the derived nucleus radii. Nevertheless, we can have a preliminary idea of its size from the data we have already available. We can see in Fig. 7.15 that the cumulative number of the discovered JF comets grows very steeply for $q > 1$ AU, while it still grows, but at a much slower pace, for $q < 1$ AU. This shows that the degree of completeness of the sample of JF comets is a strong function of their perihelion distance. The very few Earth-crossing JF comets that are being discovered, despite the large search programs of near-Earth objects carried out in the last few years, strongly suggests that this population is near completion, at least for comets brighter than absolute nuclear magnitude $H_N \sim 18.5$ (that corresponds to a nucleus radius $R_N \sim 0.7$ km for a geometric albedo $p_v = 0.04$). The larger q, the greater the incompleteness factor of the comet sample (down to a certain limiting absolute magnitude) and, thereby, the more uncertain the bias-corrected estimates of the population size. Table 7.4 shows the number of JF comets brighter than absolute nuclear magnitude $H_N = 18.5$, within intervals of $\Delta q = 0.5$ AU, as derived by Fernández et al. (1999).

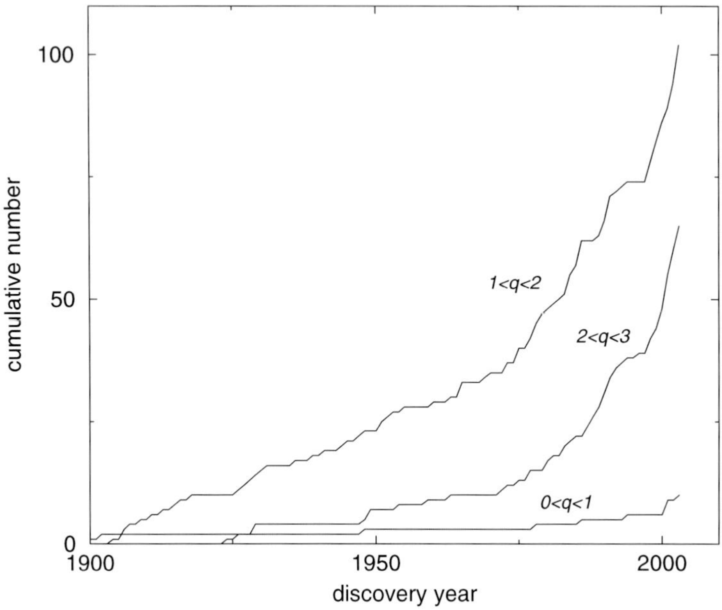

Figure 7.15. The cumulative number of JF comets discovered between 1900 and the end of 2003 in different ranges of perihelion distances. The sample has been taken from the Marsden and Williams (2003) catalogue with updates from the Minor Planet Circulars.

Table 7.4: Estimated number of JF comets brighter than $H_N = 18.5$ in different q intervals$^{(*)}$

q (AU)	Number
0.0 - 0.5	2^{+1}_{-1}
0.5 - 1.0	12^{+4}_{-4}
1.0 - 1.5	80^{+40}_{-20}
1.5 - 2.0	250^{+200}_{-100}
2.0 - 2.5	450^{+300}_{-200}

(*) From Fernández et al. (1999)

They are also plotted in Fig. 7.14 superimposed on the three theoretical q-distributions derived in the previous section.

Given the large error bars, it is difficult to conclude to which particular curve the empirical numbers of JF comets fit better. The best fit seems to be between $\bar{U} = 0.2$ and $\bar{U} = 0.3$, at least for $q < 2.5$ AU, which is again in good agreement with our discussion in the previous section. The sharp dropoff in the number of JF comets for smaller q (say $q \lesssim 1.5$ AU) may be enhanced if we also consider their shorter physical lifetimes. Levison and Duncan (1997) derived a q-distribution of fictitious JF comets with $q < 2.5$ AU from numerical simulations of thousands of objects starting in Neptune-encountering orbits and evolving by gravitational encounters with the Jovian planets. Their simulations also assume a certain physical lifetime for the JF comets with $q < 2.5$ AU. Their derived numbers (also normalized to 800 comets) are also shown in Fig. 7.14. Levison and Duncan's empirical q-distribution is somewhat steeper than that from Fernández et al. (1999), though given the different procedures, assumptions and uncertainties involved, the discrepancy cannot be taken as very significant.

The previous fit to a curve between $\bar{U} = 0.2$ and $\bar{U} = 0.3$ is relevant not only to discussions on possible source regions of JF comets, but also on the total population of JF comets. For a mean encounter velocity of, say $\bar{U} \sim 0.25$, only about 8% of the JF comets crossing Jupiter's orbit will have at a given time $q < 2.5$ AU, result in good agreement with that found by Levison and Duncan (1994). There is accordingly a steep increase in the number of JF comets with increasing q, with a concentration close to Jupiter's orbit. Thus, if the number of JF comets brighter than $H_N = 18.5$ with $q < 2.5$ AU is about 800, we can make an extrapolation to set the total number of JF comets with $H_N < 18.5$ crossing Jupiter's orbit ($q \lesssim 5.2$ AU) at about 10^4 (Fernández et al. 1999). This is consistent with the upper limit of 30,000 comets (for a similar limiting size) found by Lindgren et al. (1996) based on searches of JF comets in Jupiter's vicinity that led to negative results.

We have considered so far a limiting absolute nuclear magnitude $H_N = 18.5$. This limit was chosen because most of the comets discovered are brighter than $H_N = 18.5$. What is the situation for fainter JF comets? If the CLF derived above holds for fainter JF comets, such a population of small comets should be substantial. Yet, it is quite possible that the CLF starts to flatten as the small nuclei rapidly sublimate and disintegrate into meteoroids and interplanetary dust, thus keeping the steady-state number of small comets very low. Given the scant volume of data, we will not venture here any figure for the population of small comets and leave this issue for discussion in Chapter 9.

7.9. The source region

The study of the random-walk process of LP comets in the energy space led to the idea that JF comets could be the end products of the dynamical evolution of LP comets, involving perhaps hundreds to thousands of revolutions (e.g. Everhart 1972). Yet, despite its dynamical feasibility, the efficiency of this mechanism to maintain the steady-state population of JF comets was a matter of debate (Joss 1973, Delsemme 1973, Fernández 1980b). A more serious objection was that the inclination distribution of comets captured from an isotropic, near-parabolic flux may probably show a significant fraction of JF comets in retrograde orbits, which is in contradiction with their observed flat i-distribution (cf. Fig. 2.2). The idea that JF comets and LP comets come from different source regions is well illustrated by the plot of the energy (or reciprocal semimajor axis) versus Tisserand parameter (Fig. 7.16). We can see that the JF population occupies a different region from that occupied by LP and HT comets, with very little overlapping. In particular, we can see that JF comets have Tisserand parameters $2 < T \lesssim 3.1$.

Havnes (1970) also concluded that the direct capture of LP comets cannot be the main source of JF comets because a fraction of these would be in retrograde orbits, which is not observed. Havnes then favored a secondary source of nonparabolic comets with perihelia close to Jupiter, presumably derived from the Oort cloud, though he did not discuss how this secondary source got rid of the retrograde comets. Lowrey (1973) also favored a source of near-parabolic comets with perihelia close to Jupiter's orbit. Lowrey argued that after capture, such comets would interact with Saturn thus decreasing their initial high Jovicentric velocities ($U \gtrsim \sqrt{2} - 1$) to values more compatible with those of the JF comets. The question is if this rather complex dynamical mechanism will guarantee the required flux of low-U, trans-Jovian comets to keep the JF population, and how to avoid that a certain fraction of retrograde comets also leak to short-period orbits.

Bearing in mind the above discussion, the old hypothesis that JF comets - at least most of them - came from an isotropic source of near parabolic comets captured by Jupiter could not be supported any longer. A trans-neptunian (TN) comet belt, possibly the remnants of the formation of the outer planets, appeared as a more suitable source (Fernández 1980b). This idea was further ellaborated by Duncan et al. (1988) who showed from numerical experiments that JF comets captured from an original population of LP comets with random incli-

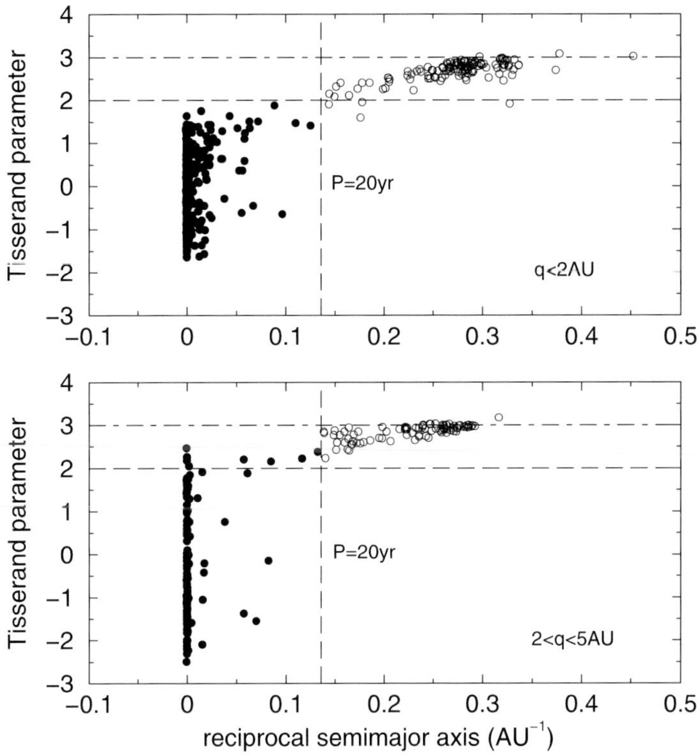

Figure 7.16. Tisserand parameter versus reciprocal semimajor axis for comets with $q < 2$ AU (upper panel), and $q > 2$ AU (lower panel). JF comets (open circles), LP and HT comets (full circles) (Fernández 2002).

nations would have a fraction of retrograde orbits, confirming Havnes's (1970) study. On the other hand, if the source was a flat TN belt, the resulting i-distribution would be in agreement with the observed one. In essence, this discussion - and the conclusion - is similar to what was presented in Section 7.7. Comets coming from the TN belt after being handed down by gravitational interactions with the Jovian planets will reach Jupiter with a mean encounter velocity $\bar{U} = 0.28$ (cf. Section 7.7) which matches quite well the U required for fitting the empirical q-distribution of JF comets. Levison and Duncan (1997) later carried out more sophisticated numerical simulations for 1 Gyr which allowed them to evaluate the efficiency of the transfer process from bodies in Neptune-encountering orbits to orbits with perihelion distances $q < 2.5$ AU. They found this to be a highly efficient process: about 30% of the original sample attained this type of final orbit, and from these 99.7% were JF comets. We will come back to the issue of the trans-neptunian population in the next chapter.

It is still possible that some JF comets come from other sources. For instance, despite what was said before, a few LP comets might trickle down to short-period orbits with Tisserand constants $T > 2$ (Levison et al. 2001). The Trojan asteroids constitute another important reservoir of, presumably, ice-rich bodies. They circle the Sun around the L4 and L5 Lagrange equilibrium points of Jupiter at the same distance as the planet, so they are locked in the 1:1 mean motion resonance with Jupiter. The Trojans around the L4 and L5 points lead and trail Jupiter respectively at an angular separation of 60°. The total number of Trojan bodies with diameters > 1 km is of about 2×10^6 (Shoemaker et al. 1989). The orbits of the Trojans are not stable indefinitely, so there is a continuous leakage of bodies from the L4 and L5 reservoirs. Rather than dynamical perturbations of Jupiter and the other Jovian planets, mutual collisions seem to be the most efficient way to knock fragments out of the L4 and L5 Lagrange points. Once they leave their safe refuges, Trojans will start to experience gently encounters with Jupiter at very low relative velocities (and thus with a Tisserand parameter $T \sim 3$). It is then possible that some comet-like bodies in the Jupiter's zone with a Tisserand constant $T \sim 3$ are indeed escaped Trojans (Rabe 1972, Marzari et al. 1995). Levison et al. (1997) estimate that about 200 escaped Trojans with diameters > 1 km are currently roaming the Solar System. In this regard, Marzari et al. (1997) estimate that Trojan collisional debris ending up in cometary orbits could account for about 10% of the current population of JF comets and Centaurs. Their identification as JF comets or asteroids depends of course on whether such Trojan collisional debris contain enough volatiles on or close to their surfaces to produce detectable gaseous activity.

Escaped Hildas, that are asteroids locked in the 3:2 mean motion resonance with Jupiter, may also provide objects with Tisserand constants $T \sim 3$. Yet, there is the question on whether escaped Hildas have enough water ice close to their surfaces to produce by sublimation detectable comae of gas and dust particles, which is an indispensable requirement to classify them as "comets". Otherwise they will be classified as "asteroids" in cometary orbits.

In summary, a source in the TN belt is favored for most JF comets, with some small contribution from other sources such as LP comets and Trojans. These interlopers may become dynamically indistinguishable from the rest coming from the TN belt, and only studies of their internal structure and chemical composition may tell us something about their different origin.

THE TRANS-NEPTUNIAN BELT

The idea of the existence of a massive planet beyond Neptune dates back more than a century when it was argued that Neptune alone could not explain the seemingly irregularities in the motion of Uranus. In 1908 the American astronomer William Pickering suggested the existence of a trans-neptunian planet with a mass about twice that of the Earth. But it was the American astronomer Percival Lowell (1855-1916), well known for his theory of intelligent life on Mars, the first to carry out an extensive search for what he called "planet X" between 1905 and 1916. The search for the elusive planet was re-started in 1929 by Clyde Tombaugh at the Lowell Observatory in Flagstaff, Arizona, the same observatory where Lowell worked until his death. Tombaugh's search led to the discovery of Pluto on the 18th of February 1930. Tombaugh continued his search for other trans-neptunian planets for another 13 years, covering the entire sky north of $-30°$ declination to apparent (blue) magnitude $m_B = 16$, but no new planets appeared (Tombaugh 1961). The inventory of trans-neptunian bodies was going to remain unchanged for as yet several decades.

Although nothing really dramatic happened in the observational front in the post-Pluto years until the discovery of Pluto's moon Charon by James Christy and Robert Harrington of the U.S. Naval Observatory (Christy and Harrington 1978), the situation was much more dynamic and stimulating in the theoretical front. Immediately after Pluto's discovery Leonard (1930) speculated that Pluto was the first of an as yet undiscovered population of "ultra-Neptunian" and "ultra-Plutonic" planets as he called them. A few years later the Irish astronomer Kenneth Essex Edgeworth (1880-1972) presented the first quantitative picture of a trans-neptunian belt based on cosmogonic principles in a treatise that has remained unpublished (Edgeworth 1938). He conjectured that the early Sun was surrounded by a vast disk of meteorites that extended far beyond the planetary orbits. While in the planetary region gravitational instabilitites led to the formation of big planets, Edgeworth argued that densities were too low in the trans-neptunian region to allow planet formation. In the outer portions the particles ceased to move in regular streams and the motion became turbulent, appearing in the fluid of particulate matter a large number of independent eddies that would further contract into comets. These comets would have been distributed in a region between $\sim 10^{10}$ km and $3-4 \times 10^{10}$ km (~ 70 to $\sim 200-300$ AU). For a total mass of one third $M_\oplus$, he estimated that about 200 million comets of an average mass of 2×10^{-9} $M_\oplus$ formed in this way, while if the total mass was 0.1 $M_\oplus$ the result would be 2000 million comets with an average mass of 5×10^{-11} $M_\oplus$. It is interesting to note that Edgeworth conjectured that this vast reservoir beyond Neptune was the source of the observed comets. He later developed further these ideas in a couple of scientific publications (Edgeworth 1943, 1949). The Dutch-American astronomer Gerard P. Kuiper (1905-1973) argued independently that icy planetesimals that formed beyond Neptune could not grow to massive bodies given the long collisional time scales at such distances. Consequently, trans-Neptunian planetesimals would have been left unaccreted in a ring stretching between $\sim 30-50$ AU with a structure resembling that of the asteroid belt (Kuiper 1951).

Whipple (1964) tried to estimate the mass of the putative trans-neptunian belt by assuming it to be the cause of the apparent irregularities in Neptune's motion. He found a mass ~ 10 $M_\oplus$ if the belt was located at 40 AU, or ~ 20 $M_\oplus$ if it was at 50 AU. Later on, Standish (1993) showed that such discrepancies in Neptune's motion were not real, so Whipple's results were spurious. Hamid et al. (1968) tried to probe the mass contained in such a belt by analyzing the motion of comet 1P/Halley whose aphelion lays beyond Neptune. The lack of measurable perturbations on this comet allowed these authors to set an upper limit of 0.5 $M_\oplus$ if the belt was at 40 AU, or 1.3 $M_\oplus$ if it was at 50 AU. Anderson and Standish (1986) used the tracking on the Pioneer 10 spacecraft to set an upper limit of ~ 5 $M_\oplus$ for a belt at ~ 35 AU.

Instead of looking into gravitational effects on known bodies, Bailey (1983) estimated the infrared flux produced by a power-law mass distribution of comets, concluding that under reasonable values of the model parameters, the IR surface brightness would dominate the extragalactic IR background. Instead of trying to detect the thermal radiation from the bodies themselves, Jackson and Killen (1988) investigated the possibility of detecting the IR flux from the dust produced by the grinding of bodies through mutual collisions in the belt. They concluded that it could be possible to detect such a belt by examining the IR flux in the ecliptic plane at about 100 μm. Under a set of plausible physical parameters, the thermal radiation from the trans-neptunian dust would show up above the noise level over the Planckian tail of the foreground zodiacal light. Even though the *Infrared Astronomical Satellite* (IRAS) did not detect the trans-neptunian disk, it did detect circumstellar dust disks around a few nearby stars. Since such disks are unstable over time scales comparable to the ages of their central stars, Weissman (1984) argued that they might be maintained through mutual collisions of comet-sized bodies.

We shall discuss in this chapter the discovery, dynamical and physical properties of the trans-neptunian belt, leaving for Chapter 10 matters concerning to its origin.

8.1. The puzzle of the Jupiter family comet population

In our brief summary of the main ideas on a trans-neptunian comet belt presented until the end of the eighties, an important point is still missing: it deals with the possible link between such a population and Jupiter-family comets. As explained before (cf. previous chapter), Jupiter family comets were assumed to arise from the capture of Oort cloud comets by Jupiter after a single encounter (the classical Laplace's hypothesis), or, more likely, through multiple moderate perturbations in successive passages (e.g. Havnes 1970, Everhart 1972). Everhart (1972) showed through a numerical study that the orbital characteristics of "short-period" comets (considered to be those with orbital periods $P < 13$ yr) could be explained, at least qualitatively, if it was assumed that it came from a population of near-parabolic comets with small inclinations ($i < 9°$) and perihelia in the range $4 < q < 6$ AU. However, from Everhart's study and assuming steady-state conditions, Joss (1973) found that the rate of captures would be 40,000 times smaller than observed. Delsemme (1973) could explain the right number of captures but only after hypothesizing the existence of a population of 30,000-100,000 intermediate-period comets with perihelion distances in the range $4 < q < 6$ AU.

Fernández (1980) found that an origin of the Jupiter family as a process of capture by Jupiter and the other Jovian planets, from a population of near-parabolic comets isotropically distributed (the Oort cloud), would imply the loss of about 300 near-parabolic comets for each one captured into a JF orbit. The total number of ejected comets would be 1.35×10^{12} throughout the solar system lifetime, i.e. about one order of magnitude greater than the number of Oort cloud comets estimated by Oort (1950). Such a enormous waste of comets led Fernández

(1980) to propose the trans-neptunian belt between $\sim 35 - 50$ AU, suggested by Edgeworth and Kuiper, as a much more efficient alternative source. He estimated that gravitational interactions among belt members could cause a diffusion of bodies towards the planetary region fast enough to keep the steady-state population of Jupiter family comets, provided that bodies of the size of Ceres or larger were present in the belt. The dynamical process was described as follows: once the bodies reached Neptune's influence zone, they were either ejected or handed down to the next planet inside (Uranus) where the process would repeat itself until a fraction managed to reach the inner planetary region where they became JF comets. If the probability of ejection and handing down to the next planet inside is evenly splitted in one half each, he estimated that the probability that a trans-neptunian body that reached Neptune's influence zone ended up as a JF comet was $(1/2)^4 = 1/16$, i.e. one order of magnitude more efficient than the capture of Oort cloud comets (in actuality, from more accurate numerical experiments Duncan et al. (1988) showed later that the process is much more efficient and amounts to ~ 0.17, and a further revision by Duncan et al. (1995) rose the ratio to 0.34).

As mentioned in the previous chapter, Duncan et al. (1988) presented another interesting piece of evidence in favor of the hypothesis of the trans-neptunian belt as the source of the JF comets. They argued that Oort cloud comets evolving toward JF orbits would tend to preserve the random inclination distribution, in contradiction with Everhart's "capture zone" for $i < 9°$. Duncan et al. found that their numerical integrations predicted a substantial number of high-inclination and retrograde JF comets. In contrast, when they started comets from low-inclination, low-eccentricity orbits with perihelia close to Neptune's orbit, they were able to reproduce the low inclination distribution of JF comets. Everhart (1972) did not find that high-inclination comets reached short-period orbits because he arbitrarily stopped the integration at 2000 revolutions. High-inclination and retrograde comets can also reach short-period orbits but on longer time scales.

8.2. The discovery

The trans-neptunian population remained in the realm of theoretical speculation for several decades. New search programs of trans-neptunian bodies were carried out during the late eighties and early nineties. Luu and Jewitt (1988) searched 200 deg^2 photographically with a Schmidt telescope to a limiting apparent visual magnitude $m_V = 20$, and another 0.34 deg^2 with a CCD camera to an apparent red magnitude $m_R = 24$ ($m_V \simeq 24.5$). Both searches led to negative results. Kowal (1989) searched 6400 deg^2 photographically to approximately $m_V = 20$, discovering 2060 Chiron in 1977, the first outer solar system, planet-crossing object after Pluto. Levison and Duncan (1990) searched 4.9 deg^2 using a CCD to $m_V = 22.5$, again with negative results.

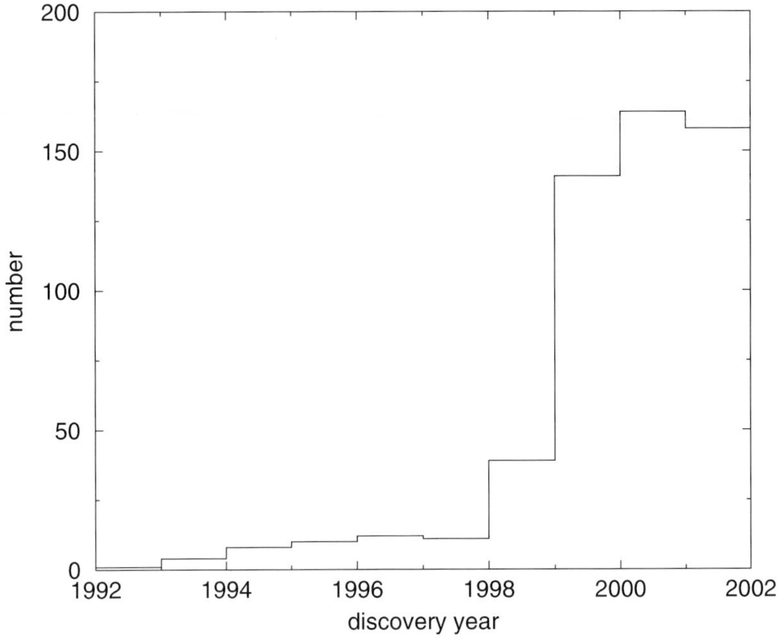

Figure 8.1. The discovery rate of trans-neptunian bodies through the first decade. Data from the Minor Planet Center's Web site: http://cfa-www.harvard.edu/iau/Ephemerides/Distant/Soft00Distant.txt

The failure of the previous sky surveys did not discourage David Jewitt and Jane Luu from pursuing the search, until their efforts were rewarded by the first successful detection of a trans-neptunian object other than Pluto and Charon - $1992QB_1$ - with the 2.2m-telescope of Mauna Kea, Hawaii (Jewitt and Luu 1993). They searched 1 deg^2 down to $m_V = 25$. This object was at its discovery at a heliocentric distance of 41.2 AU and the computation of its orbital elements yielded: $a = 43.82$ AU, $e = 0.088$, and $i = 2.21°$, i.e. a low-inclination and low-eccentricity object, in agreement with what one should have expected for bodies in a trans-neptunian belt. The object had an apparent red magnitude

$m_R = 22.8$ that, for an assumed albedo of 0.04, would correspond to a diameter ~ 250 km, i.e. a rather large body as compared with typical asteroid sizes. A second discovery of a trans-neptunian object - 1993 FW - by the same team followed a few months later. These two trans-neptunian objects were followed by a growing number of new discoveries, reaching more than 500 in the short time span of ten years (Fig. 8.1).

8.3. The naming debate

The trans-neptunian population was called the "Kuiper belt" by Duncan et al. (1988), when it was still a theoretical model, and the coined name was widely adopted in the following years. This rush to name something as yet not discovered unfairly neglected the important pioneer contribution by Edgeworth in the thirties and forties. The author of this book accepts part of the blame for this unjust situation since he himself overlooked Edgeworth's work in his 1980 paper (Fernández 1980). Bailey et al. (1990) made reference to Edgeworth's work on the trans-neptunian belt, but it was Edgeworth's countryman John McFarland who strongly argued in favor of acknowledging Edgeworth's work stating: "In order to do justice to his research efforts, and yet not detracting from Kuiper's endeavours, a suitable tribute to his memory would be to name the cometary annulus beyond the orbit of Neptune, the 'Edgeworth-Kuiper belt'." (McFarland 1996).

Paul Weissman of the Jet Propulsion Laboratory, in one of the first reviews on the trans-Neptunian belt, reflected about the naming problem. Even though he stuck to the by then widely-used Kuiper-belt denomination, because in his words "this name has already been in use for several years, and it may be confusing to try to change it now", he conceded that "Both scientists clearly suggested the existence of small objects orbiting the Sun beyond the orbit of Neptune. ... It would seem that credit should be shared." (Weissman 1995).

It was also mentioned the even earlier contribution by Frederick Leonard (see above), but despite he speculated above the possible existence of more Plutos, he never developed a self-consistent cosmogonic theory of a trans-neptunian belt as the residual leftovers of planet formation, let alone to relate it to the observed comets. Therefore, it seems clear that in the name debate Leonard should be out of the game. There is not an official denomination by the IAU, so the chosen name is a matter of author's preference. Even though, we would prefer in principle McFarland's choice, we shall adopt in the following the more neutral and somewhat simpler names: *trans-neptunian* (TN) belt and *trans-neptunian object* (TNO) to refer to all the bodies populating the region beyond Neptune up to a few hundreds AU.

8.4. Resonance dynamics

Until now we have mainly analyzed problems of non-resonant dynamics where bodies are subject to stochastic encounters (either planets, passing stars, or molecular clouds). This has been the case of Oort cloud comets or, in general, LP comets. We already saw in Section 4.4 that mean motion resonances play some role in the dynamics of comets with periods $P \lesssim 200$ yr, though even for these comets the dynamics is ruled in the long term by close interactions, mainly with Jupiter. Yet, in dealing with TNOs we face a new dynamical situation: the important role played by different types of resonances in their evolution. Resonaces are also important in other populations of the solar system, such as the Kirkwood gaps in the main asteroid belt, Trojans, the Galilean satellites and the rings of the Jovian planets. Before continuing with our study of the TN belt, we will review in this section a few basic concepts regarding resonances.

As explained in Section 4.4, *mean motion resonances* involve commensurabilities between the mean motions of two or more bodies. There are in addition *secular resonances* in which the fundamental frequencies of the reference orbits are in rational commensurability with the fundamental frequencies of the planetary system (see, e.g., Froeschlé and Scholl 1987, Knežević et al. 1991).

We will present here a heuristic approach to describe the mechanism of resonances that follows Peale's (1976) treatment. The reader interested in a more complete, mathematical description of this problem can consult, for instance, Murray and Dermott (1999). Let us consider two bodies of masses $m >> m'$ in coplanar orbits about the primary (Fig. 8.2). The inner body m is assumed to be in a circular orbit, while m' is in an eccentric orbit. The mean motions n and n' of m and m' are assumed to be near commensurable, so as to fulfill $jn \simeq (j+k)n'$, where j and k are integers. Let us assume that m is in conjunction with m' at A, a short time before reaching the apocenter. Let us consider two positions A_1, A_1' and A_2, A_2', at times $t_A \pm \Delta t$, where t_A is the time of the

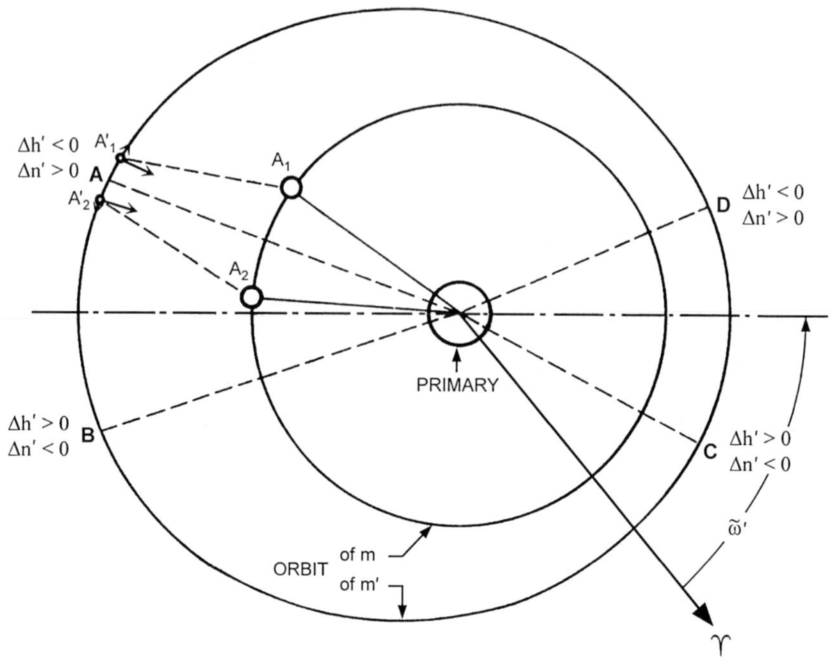

Figure 8.2. Stable and unstable configurations of a particle of mass m' in an eccentric orbit with respect to a body on a circular orbit of mass $m >> m'$, both moving around the primary. Their mean motions n and n' are near commensurable. The stable configuration is around apocenter and the unstable one is around pericenter (adapted from Peale 1976).

conjunction at A and Δt is very small as compared to the orbital period of m'. As shown in Fig. 8.2, the configurations A'_1-A_1-primary and A'_2-A_2-primary are not symmetric since the orbit of m' is eccentric, so we have: $\overline{A_2A'_2} > \overline{A_1A'_1}$. Furthermore, m' will move somewhat slower at A'_2 than at A'_1, so the relative velocity between m and m' will increase somewhat after conjunction (m keeps always the same velocity - in modulus - because it moves on a circular orbit). Because m catches up m' prior to the conjunction at A, m acting from behind brakes m'. On the other hand, after conjunction m accelerates m' because the former recedes from the latter. Therefore, the gravitational perturbation by m on m' will be somewhat stronger prior to conjunction because their relative distance and their relative velocity are both smaller. In other words, the braking effect on m' before opposition will be somewhat greater than the acceleration effect after opposition, so m' will lose some angular momentum: $\Delta h'_A < 0$. The body m' will fall somewhat closer to the primary, thus increasing its mean motion: $\Delta n'_A > 0$. By conservation of angular momentum, m will gain angular momentum, i.e. $\Delta h_A = -\Delta h'_A$, so its mean motion n will decrease.

Since n has decreased while n' has increased, the next conjunction of m with m' can occur after apocenter, let us say at B. It is easy to show that the situation reverses in this case: there is first an acceleration of m' prior to conjunction at B, and then deceleration after conjunction with a net gain of angular momentum $\Delta h'_B > 0$ and decrease of its mean motion $\Delta n'_B < 0$. Again, this means that the angular momentum of m will decrease and its mean motion increase. As a result, the next conjunction will again occur before apocenter, and so on. We will say that in this case the conjunctions of m with m' *librate* stably about the apocenter, thus preserving the commensurability. It is obvious the analogy with a pendulum that oscillates or librates back and forth around the bottom of its support. The commensurability has thus the effect of repeating the same configuration around a given position of the orbit.

We can consider two limiting situations in our example of the bodies m and m': (a) If the conjunction occurs exactly at the apocenter, there is a symmetry before and after it, so there is neither gain nor loss of angular momentum. Consequently the following conjunctions will always repeat at the apocenter. In the example of the pendulum, it would be as if we leave it hanging straight down motionless. (b) If we detach the conjunction of m with m' from the apocenter an arc sufficiently large, the restoring effect will break down, so the following conjunctions will occur at any orbital position. We will say in this case that the conjunctions *circulate*, which means that the commensurability of mean motions breaks down. In the case of the pendulum, it is as if we impart it an impulse large enough to pass from librations around the bottom to rotation around the fixed point. Note that this implies to pass from oscillations back and forth around an equilibrium point to a motion always in the same sense.

Other important question refers to stable and unstable configurations. In our previous example we saw that librations around the apocenter are stable. However, stable configurations do not occur for any point. Let us consider the case of the pericenter of m': if the conjunction of m with m' occurs exactly at pericenter it is clear that the following conjunctions will repeat themselves at pericenter, thus keeping the commensurability. Let us now assume the we displace the conjunction a bit from the pericenter, say at C (Fig. 8.2). It is easy to see that the configuration primary-m-m' prior and after conjunction at

C is just the opposite to that seen at A, so m' at C will suffer a net gain of angular momentum, i.e. $\Delta h'_C > 0$, and $\Delta n'_C < 0$. The result will be that the next conjunctions of m' will also occur before pericenter, but always receding from this point. The same reasoning can be done if we consider an opposition after pericenter, say at D. Therefore, librations around the pericenter are not possible. The pericenter conjunctions thus correspond to an unstable equilibrium configuration. In the case of the pendulum, it would be like trying to get stable librations around the top of its support.

From the previous physical example we can extract the following conclusions:

(a) A commensurability means that $jn = (j+k)n'$, though we usually have a near-commensurability in which stable librations can occur, but only around certain points like the apocenter of m'. Such librations take the ratio n/n' above and below the exact value $(j+k)/j$ but always close to it. A mean motion resonance like this is called *of order k*.

(b) The larger the amplitude of the librations, the larger the divergence from the exact commensurability, until the resonance breaks down and the conjunctions of m' start to occur at any point of the orbit.

(c) The example of the bodies m and m' shows that there is a resonant coupling between them, in which angular momentum is exchanged back and forth. This reflects in periodic variations of their eccentricities forced by the resonance. Resonances can also force periodic changes in the inclination and the other angular parameters. Yet, changes in the semimajor axis are constrained by the amplitude of the oscillations.

The example of Fig. 8.2 belongs to the eccentricity-type resonances. There are other types of mean motion resonances that involve the orbital inclination and, as we mentioned, the secular resonances. But we think that our previous heuristic example is enough to illustrate the main dynamical features of the resonant dynamics. We will next analyze a very important angle to describe the resonant mechanism. In Fig. 8.2 we assumed that the direction of the apocenter of m' is fixed in an inertial frame of reference. Yet, due to secular perturbations by m (and other massive bodies in the system) the longitude of perihelion $\tilde{\omega}'$ will actually precess at a rate

$$\dot{\tilde{\omega}}' \simeq \frac{1}{n'a'^2e'}\frac{\partial \mathcal{R}}{\partial e'}, \tag{8.1}$$

where a', and e' are the semimajor axis and the eccentricity of m', and $\mathcal{R}$ is the disturbing function (cf. eq. (4.2)). Equation (8.1) is a simplified version of one of Lagrange's planetary equations (see, e.g., Roy 1982, Murray and Dermott 1999) in which the inclination of m' is assumed to be very small and $e' << 1$. We see that the precession rate of $\tilde{\omega}'$ increases with decreasing e', so it will have to be taken into account, mainly when e' is small. Thus, if $\dot{\tilde{\omega}}'$ is non-negligible, it has to be added to the equation between the mean motions in successive conjunctions, so it now reads: $jn - (j+k)n' + \dot{\tilde{\omega}}' = 0$, i.e. n has to be somewhat smaller than that obtained for the exact commensurability $(j+k)/j$. By integrating the previous equation we get

$$\sigma = j\lambda - (j+k)\lambda' + \tilde{\omega}', \tag{8.2}$$

where λ and λ' are the mean longitudes of m and m' respectively, and σ is called the *critical angle* of the mean motion resonance.

The value of σ is associated to the amplitude of the librations. In Fig. 8.3 we show the plot of σ versus the semimajor axis a for the case

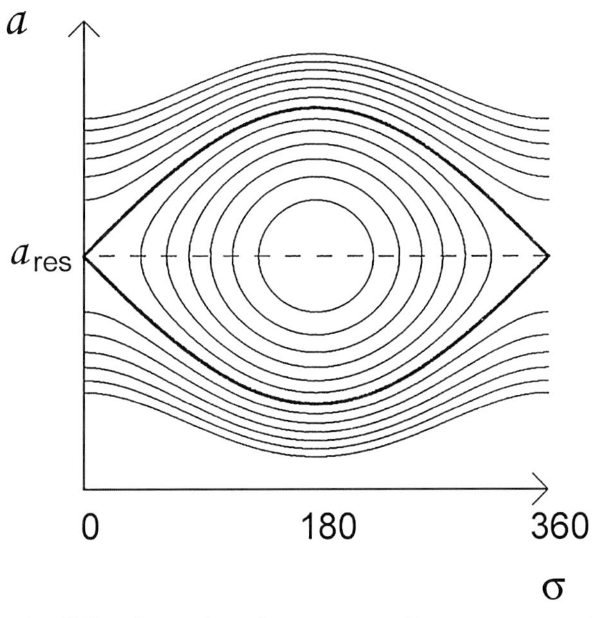

Figure 8.3. Sketch of the dynamics of a mean motion resonance of the Neptune–Pluto type (i.e. Neptune on a circular orbit and Pluto on an eccentric one). Within the thick curve σ librates, so the commensurability is restored. Outside the thick curve σ circulates. An infinitesimal change near the boundary is enough to pass from a regime to a completely different one (Morbidelli 1997).

of Neptune-Pluto, that are locked in the 2:3 mean motion resonance, as presented by Morbidelli (1997). The critical angle is $\sigma = -2\lambda_4 + 3\lambda_P - \tilde{\omega_P}$, where λ_4 is the mean longitude of Neptune, and λ_P and $\tilde{\omega_P}$ are the mean longitude and longitude of perihelion of Pluto. This is a resonance of the type shown in Fig. 8.2, so librations are stable around Pluto's aphelion, i.e. when $\lambda_P = \lambda_4 = \tilde{\omega_P} + 180°$. If we substitute this condition in the equation for the critical angle we obtain $\sigma = 180°$. We can see that $\sigma = 180°$ and $a = a_{res}$ corresponds to the center of the resonance with zero amplitude of libration. The departure from this set of values determines the amplitude of the libration in the resonance, represented in the plot by closed curves, namely σ oscillates back and forth around 180°. The close curves expand around the center until we reach the boundary or separatrix between the libration and the circulation regimes (marked by the thick curve). In the circulation regime σ can take any value between 0 and 360° and it always goes in the same sense (remember the case of a pendulum rotating around a fixed point).

Figure 8.3 also shows that the librations around the resonance can keep the semimajor axis around the narrow range: $a_{res} \pm \Delta a$, where Δa is the distance from the top (or the bottom) of the separatrix to the center of the resonance. It is then clear that if a resonant body receives and impulse that changes its semimajor axis by an amount $> \Delta a$, the resonance will break down. The width Δa will be a function of the eccentricity of the body.

We can see in Fig. 8.3 another important feature of resonances, namely the very sharp transition between the libration and the circulation regimes. This introduces the concept of *chaos*, associated to dramatic and unpredictable orbital changes that occurred when we are near the separatrix after small variations in the conditions of the problem. Chaotic motion may also be associated to more than one resonance that overlap over certain ranges of the orbital parameters. Chaotic motion can force high eccentricities and inclinations.

8.5. Dynamical structure and transfer mechanisms

The study of the dynamical structure of the trans-neptunian belt started before any body - other than Pluto and Charon - was discovered there. Torbett (1989) and Torbett and Smoluchowski (1990) showed that orbits with $a \lesssim 45$ AU would have chaotic nature because of their positive Lyapunov exponents, so they may become Neptune-crossers on short time scales as compared to the solar system age (the Lyapunov exponent measures the rate at which neighboring trajectories diverge, so it gives an idea of the degree of chaoticity of a given orbit). Levison (1991) also explored the dynamical lifetimes of TNOs by means of numerical techniques based on Markov chains. In this approach the parametric plane (q, Q) (perihelion and aphelion distances of test bodies) was divided into small bins and the probability P_{ij} that a test body leaves the bin i for the adjacent j after a time Δt was computed by integrating numerically the orbits of a sample of particles within the bin. The matrix of probabilities P_{ij} was later used to compute the diffusion speed of test bodies in the TN belt. Levison found that some of the bodies have lifetimes comparable to the solar system age, so they could be a suitable source of the observed population of JF comets. Levison and Duncan (1993) and Holman and Wisdom (1993) performed numerical integrations of a large number of particles over periods of 1 Gyr and 800 Myr respectively. They computed stability times, defined as the time required for a particle to cross Neptune's orbit, as a function of the initial semimajor axis and eccentricity of the particle. They found that for $a < 45$ AU the TN belt has a complicated dynamical structure, characterized by an alternation of stable and unstable regions. Holman and Wisdom (1993) found that the rate of TNOs encountering Neptune is roughly proportional to t^{-1}, where t is the evolution time of the population.

Preliminary computations of the orbits of the first discovered TNOs indicated that a significant fraction of them (about 40%) moved in Pluto-like orbits, i.e. in the 2:3 mean-motion resonance with Neptune and for this reason, and for being smaller than Pluto, they have been called *Plutinos*. Most of the bodies in the 2:3 resonance have in addition eccentricities and inclinations higher than the non-resonant TNOs in such a way that their perihelia approach or fall inside Neptune's orbit. These bodies are nevertheless dynamically stable, since this mean-motion resonance (and others) provide a protection mechanism from Neptune encounters. This explains why Pluto has survived throughout the solar system lifetime despite its orbit crosses that of Neptune, since this resonance prevents Pluto from getting closer than ~ 18 AU from Neptune. The first detailed characterization of the dynamical structure of mean-motion resonances was done by Morbidelli et al. (1995) who computed the borders of the resonances as the separatrices of the average planar, circular three-body problem Sun-Neptune-particle,

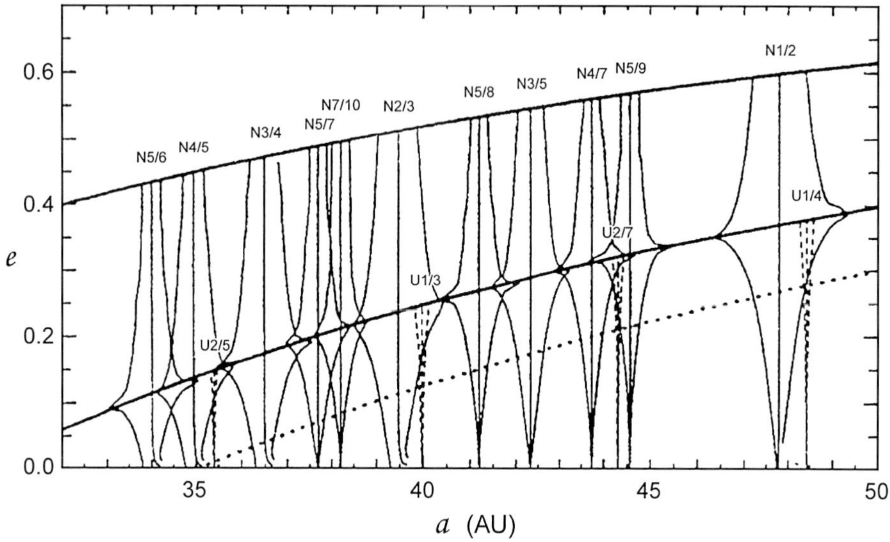

Figure 8.4. The libration widths in the (a, e) plane of mean-motion resonances with Neptune and Uranus in the trans-neptunian belt between 32 and 50 AU (Morbidelli 1999).

which is integrable. The masses of the non-resonant planets were added to that of the Sun. The location and the libration width in the plane (a, e) of the most important mean motion resonances with Neptune and Uranus are shown in Fig. 8.4. The two bold curves in the figure denote Neptune-crossing orbits and Uranus-crossing orbits. The diamond-like curves delimit the libration width of each resonance. We can see that for $e \to 0$ the libration width also tends to zero which means, as shown by the example of Fig. 8.2, that the small-mass body has to move on an eccentric orbit to experience alternate losses and gains of angular momentum before and after aphelion. As shown, the width increases with the eccentricity until the particle reaches Neptune's orbit and then it narrows for still larger eccentricities. Very eccentric orbits reaching $q < 25.5$ AU become unstable because of perturbations by Uranus (Gallardo and Ferraz-Mello 1998). Mean motion resonances with Uranus are very narrow. They are only plotted up to the Neptune-crossing limit since perturbations by this planet will destroy them very quickly.

Massive numerical computations of test bodies by Duncan et al. (1995) for the solar system age showed that for $e \lesssim 0.1$ and $a \lesssim 42$ AU, the only stable orbits were those in the mean-motion resonances 2:3, 3:4, 4:5, and 5:6, in close match with the regions in the plane (a, e) found theoretically by Morbidelli et al. (1995) and shown in Fig. 8.4.

In no case, stable orbits are found for $a \lesssim 34$ AU. The overlapping of mean-motion resonances often causes large-scale chaos, and this occurs in a zone close to Neptune ($a \lesssim 33$ AU) that is already strongly perturbed by close encounters with this planet. Thus, this inner edge of the TN belt should have been cleared up very early in the history of the solar system. Outside the stable libration regions around mean-motion resonances the motion is highly chaotic and non-resonant bodies with semimajor axes $a \lesssim 42$ AU are expected to evolve to Neptune-crossing orbits in time scales smaller than the solar system age. It is therefore expected that the primordial population of the TN belt in the range $30 \lesssim a \lesssim 42$ AU has been heavily eroded, except for a few stable zones around some mean-motion resonances with Neptune. Crossing times for test particles computed by Duncan et al. (1995) are shown in Fig. 8.5. Dark regions are particularly unstable, and their dynamical time scales are much shorter than the solar system age, so they are not a suitable source for the current population of JF comets.

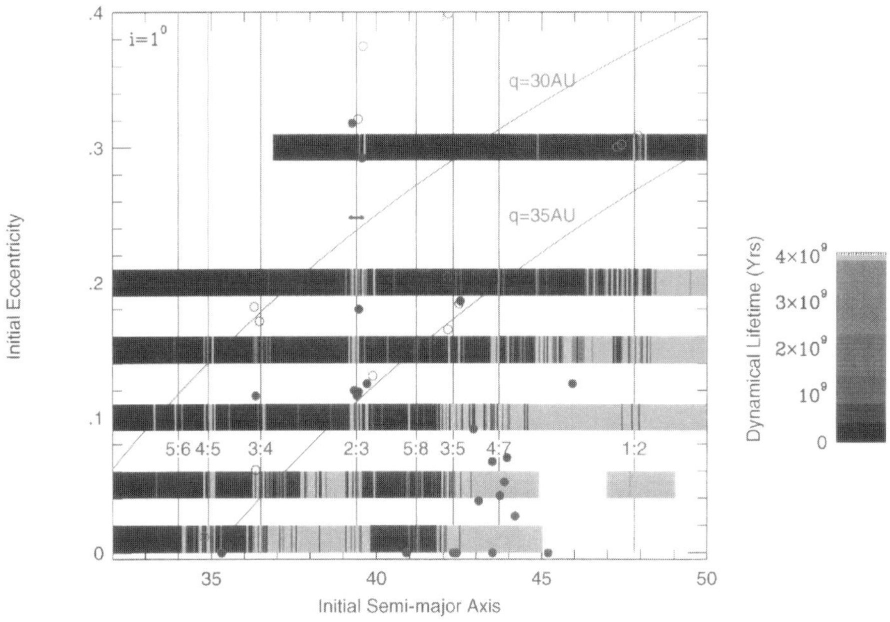

Figure 8.5. Dynamical lifetimes of test particles as a function of their initial semi-major axis and eccentricity. The initial inclination of the particles is $i = 1°$. The dynamical lifetimes are indicated with different gray tones: black for the shortest dynamical lifetimes to clear gray for the longest dynamical lifetime. The small circles are for the TNOs discovered at that moment (Duncan et al. 1995).

Ip and Fernández (1997) found that gravitational interactions of Plutinos with large TNOs (diameters > 100 km) can efficiently decouple them from the 2:3 resonance. The decoupling mechanism from the 2:3 resonance has also been analyzed by Morbidelli (1997). He showed that there is a border zone between a strongly chaotic region and a regular region, characterized by chaotic diffusion at a speed that increases monotonically as the square of the libration amplitude. It can take several billion years to reach the chaotic region, and once there the body will be subject to frequent encounters with Neptune. From numerical experiments Morbidelli could show that the escape rate from the 2:3 resonance varies with time according to the law $dN/dt \propto t^{-3/2}$. By considering only this dynamical diffusion scheme (i.e. neglecting kicks due to close encounters with massive bodies or collisions), and that 25% of the JF comet input comes from the 2:3 resonance, Morbidelli estimated that the current number of bodies in the 2:3 resonance should be about 4.5×10^8.

Figure 8.6 depicts very nicely how a TNO, initially on a near circular orbit at $a \sim 40$ AU, evolves until it becomes Neptune-crosser. As shown

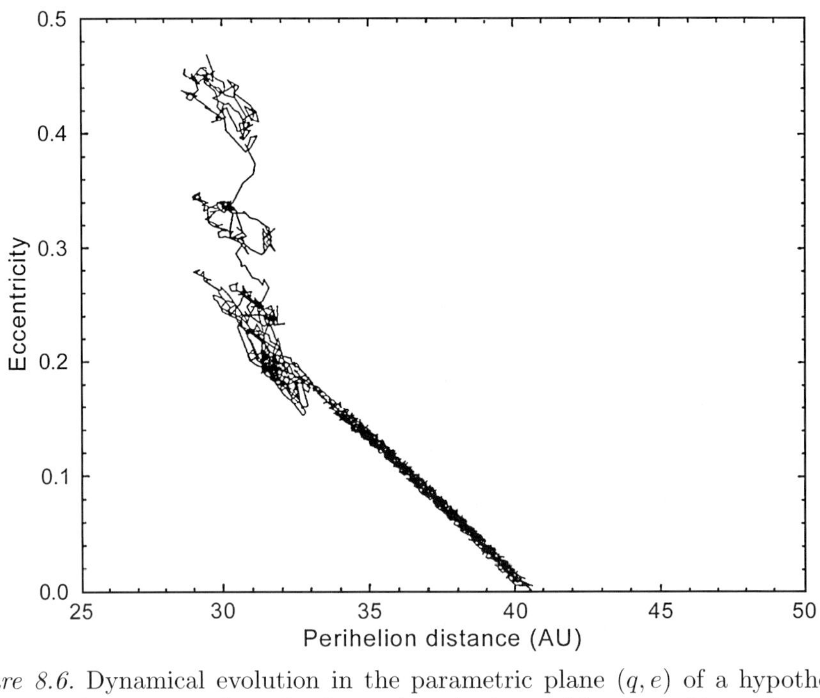

Figure 8.6. Dynamical evolution in the parametric plane (q, e) of a hypothetical TNO initially on a circular orbit located in the invariable plane. Lines of constant semimajor axis are diagonal (Holman and Wisdom 1993).

by Knežević et al. (1991), the region between 40 and 42 AU is occupied by secular resonances, in particular ν_8 in which the precession rate of the body's perihelion matches that of Neptune, and ν_{18} which involves the precession of the nodes. The ν_8 resonance excites the body's eccentricity, while the ν_{18} resonance excites the body's inclination. Therefore, the body first tends to evolve in eccentricity and in perihelion distance, keeping its semimajor axis more or less constant (represented by a diagonal in the plane (q, e)). When it is close or crosses Neptune's orbit, strong gravitational interactions with this planet produce large changes in a and e, keeping in this case q more or less constant (a vertical line in the plane (q, e)).

8.6. Transient objects

If JF comets come indeed from the TN belt, there should be a large number of stray bodies in the outer planetary region, in transit from the outer to the inner planetary region. Actually, there have already been several discoveries of this class of objects, known now as *Centaurs*, starting with 2060 Chiron (or 95P/Chiron) in 1977. They have their perihelia beyond Jupiter and semimajor axis smaller than Neptune's and move on low-inclination orbits. Their transport time from the Neptune's zone to the region interior to Jupiter's orbit is $\sim 4 \times 10^7$ yr (Levison and Duncan 1997). Chiron has a double denomination comet-minor body because it has shown gaseous activity (Hartmann et al. 1990, Luu and Jewitt 1990, Meech and Belton 1990), which is expected for bodies rich in icy material coming from the outer solar system. According to the previous definition, comet 29P/Schwassmann-Wachmann 1 and 39P/Oterma should be reclassified as Centaurs. The problem is to explain why 29P/Schwassmann-Wachmann 1 is so active while most Centaurs have so far not shown signs of gaseous activity. There is no good explanation at hand so far to explain the different behavior. It might be possible that heating by solar radiation triggers a phase transition from amorphous to cubic ice, which is exothermic, so the heat released sublimates pockets of highly volatile material like CO_2 or CO, thus producing the activity observed in 29P and other few bodies. It might be possible that once the phase transition is completed, the activity comes to an end, at least until the body reaches the inner planetary region. Another possibility is that the active/non-active behavior reflects intrinsic differences in the content of volatile material, or that radicals produced by cosmic-ray bombardment in the outer layers of the body, and stored there for several Gyr in the deep freeze of the TN belt, react exothermically once the heat pulse from the Sun reaches them (Strazzulla and Johnson 1991). Figure 8.7 shows the location of the Centaurs so far discovered in the parametric plane (a, e). Those Centaurs that have shown activity, and are then also classified as comets, are also plotted with open circles.

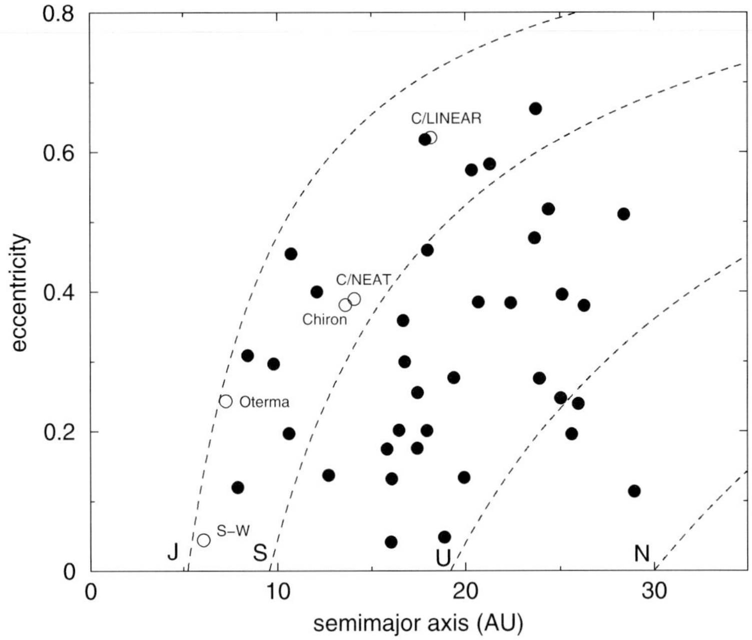

Figure 8.7. The discovered Centaurs through 2003. The dashed curves are the loci of bodies whose perihelion distance is equal to the semimajor axis of the Jovian planet indicated beside each curve. The open circles indicate the location of the active Centaurs (or comets) 29P/Schwassmann-Wachmann 1, 39P/Oterma, 2060 Chiron, C/LINEAR (2000 B4), and C/NEAT (2001 T4).

Another dynamical class of TN objects was discovered in 1996 with high-eccentricity and high-semimajor axis. The first discovered object in this class was 1996 TL_{66} with the orbital parameters: $a = 84.3$ AU, $e = 0.584$, $i = 24.°0$ (Luu et al. 1996). In general these bodies also show high inclinations. In hindsight, it was expected to find this class of objects as a result of the gravitational scattering outwards by Neptune of bodies leaving the TN belt (cf. Fig. 8.6). This class of bodies are said to belong to a *Scattered Disk* (or SD for short). *Scattered Disk Objects* (SDOs) are defined as those with perihelion distances $q > 30$ AU and semimajor axes $a > 50$ AU. The limits are somewhat arbitrary, just adopted for taxonomy purposes. They share the common property of moving on moderate to highly eccentric orbits outside Neptune's. A few SDOs have perihelion distances above 40 AU. These are said to belong to an *Extended Scattered Disk* (ESD), in the sense that it extends beyond the gravitational influence of Neptune. Duncan and Levison (1997) explain the increase in perihelion distance in many cases (though not in all of them) as due to trapping in a mean motion resonance with Neptune. Gomes et al. (2005) have shown that the Kozai mechanism (see Section 8.11 below), coupled with mean motion resonances, are responsible for raising the perihelia of SDOs above 40 AU.

8.7. Population size and size distribution

Summarizing the main dynamical features of the TNO population, Jewitt et al. (1998) characterized three different dynamical classes: (1) the classical TNOs, which occupy low-eccentricity ($e < 0.25$) orbits with semimajor axes $41 \lesssim a \lesssim 46$ AU, they are estimated to constitute $\sim$ 70% of the observed population; (2) the resonant TNOs, which occupy mean-motion resonances with Neptune, in particular the 2:3 ($a \approx 39.4$ AU) and comprise $\sim$ 20% of the known objects; and (3) the scattered TNOs, or SDOs, which possess the most extreme orbits, with median semimajor axis $a \sim 90$ AU and eccentricity $e \approx 0.6$, and comprises about 10% of the known TNOs. Since observational bias plays against the discovery of SDOs, that spend most of the time at large heliocentric distances where they are too faint to be discovered, the actual population of these bodies should be substantially larger than the observed one. The distribution of all the discovered TNOs in the parametric plane (a, e) is shown in Fig. 8.8. We can see in the figure the different dynamical groups, in particular the Plutinos, classical TNOs, SDOs and a few at the 1:2 mean-motion resonance.

All the discovered TNOs are of relatively large size, with estimated diameters greater than 100 km if low geometric albedos are assumed. The largest TNOs so far discovered -after Pluto- are 2003 VB_{12}, named Sedna after the Inuit goddess of the sea, with an upper limit for the estimated diameter of 1,600 km, and 2002 LM_{60}, named Quaoar after a creation god of the Tongva tribe of Southern California. The former moves on a very eccentric orbit with the largest perihelion distance so far known: $q = 76 \pm 4$ AU (Brown et al. 2004), whereas the latter moves on a near-circular orbit of radius $\sim$ 43 AU. Caltech astronomers

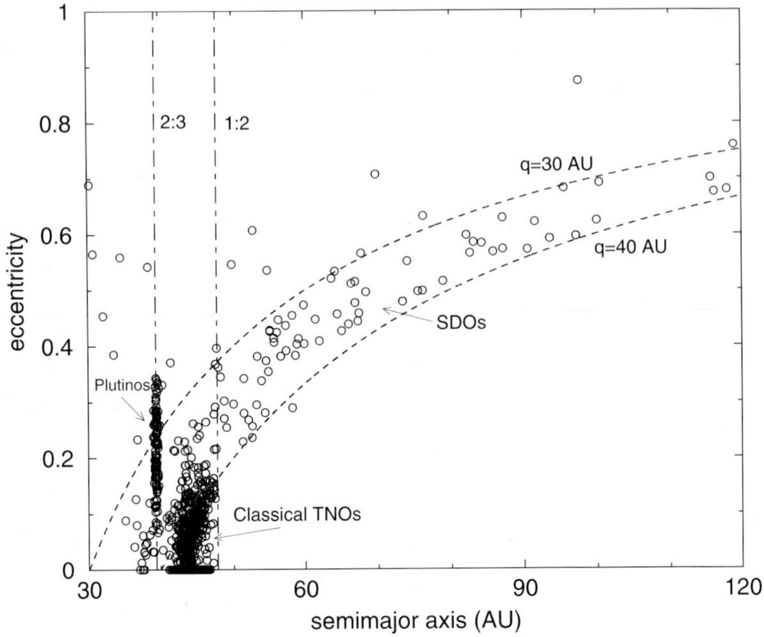

Figure 8.8. Discovered TNOs. The figure also shows the location of the 2:3 and 1:2 resonances and the dashed curves for which the perihelia of the bodies are $q = 30$ AU and $q = 40$ AU (Fernández et al. 2004).

Michael Brown and Chad Trujillo could measure an angular diameter of 40 milliarcseconds on images, which at the distance of Quaoar corresponds to a diameter of 1300 km. Another large TNO is 2001 KX_{76} whose estimated diameter lies between 960 km and 1270 km, depending on the assumed albedo. Sedna is greater than Charon, and its diameter $\sim 2/3$ of Pluto's. It is by now clear that Pluto is the largest body of a population showing a near-continuum distribution of sizes. It is possible that other Pluto-sized bodies might exist in the TN belt, and the existence of bodies even larger than Pluto cannot be ruled out at present.

Trujillo et al. (2001) modeled the cumulative luminosity function of TNOs by means of the linear relation

$$\log \Sigma(m_R) = C + \beta m_R, \tag{8.3}$$

where $\Sigma(m_R)$ gives the density of TNOs per square degree near the ecliptic brighter than m_R, and C and β are constants. The CLF can be used for determining the size distribution of TNOs, after introducing appropriate corrections for heliocentric distance, geocentric distance and detection efficiency. By doing this, Trujillo et al. (2001) obtain a power-law for the cumulative size distribution

$$N_R(R) \propto R^{-s}, \tag{8.4}$$

where $N_R(R)$ is the number of objects with radii greater than R. They obtain a slope for the CLF: $\beta = 0.63 \pm 0.06$, which corresponds to $s = 3.15 \pm 0.3$ (Fig. 8.9), and population sizes for the classical TNOs: $N_{CTNO}(D > 100km) = 3.8^{+2.0}_{-1.5} \times 10^4$, and Plutinos: $N_{Plutino}(D > 100km) \approx 1400$. This value of s turns out to be greater than those quoted in Table 7.2 for JF comets and other populations of minor bodies, though it is closer to the theoretical exponent derived by Kenyon

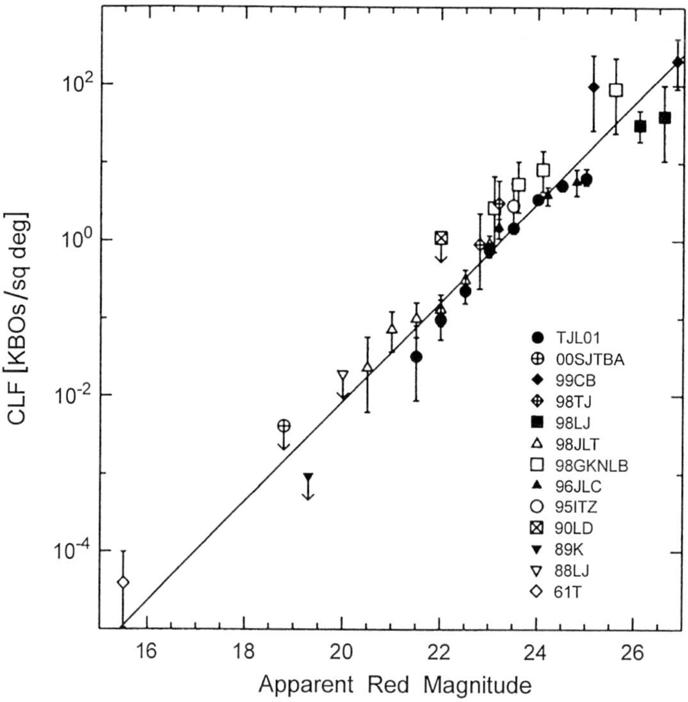

Figure 8.9. Cumulative luminosity function, which represents the number of TNOs brighter than a given apparent red magnitude. The line represents a fit to the data obtained by Trujillo et al. (2001) yielding $\beta = 0.63 \pm 0.06$. The different symbols are for the surveys: TJL01 = Trujillo et al. (2001), 00SJTBA = Sheppard et al. (2000), 99CB = Chiang and Brown (1999), 98TJ = Trujillo and Jewitt (1998), 98LJ = Luu and Jewitt (1998), 98JLT = Jewitt et al. (1998), 98GKNLB = Gladman et al. (1998), 96JLC = Jewitt et al. (1996), 95ITZ = Irwin et al. (1995), 90LD = Levison and Duncan (1990), 89K = Kowal (1989), 88LJ = Luu and Jewitt (1988), 61T = Tombaugh (1961) (Luu and Jewitt 2002).

and Bromley (2004) from collisional evolution models of icy bodies in the TN belt ($s \approx 3.5$ for bodies with radii $\gtrsim 10 - 100$ km).

Trujillo et al. (2001) estimate populations of similar size for the 2:3 and 1:2 resonances, after correcting for several observational biases. According to these authors, the size of the populations: Classical : Scattered : Plutino : Resonance 1:2 are in the respective ratios 1.0:0.8:0.04:0.07. The inferred mass of TNOs in the belt with diameters $100 < D < 2000$ km is found to be

$$M_{TNO}(D > 100km) \approx 0.03 M_{\oplus} \left(\frac{\rho_{TNO}}{1 \text{ g cm}^{-3}} \right) \left(\frac{0.04}{p_R} \right)^{1.5}, \tag{8.5}$$

where ρ_{TNO} is the mass density of the TNOs and p_R the red geometric albedo.

If we assume that the same power-law size distribution applies down to objects of radius one km, then the total mass of the TN belt would be ~ 0.1 $M_{\oplus}$ (Jewitt et al. 1998), i.e. about two orders of magnitude more massive than the asteroid belt. An independent estimate of the upper limit of the mass of the TN belt was made by Anderson et al. (1998), based on the lack of damage in the hydrazine tank of the Pioneer 10 spacecraft that passed through the TN belt in its journey out of the solar system. They set an upper limit of 0.1 $M_{\oplus}$ on low-mass (centimeter-size), low density objects within the range $\sim 35 - 65$ AU, which is consistent with Jewitt et al.'s estimate.

Larsen et al. (2001) searched a sky area of 1483.8 deg^2 which led to the discovery of five TNOs and five Centaurs or SDOs. From this survey they estimated a population of 400 TNOs, 100 Centaurs, and 70 SDOs down to $m_R = 21.5$ in the ecliptic. For a classical TNO at ~ 42 AU and albedo $p_R = 0.04$ this limiting magnitude would correspond to a diameter $D = 460$ km. Extrapolating down to $D = 100$ km it would give a total population of classical TNOs of $\sim 40,000$, in good agreement with the value derived by Trujillo et al. (2001).

Duncan et al. (1995) combined the results for the transfer efficiency of bodies from the TN belt to the JF population , derived from their simulations, with the estimated total number of JF comets (cf. previous chapter) to estimate the total number of comet-sized objects in the TN belt. The total number of JF comets (both active and extinct) can be expressed as

$$N_{JFC} = N_{TNO} \times \nu \times f_{JFC} \times L_{JFC}, \tag{8.6}$$

where N_{TNO} is the current number of comet-sized bodies in the TNO belt; ν is the fraction of bodies that leave the TN belt per year; f_{JFC} is the fraction of bodies that become JF comets once they leave the TNO belt; and L_{JFC} is the median dynamical lifetime of JF comets. From numerical simulations, Duncan et al. estimated the following numerical values: $L_{JFC} = 3.3 \times 10^5$ yr, $f_{JFC} = 0.34$, and $N_{JFC} = 21,000$. Furthermore, $\nu = 3 \times 10^{-11}$ for a model of TNOs with $e = 0.05$; or $\nu = 5 \times 10^{-11}$ for $e = 0.15$. By introducing these numerical values in eq. (8.6) they derived a population of comet-sized bodies in the TN belt interior to 50 AU of $N_{TNO} = 6 \times 10^9$ for $e = 0.05$, or $N_{TNO} = 3 \times 10^9$ for $e = 0.15$. One should note that the actual lifetime of JF comets may be shorter than L_{JFC} since factors other than dynamical ejection may be at work, as for instance sublimation/disintegration, or collision with the Sun or any of the planets. Their derived TNO population can then be taken as a lower limit. Duncan et al.'s results are in good agreement with the population $N(R > 1km) \sim 10^{10}$ estimated by Luu and Jewitt (2002) from the extrapolation of the estimated population of bodies in the classical belt with $D > 100$ km by means of the power-law mass distribution given by eq. (8.4). There is also a fairly good agreement with Morbidelli's (1997) estimate of 4.5×10^8 bodies in the 2:3 resonance (cf. Section 8.4) that, assuming that they represent 25% of the total population, would be 1.8×10^9 bodies. If Plutinos represent much less than 25% of the total population, as Trujillo et al. estimate, a total TNO population of $\sim 10^{10}$ with $R > 1$ km seems to be more reasonable, in line with the above results.

Bernstein et al. (2004) have searched 0.02 deg^2 of the invariable plane for TNOs using the Advanced Camera for Surveys aboard the Hubble Space Telescope. They could search for objects as faint as $m_R \simeq 29.2$ which, at the distance of the TN belt, means objects of a few tens km diameter. Bernstein et al. discovered only three objects in this size range, which turns out to be about 25 times smaller than would be expected, if the population followed the power-law given by eq. (8.4) with $s = 3.15$ down to $R \sim 10$ km. These survey results, if confirmed, would imply a heavy depletion of comet-sized bodies for which present-day collisions are expected to be disruptive. Because of this depletion, Bernstein et al. estimate that the classical belt and Plutinos fall at least 10 times short of the required reservoir to supply the JF population. We have then here a source of potential conflicts. The authors argue that the scattered disk might be a more suitable precursor population of JF comets. We will come back to this point in Section 8.11.

8.8. Binary TNOs

Another interesting feature is the discovery of a few percent of binaries among the TNOs, which show wide separations of about 100 to 1,000 times their radii (for instance, the Moon is at about 60 Earth's radii), and mass ratios of order unity (Fig. 8.10). Pluto and Charon may be taken as the first example of binary TNOs of comparable mass, though their separation is only $\sim$ 17 times Pluto's radius. The determination of the orbital period of binary systems will permit to derive their total mass and the individual component masses, and hence their bulk densities, a key parameter to gain insight into the composition and internal structure of these distant bodies. It is still too premature to assess what is the fraction of TNOs with companions, and what is the debiased

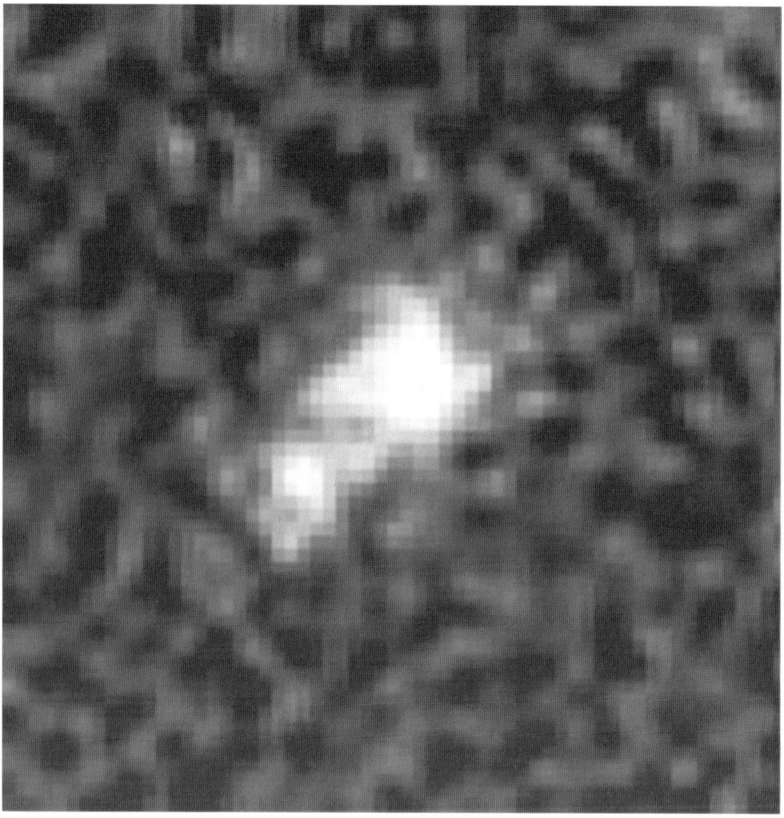

Figure 8.10. The binary TNO 1998 WW_{31} as observed with the wide-field imaging camera mounted at the prime focus of the Canada-France-Hawaii 3.6-m telescope (Christian Veillet, CFHT).

distribution of sizes and separations. From the rate of binaries detected among TNOs, Noll et al. (2002) estimate a frequency of binaries of $\approx 4\% \pm 2\%$ for separations greater than $0.15''$ (or ≈ 4500 km at the distance of 40 AU), and less than one mag difference (in the V-band) between the components. This separation marks an observational limit for discriminating the two components of the binary on high-resolution images.

Of course, the frequency of binaries among TNOs may greatly increase if we consider closer binary systems, or binaries in which the difference between the primary and secondary exceeds 1 mag. Unfortunately, these close pairs cannot be resolved on images, so their detection requires other techniques as, for instance, from their lightcurves. Several lightcurves have been determined photometrically by Sheppard and Jewitt (2004) with the aim to derive rotation periods and shapes of TNOs. During the course of their research, they found the anomalous case of the TNO 2001 QG_{298} with a peak-to-peak lightcurve of 1.14 ± 0.04 mag. If this magnitude variation were attributed to shape, it would correspond to a very elongated body of axial ratio $\approx 3:1$. Such cigar-shaped bodies are physically unsound, since gravitational compression acting on bodies with radii $\gtrsim 50-75$ km would lead to more spherical shapes (2001 QC_{298} has an estimated mean effective radius of $R = 122$ km if a geometric albedo 0.04 is assumed). Sheppard and Jewitt rule out albedo differences as the cause of the large magnitude variation since the BVR colors show no variation from minimum to maximum brightness. They also rule out rotational deformation because of its rather long rotation period (13.77 hr). After discarding the previous effects, they argue in favor of 2001 QG_{298} being a contact binary or a near-contact binary viewed from a nearly equatorial perspective, so the components are subject to mutual occultations as seen from the Earth. To produce such a large brightness variation, the components should be of similar size and distorted by their mutual tidal interactions. Sheppard and Jewitt then conclude that the frequency of contact or near-contact binaries with components of similar size could be at least 10% to 20% of the TNO population.

Tancredi and Fernández (1991) explained the high content of specific angular momentum of the Pluto-Charon system, that would make it rotationally unstable if Pluto and Charon coalesced into a single body, as the result of the accretion of planetesimals with masses up to a few tenths the mass of the proto-Pluto itself. It is possible that this mechanism also worked for other massive TNOs, that increased their specific angular momentum as they grew more massive, until they fissioned or were disrupted and reassembled in two or more components. This process would have led to close binaries, though their mutual separation could have increased with time under the action of mutual tides as happened with the Pluto-Charon system. Weidenschilling (2002) offers a different explanation for the origin of TN binaries in terms of two bodies that collide within the Hill's sphere of influence of a third massive body. The loss of energy upon impact would have led to the permanent capture of one or the two bodies by the massive body. Three-body encounters would have occurred with a frequency high enough to justify the fraction of binary systems if the number density of massive bodies was much higher than the current one ($\sim$ 100 times), which is consistent with some estimates of the primordial mass of the TN disk. Goldreich et al. (2002) offer an interesting alternative scenario in which two bodies enter each other's Hill's sphere, becoming afterwards a bound system as the result of energy loss by dynamical friction (this is a kind of equipartition of energy in which massive bodies exchange energy with smaller ones after gravitational encounters). These authors argue that the loss of energy could have also occurred after a gravitational interaction with a third massive body, though they estimate this process to be somewhat less efficient than the previous one. They predict that five percent of the TNOs are binaries with separations greater than 0.2 arcsec, and that an even larger percentage are in more tightly bound binaries. These percentages are in rather good agreement with observations. From the previous discussion, we can conclude that the high binary frequency among TNOs may be a consequence of the early intense collisional environment in the TN belt.

8.9. The outer edge of the belt

Large-scale ecliptic surveys of bright TNOs uncovered an interesting feature of the TN belt: if the belt extended to large heliocentric distances, then one should expect to find some of the brightest members of the classical belt with semimajor axes $a > 50$ AU. However, in one of this deep surveys carried out by Jewitt et al. (1998) none showed up, despite many classical TNOs were detected with $a \lesssim 47$ AU. Confirming the previous result, Trujillo et al. (2001) found no objects beyond $\sim$ 49 AU among the 86 TNOs detected in another survey of 73 deg^2 to a limiting red magnitude of 23.7, which led the authors to conclude that the outer edge of the classical TNO belt is at about 50 AU.

Allen et al. (2002) searched for a dynamically cold distant TN belt near the invariable plane of the solar system. They argued that the early protoplanetary disk, in which planetesimals formed, was located in the invariable plane of the solar system, i.e. the plane normal to the total angular momentum vector of the solar system, not in the ecliptic plane which is inclined 1.°6 to it, so the searches should concentrate on the invariable plane, not on the ecliptic. They surveyed a total of 2.3 deg^2 in which they discovered 33 TNOs and one Centaur, but no objects in circular orbit beyond 50 AU. This result allowed them to set an upper limit to the population of TNOs between 50 AU and 60 AU, at the 95% confidence level, at no more of 1.2 times the observed population of TNOs inside 50 AU for diameters $D > 185$ km, if the distant disk is thinner than 1.°75.

What could be the causes for a sharp drop in the TN population at ≈ 50 AU? It might be due to cosmogonic reasons, following to the longer time scales required to form bodies of a certain size at larger heliocentric distances (see, e.g., discussion by Luu and Jewitt 2002). As it will be shown in Chapter 10, the mass growth rate of grains in the protoplanetary disk decreases as $r^{-11/4}$. Perhaps, dust grains beyond ~ 50 AU did not have the time to settle in a thin disk before the dispersal of the gaseous component of the nebula, requirement that seems indispensable to proceed further with their growth into planetesimals. Brunini and Melita (2002) have argued that a Mars-size planetoid with semimajor axis $a \approx 60$ AU and perihelion distance ≈ 49 AU could have cleared a gap in the TN belt between $\sim 49 - 78$ AU. According to the authors, such a large body could have escaped detection until now because, among other reasons, a rather high inclination ($i \gtrsim 10°$) or a low albedo. The authors argue that such a massive body could have been scattered by Neptune from its accretion zone in the early solar system, as has been discussed by, e.g. Ip (1989), as a consequence of the macro-accretion process of Uranus and Neptune. The scattered massive body would have passed through the TN belt where exchange of orbital angular momentum with TNOs rose its perihelion, thus decoupling the body from Neptune's gravitational influence. Yet, new numerical simulations by Melita et al. (2004) of samples of fictitious TNOs with initial semimajor axes in the range $35 < a < 80$ AU and a planet X with a mass between $0.1 - 3.3$ $M_{\oplus}$, were unable to reproduce the observed dynamical features of the TN belt, in particular the sharp edge at

$a \approx 49$ AU. Levison and Morbidelli (2003) have offered a different explanation in terms of a TNO population formed closer to the Sun and subsequently drove outwards through resonant interactions with Neptune. We shall come back to this theory below and in Chapter 10.

8.10. Dynamical and physical erosion of the belt

The belt is dynamically excited and appears to have a sharp transition at 40 - 42 AU, allowing a natural distinction between the inner portions for $a \lesssim 40$ AU and the *classical belt* ($a > 41$ AU and $q > 35$ AU) (Jewitt et al. 1998, Petit et al. 1999). The classical TNO population has a median eccentricity ~ 0.07 and inclination $\sim 4.^\circ 0$. Furthermore, its current estimated mass of $\sim 0.1\ M_\oplus$ is about two orders of magnitude smaller than the primordial mass estimated on cosmogonic grounds (see Chapter 10). A minimum mass solar nebula with a radial distribution of mass surface density $\Sigma \propto r^{-3/2}$ would give $\sim 7-15\ M_\oplus$ of solid matter in a 6 AU wide annulus centered at 35 AU. Kenyon and Luu (1999) found that kilometer-sized planetesimals in such a massive belt would experience runaway growth to a Pluto-sized body on a time scale of 100 Myr, provided that the belt was "cold" (initial eccentricities $\sim 10^{-3}$). If the initial mass in the annulus was 100 $M_\oplus$, the runaway time scale would reduce to ~ 10 Myr. Stern (1996) found that the required mass of the primordial belt to form QB_1-sized objects within 10^9 yr by binary accretion from kilometer-sized planetesimals would be between 35-50 $M_\oplus$, in agreement with Kenyon and Luu's result. Stern also found that a runaway growth would have proceeded in a dynamically cold disk with initial eccentricities $\lesssim 5 \times 10^{-3}$.

There are several possible mechanisms that can explain the passage from a massive (several tens $M_\oplus$), dynamically cold ($e << 0.01$) TN belt, to the current low-mass, dynamically hot belt. We can mention among them:

8.10.1. Dynamical erosion

As seen, the inner portions of the TN belt have been severely depleted by chaos induced by overlapping mean-motion and secular resonances. Figure 8.11 shows the results of the numerical integration of thousands of test bodies by Levison et al. (1995). The bodies follow initially a radial distribution $\propto r^{-2}$, simulating the primordial radial distribution of

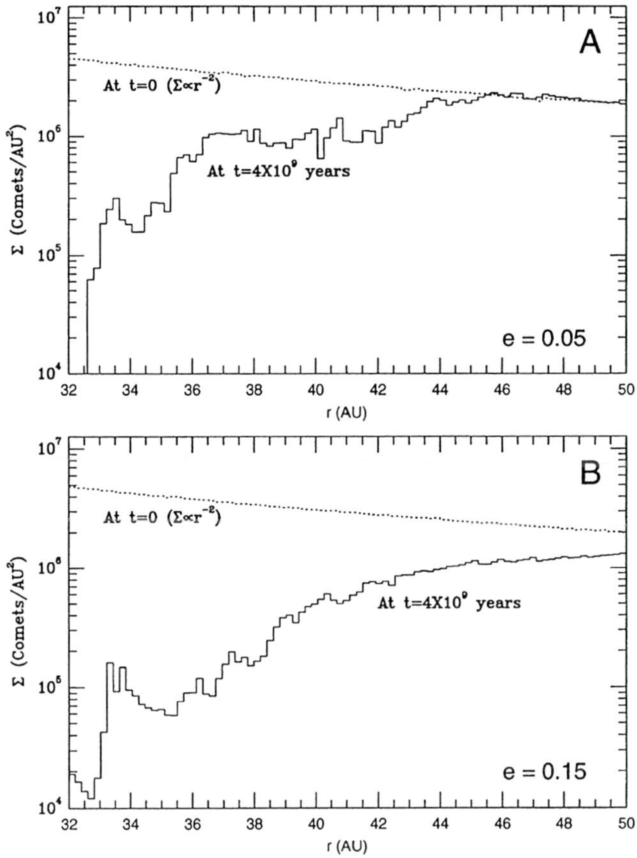

Figure 8.11. The current surface density distribution of bodies in the TN belt as determined by Monte Carlo simulations. These models assume that the initial radial distribution of bodies follows the law r^{-2}, that the initial inclination is $i = 1°$, and that the initial eccentricities are 0.05 (A) and 0.15 (B) (Duncan et al. 1995).

matter in the protoplanetary disk (different cosmogonic models adopt exponents generally between -3/2 and -2 for the radial distribution of matter in the protoplanetary disk). All the test bodies are initially in low-inclination orbits ($i = 1°$) with two different eccentricities: $e = 0.05$ and $e = 0.15$. We can see a heavy depletion for semimajor axes $a \lesssim 36$ AU, and still a substantial depletion for $36 \lesssim a \lesssim 45$ AU. For $a \gtrsim 45$ AU the disk remains practically unperturbed. As expected, the hotter disk leads to a much heavier depletion. Despite the crucial role played by dynamical erosion in depleting the TN disk in the region $30 \lesssim a \lesssim 40$ AU, dynamical causes alone cannot explain the heavy depletion and dynamical heating for $a \gtrsim 42$ AU, mainly considering that the primordial disk was possibly "cold" ($e << 0.05$).

8.10.2. Mutual collisions

From numerical simulations, Stern (1996) finds that binary accretion (i.e. impactor + target pair) in the present day TN disk cannot explain the growth of QB_1-class bodies on time scales comparable to the solar system age. From more sophisticated accretion models, Stern and Colwell (1997) later found that, in order to achieve the current population of QB_1-class objects, it could have been required an initial solid mass $\gtrsim 20$ $M_\oplus$ in the 30-50 AU zone and mean orbital eccentricities $\lesssim 0.0025$. The time scale for accretion would have been $\sim 10^8 - 10^9$ yr. The growth of Neptune would have inhibited accretion by dynamically heating the TN population. Afterwards, accretion turned into an erosion process due to the energetics of collisions, with a heavy mass loss via radiation-transport of the dust produced in the collisions.

8.10.3. Perturbations from massive bodies ($M \sim 0.1 - 1$ $M_\oplus$) scattered by Neptune and the other Jovian planets

The scattering of massive bodies by the accreting Jovian planets was already anticipated before the discovery of TNOs (e.g. Ip 1989). Cosmogonic models show that in order to form Uranus and Neptune, it would have been required a mass ~ 100 $M_\oplus$ in their accretion zones (e.g. Fernández and Ip 1996, Brunini and Fernández 1999). With such a large mass, the formation of several Earth-sized planetoids - besides those that originated Uranus and Neptune - would have been quite possible. The massive planetoids were ultimately ejected by gravitational interactions with Uranus and Neptune, once these planets grew to nearly its current sizes. Morbidelli and Valsecchi (1997) showed that a small number of Neptune-scattered planetesimals of 1-5 $M_\oplus$ could have excited the eccentricities of most primordials TNOs through gravitational interactions. From more refined numerical simulations, Petit et al. (1999) found that a planetesimal of 1 $M_\oplus$ acting for 100 Myr could have ejected most of the mass in the region $35 - 47.7$ AU.

8.10.4. Collisions with planetesimals scattered by the Jovian planets

As mentined before, models of the accretion process of Uranus and Neptune suggest a high inefficiency, with the consequent loss of several tens $M_\oplus$ during the last stages of formation of these planets. Thus, a primordial heavy bombardment of the TN belt may be expected from this massive scattering of planetesimals. The first consequence could have been an attrition of the TNO population down to dust that was subsequently removed by Poynting-Robertson drag, and the dynamical heating of the remaining population that made it more prone to fall under the gravitational influence of Neptune, thus further increasing the loss of matter.

8.10.5. Perturbations by massive stars in an early dense galactic environment

It is possible that the Sun formed together with other stars in a cluster, as most stars are observed to form. Under such conditions, close encounters with nearby stars could have destroyed the regular structure of the outer TN belt. Ida et al. (2000) found that a stellar encounter at about 100 AU could have excited the high eccentricities and inclinations of the TN belt. Yet, it could have been quite fortuitous such a fine tuning in the stellar encounter that strongly excited the orbits of TNOs but not Neptune's. A stellar encounter at $\sim$ 100 AU in $\sim 10^8$ yr could be highly probable only in an extremely dense cluster. Simulations carried out by Fernández and Brunini (2000) show that even if the Sun formed in a very dense star cluster ($\sim$ 100 stars pc^{-3}), the frequency of close stellar encounters would not be high enough to perturb significantly TNOs with $a \lesssim 10^2$ AU. As shown in Fig. 8.12, the expected energy change is only $\Delta x/x \sim 10^{-3}$ for a semimajor axis $a = 250$ AU. It is therefore very unlikely that stellar perturbations would have had an appreciable influence on bodies at $a \sim 50$ AU. Furthermore, Levison et al. (2004) found that a close stellar encounter (for a perihelion distance of the intruder star $\leq$ 200 AU), at a time between $\approx 10 - 100$ Myr (namely, between the time when TNOs formed and the dispersal of the dense star cluster) would have produced an extended scattered disk too massive to be reconciled with observations. These studies seem to rule out a close stellar encounter as an excitation mechanism of the early TN belt.

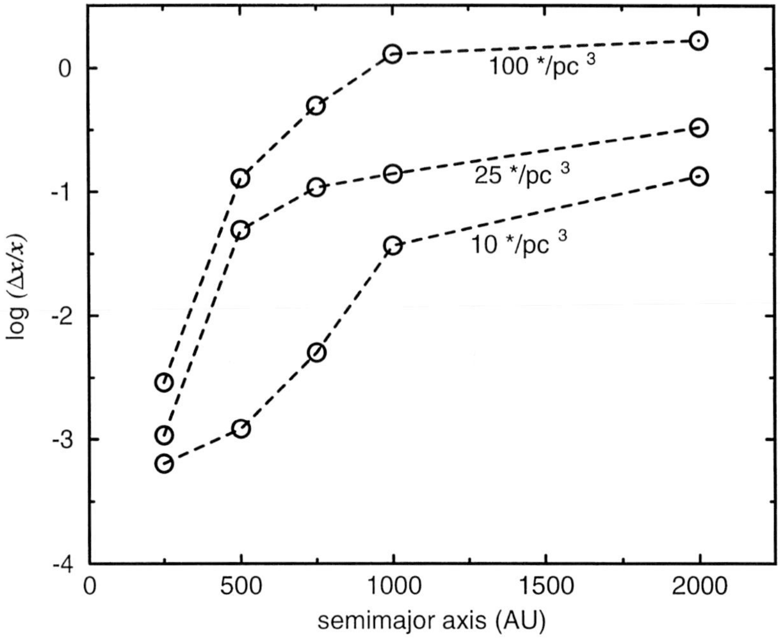

Figure 8.12. The r.m.s. change in the relative energy of TNOs by perturbations of cluster stars as a function of the semimajor axis, and for the three densities of star clusters indicated beside each curve (Fernández and Brunini 2000).

8.10.6. Cosmogonic causes

As mentioned above, Levison and Morbidelli (2003) argue that the TNO population formed closer to the Sun where the density of solid matter was higher, thus explaining the formation of Pluto-sized bodies on reasonably short time scales ($\sim 10^8$ yr or less). This population was subsequently pushed outwards through resonance capture by the proto-Neptune, which itself was driven outwards through exchange of angular momentum with the interacting planetesimals. This is a radically different view on how the TN belt formed, and why is dynamically hot, from the previous theories that implicitly assume *in situ* formation. We will come back to this theory in Chapter 10 when we discuss the conditions in the early solar system.

8.11. The trans-neptunian belt as a replenishment source of the Oort cloud

We have seen that the TN belt is a suitable source of JF comets and Centaurs. TNOs can reach Neptune after being decoupled from the 2:3 mean motion resonance (e.g. Morbidelli 1997, Ip and Fernández 1997), or from the scattered disk (Duncan and Levison 1997, Bernstein et al. 2004). Yet, not all SDOs will end up entering the planetary region as Centaurs; a fraction of them - actually the majority, as we shall see below - are scattered to the Oort cloud. We will analyze next this important point.

Observed SDOs have semimajor axes of several tens to hundreds AU. It is understandable that under the weak perturbations of Neptune, SDOs will random-walk in the energy space until they reach the Oort cloud, are hyperbolically ejected, or transferred to the planetary region. From numerical computations of samples of massless test particles in Neptune-encountering orbits, Duncan and Levison (1997) found that about 1% still remained after 4 Gyr in eccentric orbits forming the Scattered Disk, and that a fraction of them eventually evolved to near-parabolic orbits. Fernández et al. (2004) have carried out numerical simulations of the orbital evolution of a sample of 76 observed SDOs and 399 clones for 5 Gyr. Figure 8.13 shows the example of the SDO 1999 DP_8 that reaches the Oort cloud after 3.35 Gyr. It is also noteworthy to see that the argument of perihelion of this body librates during a time at $\sim 0.9 - 1.2$ Gyr around $\omega \sim 180°$. This is what we call a Kozai resonance that is responsible for raising its perihelion to $q \sim 50$ AU, namely 1999 DP_8 becomes for a time a member of the Extended Scattered Disk quite detached from the planetary region. The body suffers some close encounters with Neptune during the course of its evolution, though most of the time it remains with a perihelion outside Neptune's orbit. Actually, this is a dynamical behavior that we observed in many test SDOs: they get scattered to the Oort cloud from orbits with perihelion distances $q > 30$ AU. Because planetary perturbations at such distances are very weak (cf. Section 4.1), the random-walk in the energy space is very smooth, which warrants that most bodies diffusing outward will be trapped in the Oort cloud before being kicked to interstellar space. In effect, the typical energy change per orbital revolution of low-inclination bodies (such as SDOs) scattered by Neptune is $\sim 4 \times 10^{-5}$ AU^{-1}, while the Oort cloud population covers an energy range $\sim 10^{-4}$ AU^{-1}.

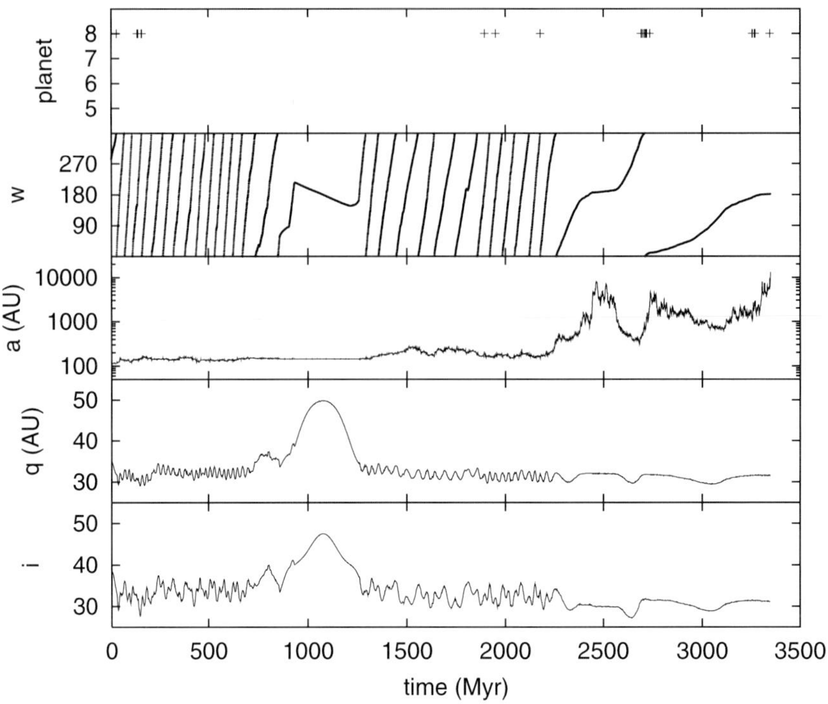

Figure 8.13. Dynamical evolution of the TNO 1999 DP_8. The following parameters are represented: close encounters with Jupiter (5), Saturn (6), Uranus (7) and Neptune (8) (top panel); argument of perihelion (second panel); semimajor axis (third panel); perihelion distance (fourth panel); and inclination (bottom panel) (Fernández et al. 2004).

From their numerical simulations, Fernández et al. (2004) estimate a dynamical half-life for SDOs given by

$$t_{dyn} = 10^{(q-33.5)/4.7} \text{ Gyr}, \tag{8.7}$$

where the perihelion distance q is expressed in AU.

From eq. (8.7) we find that SDOs with $q \sim 30 - 40$ AU have dynamical half-lives between $10^8 - 10^{10}$ yr, i.e. some of them may be in the SD over time scales exceeding the solar system age. Therefore, the escape process of bodies from the SD is still taking place, even for those bodies that have been there since primordial times. What are the end states of the bodies that leave the SD?. Fernández et al. (2004) find that the fraction that reach the Oort cloud, that are ejected hyperbolically, or that reach Jupiter's region are: $f_{oort} = 0.47 \pm 0.04$, $f_{hyp} = 0.25 \pm 0.03$, and $f_{jup} = 0.28 \pm 0.03$, respectively.

The previous results tell us that close to 30% of SDOs will become Centaurs that reach Jupiter, and a fraction of these will ultimately become JF comets. This agrees with Bernstein et al.'s (2004) suggestion that the scattered disk may be the precursor source of most JF comets, in addition to the Plutinos. Even more important is to highlight that nearly 50% of the SDOs are incorporated into the Oort cloud. As said before, the random-walk in the energy space of bodies scattered by Neptune is very smooth, so most of them get trapped in the Oort cloud before being ejected. This explains why for every SDO ejected to interstellar space, about two are incorporated into the Oort cloud. Given the long dynamical half-lives of SDOs, the transfer of bodies to the Oort cloud should be still a very active process. Fernández et al. (2004) estimate a transfer rate of bodies with radii $R > 1$ km from the SD to the Oort cloud of 5 yr^{-1}. This rate is comparable, at least within the order of magnitude, with that of new comets entering the Jupiter-Saturn region per year. If this constitutes the main comet loss mechanism from the Oort cloud, then the transfer of bodies from the SD to the Oort cloud may play a very important role as a replenishment source. Therefore, far from considering the Oort cloud as a fossil structure, dating back to the very beginning of the solar system lifetime, we can imagine it as a very dynamic environment where comet losses are made up - at least partially - with bodies coming from the scattered disk or the inner core.

Levison et al. (2005) also discuss the transfer of SDOs to very eccentric orbits where galactic tides act on the bodies, driving the perihelia of a fraction of them into the planetary region. From numerical simulations, they find that about 0.01% of the SDOs that reach the Oort cloud ($a > 10^4$ AU) evolve into Halley-type orbits after being perturbed by the giant planets. Since the residence time in the Oort cloud is very short, external perturbers do not have the time to randomize the orbital planes of these bodies, so most Halley-type comets will preserve the low inclinations they had in the SD. Levison et al. explain the retrograde orbits of some HT comets as due to the precession of the SDO's ascending node in the galactic coordinate system by tides of the galactic disk. This precession makes the orbit to flip over from a direct to a retrograde orbit, as explained in Section 5.5. Although Levison et al.'s (2005) study focuses on the origin of HT comets, it essentially confirms Fernández et al.'s (2004) results in that a substantial number of bodies is being incorporated into the Oort cloud, at least for a short time. Levison et al. found a probability that a body will leave the SD for the Oort cloud of $\approx 1.8 \times 10^{-10}$ yr^{-1}. For a population of $\sim 10^{10}$ bodies with radii $R > 1$ km (cf. Section 8.7), the transfer rate would give ≈ 1.8 yr^{-1}, in fairly good agreement with that found by Fernández et al.

8.12. Physical properties of trans-neptunian objects and Centaurs

There is already a good wealth of photometric data in the visible and near-infrared of several TNOs and Centaurs. The observations have been carried out mainly in the broad band filters BVRIJHK and have permitted a preliminary broad spectral characterization of these populations. Near-infrared reflection spectra of 1996 TO_{66} (Brown et al. 1999), 20000 Varuna (Licandro et al. 2001), and 1999 DE_9 (Jewitt and Luu 2001) show absorption features corresponding to water ice. Yet other TNOs like 1993 SC (Jewitt and Luu 2001) or 2000 EB_{173} (Brown et al. 2000) show featureless spectra in the range 1.0-2.5 μm (Fig. 8.14). Luu et al. (2000) argue that water ice is ubiquitous in TNOs but it may not be detectable in some objects due to its high degree of contamination or low abundance. Jewitt and Luu (2001) further discuss that highly carbonized material is commonly neutral and featureless, reflecting a deficiency of hydrogen bonds. In this regard cosmic-ray bombardment would be responsible for the hydrogen depletion of the surface layers of TNOs because of sputtering and chemical alteration, involving the dissociation of water molecules: $H_2O \rightarrow H_2 + 1/2O_2$ (Strazzula and Johnson 1991). Even the H_2O features observed in some objects are severely weakened by contamination. For instance, the reflection spectrum of 1999 DE_9 can be matched by that obtained from a mixture of red cinder and 1% (by mass) water ice (Fig. 8.14). For the Centaurs Chiron and Chariklo (1997 CU_{26}) Groussin et al. (2004) could also detect the water absorption feature in their near-IR spectra. They found that their spectra can be well fitted to laboratory spectra produced by mixtures of refractory grains ($\sim 70\%$) and water ice ($\sim 30\%$) in the case of Chiron, and $\sim 80\%$ to $\sim 20\%$ in the case of Chariklo.

Cruikshank et al. (1998) present a detailed model of the composition of one of the reddest objects so far known: the Centaur 5145 Pholus ($V - R = 0.78 \pm 0.05$, according to Bauer et al. 2003). Pholus is still moving on a quite distant orbit: $q = 10.6$ AU, $a = 24.7$ AU, $i = 20.3°$.

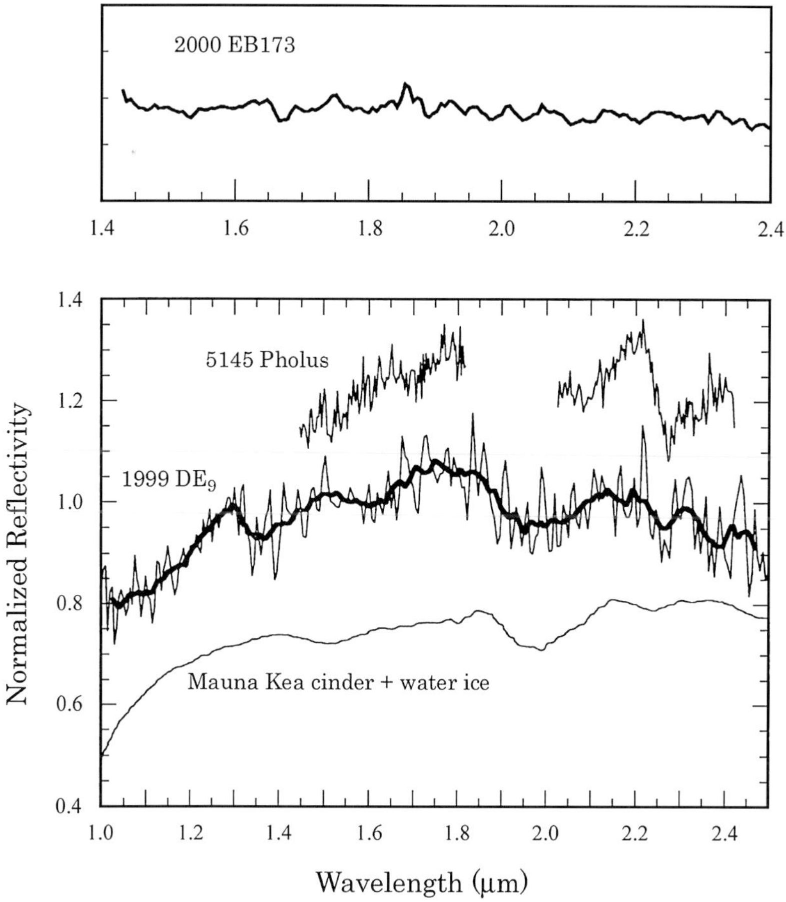

Figure 8.14. Near-infrared reflection spectra of TNOs. (a) 2000 EB173 showing a featureless spectrum (Brown et al. 2000). (b) 1999 DE_9 compared with the spectrum of a mixture of Mauna Kea cinder + 1% H_2O ice (by mass). The water absoption feature at ~ 2.0 μm is clearly seen (Jewitt and Luu 2001).

Its reflection spectrum shows the absorption band of water ice at 2.04 μm and a strong band at 2.27 μm, that the authors interpret as due to methanol ice. They found that a mixture of $(61.5 \pm 5)\%$ carbon black and $(38.5 \pm 5)\%$ of an olivine-tholin-water-methanol mixture produces the best fit to the observed spectrum. The tholin is a carbon-rich refractory compound produced by plasma irradiation of a gas mixture of N_2+CH_4 in the proportions of 0.9 to 0.1 as observed in Titan's atmosphere. Thus, Cruikshank et al.'s model can be the best hint until now about the nature of the ultrared matter in the outer solar system.

Jewitt and Luu (2001) find no evidence of a correlation between color and semimajor axis for the TNOs, in sharp contrast with the main-belt asteroids in which strong radial color gradients exist. In particular, the mean colors of the classical and resonant TNOs are indistinguishable. Relative to the Sun, TNOs exhibit reflected colors from nearly neutral ($V - R \approx 0.3$) to very red ($V - R \approx 0.7 - 0.8$). The very red material, present in some TNOs and Centaurs, is thought to arise from prolonged exposure of organics to cosmic ray bombardment that form complex carbon compounds. This process competes with resurfacing, i.e. the ejecta blanket deposited after an impact, which will tend to bury the ultra-red material with fresh sub-surface icy material. Yet, Jewitt and Luu (2001) found that resurfacing does not seem to play a fundamental role in the alteration of colors, based on the lack of observed rotational color variations in individual TNOs. If resurfacing was the cause of the color variations, one would expect to find the same color variation in individual TNOs with the rotational phase, as the broad range of colors found in different TNOs that go from $V - R = 0.35$ to $V - R = 0.80$. However, the color variations with the rotational phase are in general several times smaller. It is then possible that real compositional differences among different TNOs also plays a role in the color diversity.

Licandro and Pinilla-Alonso (2005) have taken low resolution spectra of the Centaur (32522) Thereus over more than half its rotation period (8.3 hr). They note that whereas the near-infrared spectra for different rotation phases are all very red with similar slopes in the range 0.9-1.8 μm, the absorption band of water at $\lambda \approx 2$ μm is clearly shown during part of the rotation phase while it is absent in the rest (Fig. 8.15). This study then provides the first direct evidence of inhomogeneities on the surfaces of Centaurs/TNOs, perhaps due to a recent impact that left exposed fresh icy material. The interesting thing is that this feature does not affect the global color, which poses the question as to what extent a similar color independent of the rotation phase reflects a homogeneous surface.

Jewitt (2002) compares the color properties of TNOs and of cometary nuclei. He finds that the cometary nuclei are on average bluer than the TNOs, indicating chemical and/or physical differences in the surfaces of both groups. He parameterizes the spectra using the normalized reflectivity gradient $S'[\%/1000\text{Å}]$ defined as

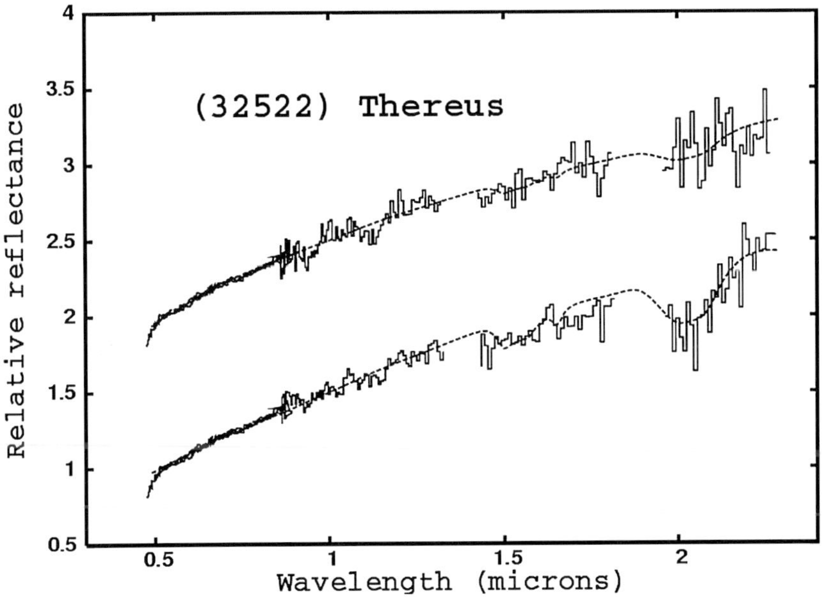

Figure 8.15. Visible and near-infrared spectra of (32522) Thereus taken at two different rotational phases. The upper spectrum is nearly featureless, while the lower spectrum clearly shows the absorption feature of water around ~ 2 μm. For clarity, both spectra are shifted by 1.0 in the relative reflectance (Licandro and Pinilla-Alonso 2005).

$$S' = \frac{dS/d\lambda}{\bar{S}} \tag{8.8}$$

where S is the reflectivity, and $\bar{S}$ is the mean value of the reflectivity in the wavelength range over which $dS/d\lambda$ is computed. The S' distributions of comets and TNOs are shown in Fig. 8.16, and it is readily seen that they are different. Furthermore, Jewitt finds that Centaurs have optical colors like those of TNOs and, therefore, different from those of comet nuclei. He argues that the color change from Centaurs to comet nuclei is due to the rapid burial of ultra-red material under a debris mantle, once the body reaches the water sublimation zone ($r \lesssim 6$ AU). He also argues that sublimation of a very volatile substance from a few active areas, like CO, has to do with the strong color change. Thus, dust grains are carried away with the sublimating gases, but a fraction of them will fall back over the surface burying the ultra-red material. Jewitt estimates that a centimeter-thick dust mantle can grow on a 100-km size body on a time scale of $\sim 10^3$ yr, much less than the dynamical

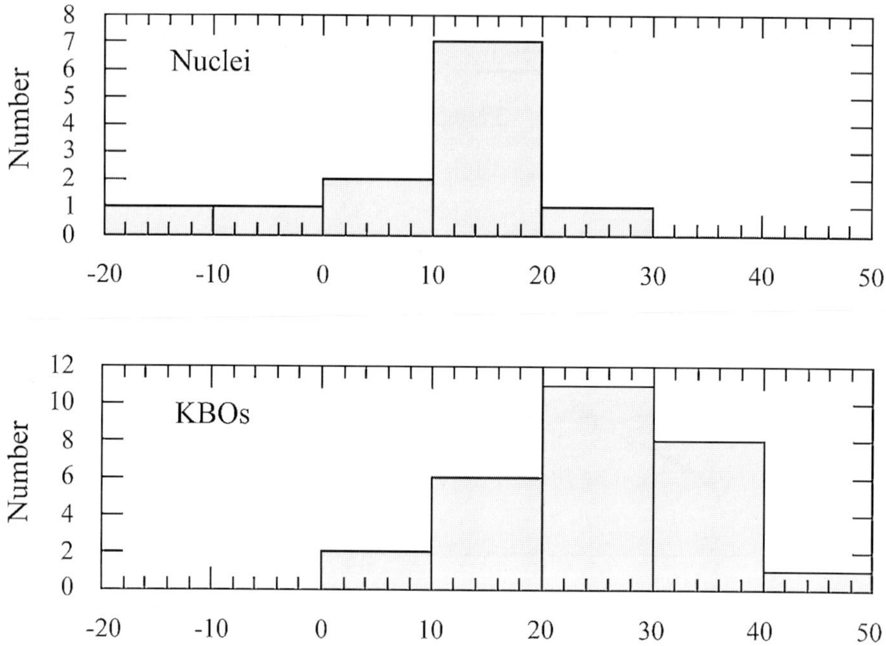

Figure 8.16. Distribution of the normalized reflectivity gradient, S' [%/1000 Å] for cometary nuclei and TNOs (Jewitt 2002).

time scale for transfer from Neptune's zone to a JF orbit of $\sim 4 \times 10^7$ yr (cf. Section 8.6). In particular, Jewitt finds that ultra-red matter, defined as material having a spectral gradient $S' > 25$ %/1000 Å in the VR region of the spectrum, is present in some TNOs and Centaurs but is missing in the nuclei of JF comets and candidate dead comets. He then concludes that this material is destroyed or removed during the journey of the body from the TN belt to the inner planetary region.

As for the reflectivity gradient, the colors of the different classes of minor bodies of the solar system also show striking differences (Fig. 8.17), which is not surprising since both physical parameters are closely related. We can clearly see a trend in the (V-R) color distribution from ultrared to gray or neutral, as the heliocentric distance decreases. We already note a shift from very red to less red colors when we pass from TNOs to Centaurs. Cruikshank et al. (1998) argue that the exposure to more intense solar UV radiation and heat in the planetary region can remove hydrogen from the small organic molecules, transforming the surface into a less red-colored and spectrally featureless macromolecular carbonaceous mass. This may explain the shift from ultrared to somewhat less red colors from TNOs to Centaurs. Comet nuclei and Trojans

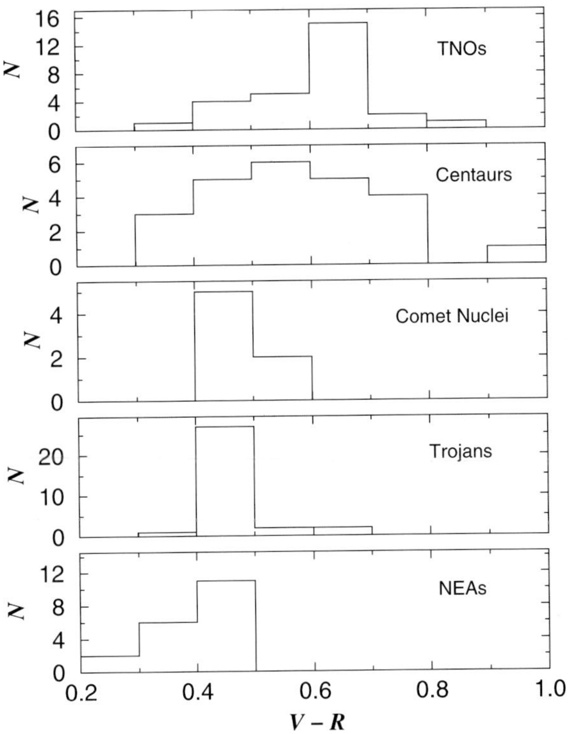

Figure 8.17. (V-R) color distribution of different classes of minor bodies. The Centaur data have been taken from Bauer et al. (2003), whereas the rest from Luu and Jewitt (2002).

are less red than TNOs and Centaurs but their color distributions are very similar to each other, which suggests that escaped Trojans might be intermingled with JF comets without chance to distinguish them, at least from their colors. We can see that NEAs are the bodies with the most gray color distribution.

Trujillo and Brown (2002) have found a correlation between color and inclination among the classical TNOs (but not among the Plutinos) which shows that the redder the TNO, the smaller its inclination on average. There is not a convincing physical argument to explain such a correlation, so one has to ask first whether unaccounted observational biases or a statistical fluke due to the smallness of the sample may be producing such a spurious correlation.

Table 8.1: Albedos of outer solar system bodies

object	class	radius (km)	albedo$^{(*)}$	source
(8405) Asbolus	Centaur	33 ± 2	0.12 ± 0.03	(1)
(2060) Chiron	Centaur	74 ± 4	0.17 ± 0.02	(1)
(5145) Pholus	Centaur	95 ± 13	0.044 ± 0.013	(2)
(10199) Chariklo	Centaur	151 ± 15	0.045 ± 0.010	(3)
1993 SC	TNO	164^{+29}_{-33}	$0.022^{+0.013}_{-0.006}$	(4)
(20000) Varuna	TNO	450^{+65}_{-73}	$0.070^{+0.030}_{-0.017}$	(4)
(50000) Quaoar	TNO	630 ± 95	$0.092^{+0.036}_{-0.023}$	(5)
Pluto	Planet/TNO	$1145 - 1200$	$0.44 - 0.61$	(6)
Charon	Satellite/TNO	$600 - 650$	0.38	(6)

(1) Fernández et al. (2002b)
(2) Davies et al. (1993)
(3) Jewitt and Kalas (1998)
(4) Jewitt et al. (2001)
(5) Brown and Trujillo (2004)
(6) Tholen and Buie (1997)
(*) Red geometric albedo for the first seven objects, blue albedo for Pluto and Charon

It has been possible to determine the albedos of only a handful of TNOs and Centaurs. In most cases the albedos have been derived from combined observations of the thermal emission and reflected optical light. The exception is the TNO (50000) Quaoar for which it was possible to measure its apparent diameter (Brown and Trujillo 2004). Table 8.1 lists the few Centaurs and TNOs whose geometric albedos have been determined. Their albedos tend to be somewhat higher than those found for JF comets, but this might be due to their different sizes rather than different compositions. Pluto and Charon have quite remarkable high albedos, which is due to their capability to retain their sublimating gases (like CH_4 or CO) in a transient atmosphere, which later condense onto the surface in a thin layer of high-albedo frost. Resurfacing by impact gardening or cryovolcanism, that leave exposed fresh unirradiated ice (Jewitt and Luu 2004), might be other processes at work that contribute to raise the albedos of outer solar system bodies.

There are still too few time-series photometric measurements of TNOs to determine rotation periods and lightcurve amplitudes. We have already seen the case of 2001 QG_{298} whose lightcurve has been interpreted as due to a contact or near-contact binary. 20000 Varuna is another well-studied case for which a reliable period of 6.34 hr and amplitude 0.42 mag were derived (Jewitt and Sheppard 2002). The large amplitude and rather short rotation period suggest that this body is rotationally distorted from spherical shape. Jewitt and Sheppard find that Varuna's high angular momentum is consistent with a triaxial Jacobi ellipsoid of axis ratio 3:2:1.4 and bulk density $\rho \approx 1$ g cm^{-3}. For a typical ice/rock mass ratio ~ 1, such a low density would imply porosities up to several tens percent, consistent with a granular structure or a loose reassemblage of fragments after a collisional disruption. Varuna's internal pressure is not large enough to compact the material. The authors argue that the high specific angular momentum of Varuna cannot have been supplied by collisions in the present-day environment. However, if the mass in the early TN belt was 10^2 times larger, collisions were frequent enough to produce highly rotationally deformed bodies and close binaries formed by fission. As we discussed in Section 8.8, Pluto and Charon, whose total angular momentum exceeds the limit for rotational stability of the putative parent body, and 2001 QG_{298}, may be other examples of such an early intense collisional environment.

PHYSICAL END STATES OF COMETS

Comets we observe today have very likely been originated with the rest of the solar system, about 4.6×10^9 yr ago. Their lifetimes are however much shorter than the solar system age, whereby they should be newcomers in the Sun's neighborhood from the deep freeze of the outer space where they have been stored for several aeons. Dynamical ejection is the ultimate fate of most comets reaching the inner planetary region (cf. Chapter 4). About 90% of the LP comets crossing Jupiter's orbit will be lost after 25 revolutions. For the small fraction of LP comets that remain gravitationally bound for at least several hundreds revolutions, as well as for a large fraction of JF comets, their demise will be more likely due to physical causes rather than dynamical ones. Among the former we can mention: (1) sublimation of volatiles and final disintegration into meteoroid streams and zodiacal dust; (2) repeated outbursts and splittings, again leading to their final disintegration; (3) collision with the Sun; (4) collision with any of the planets; and (5) collision with asteroids and interplanetary boulders. It is also possible that a comet transits through stages of total inactivity getting an asteroid-looking appearance. This is not strictly speaking an end state since the body survives and it may be reactivated if it keeps volatile material in its interior but, if discovered, it may be misclassified as an asteroid. We shall next review the previously quoted physical mechanisms leading to the destruction - or deactivation - of a comet.

9.1. The sublimation of the volatile material

The steady mass loss by sublimation sets a sublimation lifetime for a comet of a given size. This should be taken as an average value, since effects like outbursts and splittings or formation of an insulating dust mantle can accelerate or retard its final demise. Let $Z = Z(r)$ be the gas production rate (in number of molecules per unit area and unit time), which is a function of the heliocentric distance r. When the comet is close to the Sun (say $r \lesssim 3$ AU), the gas production rate is controlled by the sublimation of water ice. Values of Z are shown in Fig. 9.1 which were computed by means of the energy balance equation (cf. eq. (3.3)) for a fictitious spherical comet nucleus of (visual) Bond albedo $A_v = 0.04$ and infrared albedo $A_{IR} = 0$, and for two extreme cases: a) isothermal, and b) with a sunward-oriented hemisphere. The model results are not very sensitive to A_v, provided $A_v << 1$. Note that due to the particular spin configurations, no seasonal effects occur and the water flux is only a function of the heliocentric distance. There is not much difference between the two extreme cases up to $r \sim 2$ AU. However, the solutions start to diverge quickly for larger r, so large errors bars should be attached to Z when $r \gtrsim 2$ AU. We can also see that Z drops quickly to negligible values for $r \gtrsim 2.5 - 3$ AU.

The consideration of a spherical comet nucleus may be an oversimplification bearing in mind that most comet nuclei seem to be of very irregular shape (cf. Section 7.4). Gutiérrez et al. (2003) have computed gas production rates of fictitious bodies with different degrees of elongation and surface irregularity. They find that the fast rotation and the subsolar point approximations plotted in Fig. 9.1 generally yield large

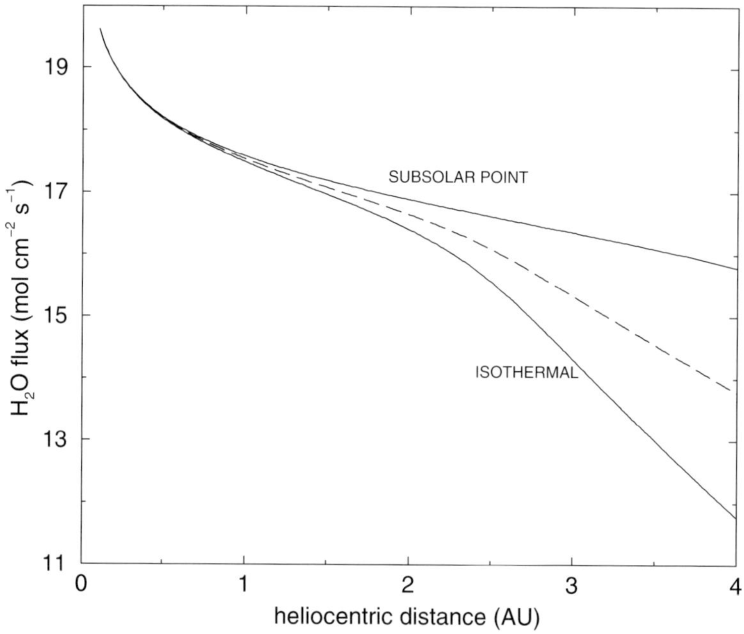

Figure 9.1. Theoretical curves of the gas production rate per unit area of water ice (in logarithmic scale) as a function of the heliocentric distance for a body with $A_v = 0.04$ and $A_{IR} = 0$. Two extremes cases are considered: isothermal and a hemisphere always oriented toward the Sun (based on energy balance at the subsolar point). The average between both extremes is represented by the dashed curve (Tancredi et al. 2005).

over- and under-estimates of the active area fractions, as compared to their more realistic models. Nevertheless, acceptable relative errors ($< 100\%$) are found for small heliocentric distances ($r \lesssim 2$ AU), so we can still use the approximation of a spherical nucleus with some confidence.

Let us assume that Δm is the mass loss per unit area and per orbital revolution of the nucleus. We can compute Δm by integrating Z along the orbit, namely

$$\Delta m = \int_0^P Z dt. \tag{9.1}$$

If ρ_N is the bulk density of the comet nucleus, the thickness Δh of the outer layer lost by sublimation during a revolution is

$$\Delta h = \frac{\Delta m}{\rho_N}. \tag{9.2}$$

strictly speaking, eq. (9.2) does not take into account the presence of other components (refractory dust and other volatiles) mixed with water ice in the cometary material. Nevertheless, a pure water ice model is enough for our order-of-magnitude estimate of sublimation lifetimes. Therefore, the sublimation lifetime N_{subl} (in number of revolutions) of a comet nucleus of radius R_N is

$$N_{subl} \approx \frac{R_N}{\Delta h}. \tag{9.3}$$

Computed values of Δm and N_{subl} are shown in Fig. 9.2. The sublimation lifetimes were computed for a fictitious comet nucleus of $R_N = 1$ km and $\rho_N = 0.5$ g cm^{-3}. We can see that N_{subl} is less than a few hundreds revolutions for an Earth-crossing comet, but it rapidly increases to more than several thousands revolutions for comets with perihelia $q \gtrsim 2.5$ AU. These results of course assume a free-sublimating comet nucleus. The formation of a dust mantle chokes off sublimation, so eq. (9.3) must be taken as a lower limit. This point will be further analyzed in the next section.

Based on the mean rate of secular brightness decrease, Kresák and Kresáková (1990) have estimated a mean active lifetime of ~ 300 revolutions for JF comets with $q < 1.5$ AU, while Fernández (1985b) found a lifetime of ~ 1000 revolutions based on his analysis of the population of JF comets that enter and leave the region $q < 1.5$ AU during ~ 400 yr in the past and in the future. Given the quite different methods employed,

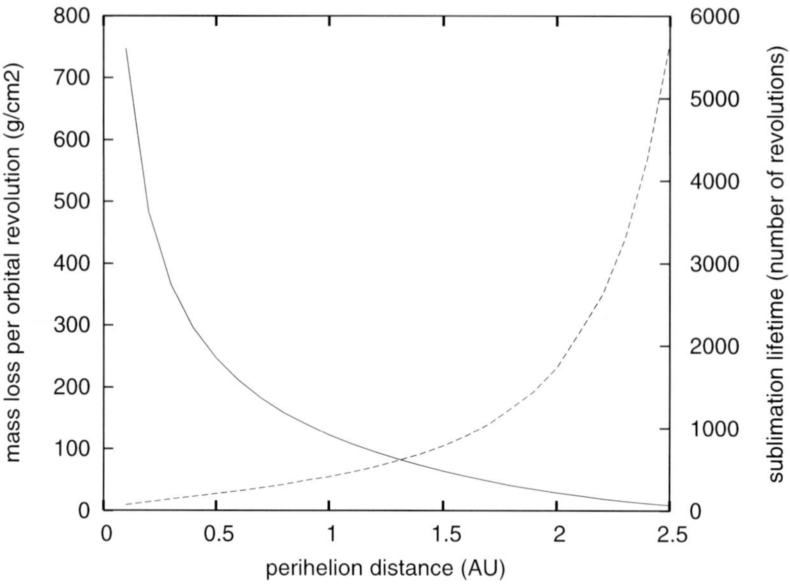

Figure 9.2. Mass loss by sublimation per orbital revolution of a comet nucleus composed of water ice of visual albedo $A_v = 0.04$ (solid curve) and the sublimation lifetime if the nucleus radius is $R_N = 1$ km and the bulk density $\rho_N = 0.5$ g cm^{-3} (dashed curve).

both results can be considered in reasonable agreement, and consistent with the sublimation lifetimes derived before. We stress that individual lifetimes may show a high dispersion due to the different sizes of comet nuclei, shapes, and/or possible variations in their chemical composition and physical structure, and the incidence of other physical processes like dust mantle buildup, outbursts and splittings.

9.2. Formation of a dust mantle

Once a comet enters the inner planetary region a dust mantle may develop as a result of the sublimation of its ices. The ices are mixed with dust particles of different sizes and not all of them are carried off by the sublimating gases. Thus, the heavier particles will stay on the nucleus surface contributing to the buildup of a dust mantle. Let us consider a dust particle of radius a and density ρ_p, it will be dragged away by the sublimating gases if the drag force (F_D) overcomes the gravitational attraction of the nucleus (F_G), i.e. if

$$F_D - F_G = \pi a^2 u_g Z m - \frac{GM_N}{R_N^2}\frac{4}{3}\pi\rho_p a^3 > 0$$

$$\pi a^2 u_g Z m - \frac{16}{9}\pi^2 \rho_N \rho_p a^3 G R_N > 0 \tag{9.4}$$

where u_g is the thermal expansion velocity of the sublimating gases at the nucleus surface. From the kinetic theory of gases we have $u_g = (3kT/m)^{1/2}$ where k is the Boltzmann's constant, m is the molecular mass and T is the temperature of the sublimating gases. $M_N = 4/3\pi\rho_N R_N^3$ is the mass of the cometary nucleus. For the molecular mass m we can take the mass of the water molecule ($m_{H2O} \simeq 3 \times 10^{22}$ g) since, as discussed, it is by far the most abundant molecular species. From the balance $F_D = F_G$ in eq. (9.4) we obtain the maximum radius a_M of the escaping dust particles, namely

$$a_M = \frac{9}{16\pi}\frac{u_g Z m_{H2O}}{\rho_N \rho_p G R_N}. \tag{9.5}$$

In the mathematical expression for the drag force of eqs. (9.4) and (9.5) we assume free molecular flow and ellastic collisions between molecules and dust particles. However, if the mean free path of the gas molecules is much shorter than the dust particle size, which may happen at heliocentric distances smaller than a few tenths AU, the expression for the drag force must be replaced by Stokes law (Huebner 1970), namely: $F_D = 6\pi\eta a u$, where the coefficient of viscosity η depends only on the gas temperature and can be expressed approximately by (e.g. Keller 1990)

$$\eta \simeq \frac{1.85 \times 10^{-6} T^{1/2}}{1 + 680/T},$$

where T is expressed in Kelvin degrees.

The maximum radius of the escaping dust particles in this case becomes

$$a_M = \left(\frac{27\eta u_g}{8\pi\rho_N\rho_p G R_N}\right)^{1/2}. \tag{9.6}$$

We note that a_M in the latter case is no longer dependent on the gas production rate Z. Furthermore, the temperature of the sublimating gases is almost independent of r for small heliocentric distances, because most of the incident solar radiation is spent in sublimating the ices. Consequently, η and u_g, and then a_M, in eq. (9.6) become independent of r; in other words, the maximum size of the particles that can be dragged away reach a ceiling that cannot be overcome no matter how large the gas production rate is.

Particles with radii $a > a_M$ are too heavy to be carried off by the sublimating gases, so they will stay on the nucleus surface contributing to the formation of the dust mantle (e.g. Brin and Mendis 1979, Prialnik and Bar-Nun 1988, Rickman et al. 1990). The buildup of the dust mantle stops when the decrease of the Sun's radiation with the increasing heliocentric distance and the insulation of the formed mantle itself prevent the underlying ices from sublimating. When the comet approaches the Sun in the next return, larger and larger dust particles will be removed following the increase in the gas production rate as the heat wave reaches the underlying ices, so the dust mantle grows thinner and may even be partially or entirely removed. Therefore, the buildup and purge of a dust mantle may be cyclic, as the comet approaches and recedes from the Sun. Yet, effects like the increase of the perihelion distance may help to consolidate a permanent dust mantle, as well as the action of cohesive forces (Kührt and Keller 1994, Möhlmann 1995), leading to a dramatic decrease of the sublimation of ices which may be constrained to small active areas, or even depend entirely on the diffusion through the porous mantle.

It is therefore very likely that not all the nucleus surface will be free-sublimating, and this will be specially true for aging comets that have endured many passages through the inner planetary region. The cases of 1P/Halley, 19P/Borrelly and 81P/Wild 2, imaged from spacecrafts during flybys, confirm this conjecture. In principle, it will be possible to estimate the fraction of free-sublimating area of a certain comet by combining observational data of the water production rate at a certain heliocentric distance r (in general close to its perihelion passage), with its nuclear magnitude determined in general at large r. Let us consider a comet nucleus of radius R_N, the total water mass production rate is

$$Q_{H2O} = 4\pi R_N^2 f Z m_{H_2O}, \tag{9.7}$$

where f is the fraction of active surface area of the comet nucleus.

The photometric cross-section S can be obtained from eqs. (2.6)-(2.8)

$$\log (p_v S) = 16.85 + 0.4 \times [m_{\odot} - H_N], \tag{9.8}$$

where $S = \pi R_N^2$ is expressed in km^2, $m_\odot = -26.77$ is the apparent (visual) magnitude of the Sun, and p_v and H_N are the geometric albedo and the absolute nuclear magnitude of the comet.

Delsemme and Rud (1973) used for the first time a combination of data of gas production rates with nuclear magnitudes to determine the radii and albedos of comets C/1969 T1 (Tago-Sato-Kosaka) and C/1969 Y1 (Bennett). They solved eqs. (9.7) and (9.8) by adopting $f = 1$, namely that all the surface area was freely sublimating. The computed Bond albedos for Tago-Sato-Kosaka and Bennett were 0.63 and 0.66 respectively, presumably too high for what was later found to be typical cometary albedos. The large errors in the computed albedos were undoubtedly due to the assumption of $f = 1$ and the large intrinsic errors of the adopted nuclear magnitudes which came from the unreliable photographic measurements by Elizabeth Roemer. Nevertheless, Delsemme and Rud's technique proved to be very useful as we will see below.

The gas production rate Q_g (mol/s) of a comet can be estimated from visual, radio and UV observations of hydrogen, OH and other radicals such as CN, C_2 and C_3 (cf. Section 3.10). As mentioned above, Q_g basically represents the production rate of water molecules, at least when comets are close to the Sun, say $r \lesssim 3$ AU, in which case $Q_g \sim Q_{H2O}$. Most water production rates of JF comets have been taken from A'Hearn et al. (1995), complemented with some spectrophotometric results from Fink and Hicks (1996) and Newburn and Spinrad (1989), some Lyman-α observations from Mäkinen et al. (2001), and some radio observations of the OH 18-cm line from Crovisier et al. (2002). We constrain the use of eq. (9.7) to distances $r < 2$ AU, for which the computed values of the gas production rate Z shown in Fig. 9.1 do not strongly depend on the adopted model (isothermal or sub-solar point).

By combining eqs. (9.7) and (9.8) for the sample of comets with derived values of both Q_{H2O} and H_N, we can get relationships among R_N, p_v and f. We note that whereas the computed Z in eq. (9.7) uses the Bond albedo A_v, the geometric albedo p_v is used in eq. (9.8). As shown before, there is a simple relation between both albedos: $A_v = p_v q$ (cf. eq. (3.1)), where q is the phase integral. The phase integral depends on how the surface of the body reflects the light in different directions. For a dark surface like that of asteroid 253 Mathilde we have $q = 0.28$ (Clark et al. 1999) and, as we shown before (cf. Section 7.6), 19P/Borrelly has also a phase integral of this order, so if the dark surfaces of comet nuclei have phase integrals around this value, their

Bond albedos should be around one third of their geometric albedos. Fortunately, some difference between the values of A_v and p_v will have little effect on the computation of Z, so this point is of little concern for us.

If we assume a standard geometric albedo of $p_v = 0.04$, we can compute the nuclear radii and fraction of active surface areas for the sample of JF comets with measurements of their gas production rates and estimates of their nuclear magnitudes as presented in Table A2.1 of Appendix 2. Results are shown in Table 9.1 and plotted in Fig. 9.3. Most of the studied comets give consistent results for f within the range (0,1), although one of the measurements of the water production rate of 73P/Schwassmann-Wachmann 3 right after the comet split leads to an extremely high value of f. This can be explained in terms of a sudden increase in the gas and dust production rate after the splitting event, leading to a spurious fraction $f >> 1$. Such "hyper-active" cases for which $f > 1$ can thus be expected if, for instance, part of

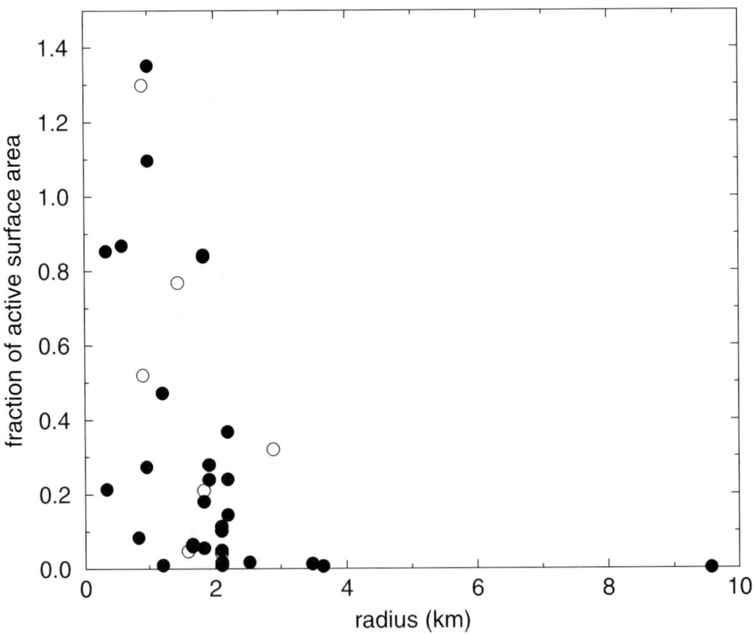

Figure 9.3. Computed fraction of free-sublimating area versus the radius of the comet nucleus for an assumed geometric albedo $p_v = 0.04$. Filled circles are for JF comets whose estimated absolute nuclear magnitudes are of quality classes 1-3 (see definition in Table A2), while open circles are for comets of quality class 4 (Tancredi et al. 2005).

the sublimating water ice comes from icy grains in the coma, or from rapidly evaporating fragments released during a splitting (Rickman and Jorda 1998).

Table 9.1: Measured gas production rates and computed fractions of active surface area$^{(*)}$

Comet	$\log Q_{H2O}(r)$	r (AU)	Ref.	f
2P/Encke	27.87, 28.30, 28.11	0.71, 0.33, 0.50	AH, Cr, Ma	0.018, 0.010, 0.015
4P/Faye	27.70	1.78	AH	0.180
6P/d'Arrest	27.51, 27.48	1.41, 1.40	AH, Cr	0.066, 0.060
7P/Pons-Winnecke	27.51	1.42	FH	0.056
9P/Tempel 1	28.09	1.41	AH	0.145
10P/Tempel 2	27.12	1.78	AH	0.013
16P/Brooks 2	27.01	1.78	AH	0.048
19P/Borrelly	28.31, 28.48	1.41, 1.43	AH, Cr	0.240, 0.369
21P/Giacobini-Zinner	28.55, 28.78, 28.71	1.12, 1.04, 1.05	AH, Cr, Cr	0.627, 0.887, 0.773
22P/Kopff	28.37, 28.46	1.78, 1.68	AH, Cr	0.843, 0.839
24P/Schaumasse	28.46, 28.00	1.21, 1.28	FH, Cr	1.299, 0.521
26P/Grigg-Skjellerup	26.67	1.12	AH	0.010
28P/Neujmin 1	27.20	1.41	AH	0.001
43P/Wolf-Harrington	27.62	1.78	AH	0.114
45P/Honda-Mrkos-Pajdušáková	26.88, 28.18	1.12, 0.55	AH, Cr	0.213, 0.854
46P/Wirtanen	27.97, < 28.18	1.12, 1.12	AH, Cr	0.869
49P/Arend-Rigaux	27.23	1.41	AH	0.007
64P/Swift-Gehrels	27.99	1.53	NS	0.210
67P/Churyumov-Gerasimenko	27.59, 27.95	1.41, 1.35	AH, Cr	0.050, 0.102
68P/Klemola	26.95	1.78	AH	0.017
69P/Taylor	27.19	1.78	AH	0.042
73P/Schwassmann-Wachmann 3	27.62, 29.35$^{(*)}$	1.44, 0.98	FH, Cr	0.273, 5.456$^{(1)}$
81P/Wild 2	28.11, 27.90	1.58, 1.74	Ma, Cr	0.279, 0.239
97P/Metcalf-Brewington	28.13	1.78	AH	0.769
98P/Takamizawa	28.35	1.78	AH	0.321
103P/Hartley 2	28.35	1.12	AH	0.473
108/Ciffréo	26.69	1.78	AH	0.084

AH: A'Hearn *et al.* (1995)
Cr: Crovisier et al. (2002)
Ma: Makinen et al. (2001)
FH: Fink and Hicks (1996)
NS: Newburn and Spinrad (1989)
$^{(1)}$Observed after the comet split
$^{(*)}$Source: Tancredi et al. (2005)

Most of the comets have fractions $f < 0.5$ and, in particular, 15 out of the 27 studied comets have $f < 0.2$. The largest comet in the sample, 28P/Neujmin 1 ($R_N = 9.58$ km), shows a very small fraction of active surface area (0.1%), though the sample is still too small to assess whether this is a general property of large comet nuclei. Nevertheless, it is suggestive that the few comets in our sample larger than $R_N \sim 3$ km have fractions $f \lesssim 0.01$. On theoretical grounds (e.g. Rickman et al. 1990), we should expect that large comet nuclei are more capable of building insulating dust mantles that choke off gas sublimation to very low levels.

In summary, the computation of the fraction of active area for the sample of JF comets strongly suggests that active areas occupy in general only a minor fraction of the total surface area, and that this may be a general property of aging comets. This agrees with what was found in 1P/Halley, 19P/Borrelly and 81P/Wild 2 from *in situ* imaging. In some cases, values $f \sim 0.01$ or even smaller are derived, which suggests that the sublimating gases might come from diffusion through a dust mantle that already covers the entire surface, but it is still not insulating enough to prevent the sublimation of the ices underneath. The very small fractions of active areas computed for many JF comets suggests that they might pass through stages of complete inactivity. Kresák (1987) has argued that active phases of periodic comets are intermitted with phases of very low activity or even dormancy, based on the lack of observations of some periodic comets at very favorable apparitions. From numerical simulations of the physical and dynamical evolution of JF comets, Tancredi (1994) also found that periods of dormancy should alternate with stages of activity, the stage of cometary activity being about half the dynamical lifetime. It is quite possible that large comets are more prone to pass through stages of dormancy as they are more capable of building insulating dust mantles. The buildup of a dust mantle will prolong the sublimation lifetime computed from eq. (9.3) in such a way that if f is the time-average fraction of active (free-sublimating) surface area, the corrected lifetime is $N'_{subl} = N_{subl} \times 1/f$.

9.3. Can defunct or dormant comets be disguised as asteroids?

As discussed, the dust mantle may be completely sealed, thus stopping all the gaseous activity. As a consequence, comets may go into stages of dormancy or become extinct altogether looking like asteroids. It has long been suggested that most or at least some Near-Earth Asteroids (NEAs) are of cometary origin (e.g. Öpik 1963, Wetherill 1988). There are several objects whose real nature has been a matter of debate, among them 107P/Wilson-Harrington (also cataloged as asteroid 4015 Wilson-Harrington), 133P/Elst-Pizarro (also cataloged as asteroid 7968 Elst-Pizarro), 2201 Oljato and 3200 Phaethon. 107P/Wilson-Harrington showed some activity when it was discovered in 1949, but it has remained inactive in following returns since its rediscovery in 1979. With regard to Elst-Pizarro, it was discovered in 1979 as an ordinary main-belt asteroid (orbital elements $q = 2.63$ AU, $a = 3.15$ AU, $i = 1^\circ.38$), but it showed a tail in a photograph taken in 1996 by Guido Pizarro at the European Southern Observatory, which led Belgium Royal Observatory's astronomer Eric Elst to classify it as a comet. Yet, the object showed no coma nor spectral features of gaseous species characteristic of comets during its whole active period, so the activity could have been induced by an impact (Toth 2000). As regards to 2201 Oljato, McFadden et al. (1993) reported the observation of a high ultraviolet reflectance during its 1979 and 1983 apparitions, which they suggested to be related to fluorescent emission from neutral species found in comets, such as CN or OH. The orbit of 3200 Phaethon was found to match the mean orbit of the Geminid meteoroid stream (Williams and Wu 1993), and since meteoroid streams are generally associated to the debris of comets in their disintegration process (see next section), it again would suggest a cometary nature of the object. Jenniskens (2003) suggests that the NEA 2003 EH_1, that moves on a cometary orbit: $q = 1.192$ AU, $Q = 5.058$ AU, $i = 70^\circ.8$, is the parent of the Quadrantid meteor stream which, if confirmed, would put this body on the list of candidates to be deactivated comets.

The search for residual activity in some of the above mentioned bodies led to negative results, giving upper limits for the possible active areas of 0.02% for Wilson-Harrington and 0.01% for 3200 Phaethon (Chamberlin et al. 1996). Fernández et al. (2001, 2005) have recently argued that some extinct comet candidates among the asteroids show very low geometric albedos ($p_v \sim 0.02 - 0.03$), similar to those found for comets, but significantly lower than the geometric albedos of NEAs with Tisserand constant > 3 ($p_v \sim 0.1 - 0.6$). Yet, the spectral types P and D predominate among the asteroids of the outer belt ($a \gtrsim 3.5$ AU) (e.g., Hartmann et al. 1987), whose surfaces are dark, red to very red, probably of a very low albedo. Therefore, a low albedo does not necessarily mean a cometary origin, since the outer belt is also a possible source of dark objects.

Figure 9.4 plots the aphelion distance Q versus the Tisserand parameter T of all cataloged NEAs with $Q > 3.5$ AU and Earth-approaching JF comets with $q < 1.3$ AU taken from Fernández et al. (2002a). We can see that both populations tend to occupy different regions in the parametric plane (Q, T): most NEAs have $T > 3$ so encounters with Jupiter are not possible at present (cf. Section 7.5), while most JF comets have $T < 3$ indicating that they are subject to close interactions with Jupiter. Notwithstanding the segregation of both populations, a

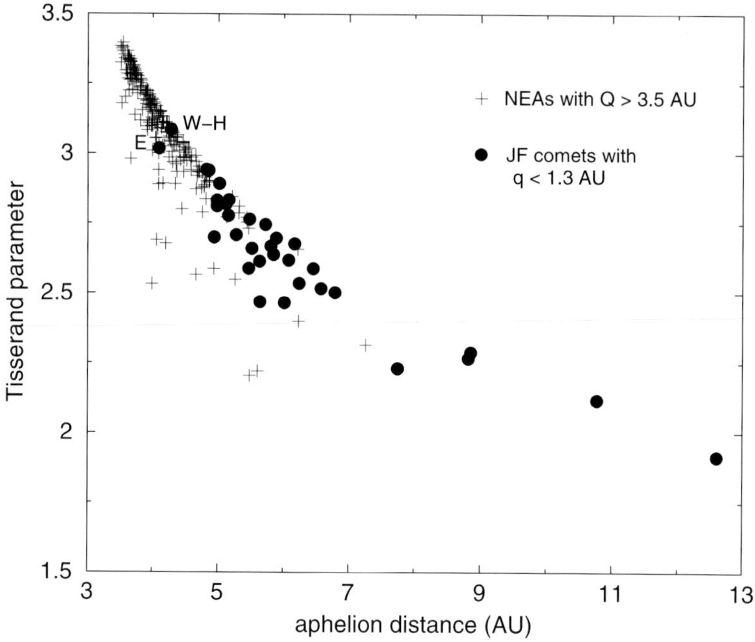

Figure 9.4. Aphelion distance versus Tisserand parameter of NEAs with $Q > 3.5$ AU and JF comets with $q < 1.3$ AU. Only two comets: 2P/Encke and 107P/Wilson-Harrington, shared the NEA space ($Q < 4.5$ AU, $T > 3$) clearly detached from the rest of the JF comets (Fernández et al. 2002a).

small fraction of NEAs extends and overlaps the JF comets zone. It is interesting to note that two comets: 2P/Encke and 107P/Wilson-Harrington are clearly detached from the rest of the JF population, deep inside the NEA zone with $T > 3$. Wetherill (1991) argues that several dynamical mechanisms can decouple JF comets from Jupiter into Encke-type orbits on time scales of $10^5 - 10^6$ yr. He quotes: (a) secular and resonant perturbations by Jupiter and the other Jovian planets; (b) perturbations by the terrestrial planets; and (c) nongravitational forces. Fernández et al. (2002a) found that only very strong nongravitational forces acting for at least several tens of revolutions can produce Encke-type orbits so, although possible, only in very favorable circumstances could nongravitational forces decouple orbits from Jupiter.

Are some NEAs, and in particular those on "cometary" orbits (i.e. those that approach Jupiter, say with aphelion distances $Q > 4.5$ AU) of cometary origin?, or they come from the main asteroid belt? The transport of bodies from the main-belt to NEA-type orbits was regarded as very slow and inefficient. It was also noted that perturbations by Mars on approaching asteroids were not efficient enough to produce the right number of Earth-crossers from Mars-crossers. This dynamical argument was raised by Öpik (1963) in support of a cometary origin for most NEAs. However, it was shown later on that there exist mechanisms that could provide efficient dynamical routes of escape from the main belt to NEA-type orbits (e.g., Wisdom 1983, Gladman et al. 2000). Mutual collisions among main-belt asteroids can inject fragments into either mean motion resonances with Jupiter (e.g. 3:1, 5:2), or the ν_6 secular resonance, from which they are quickly transferred to NEA-type orbits on a time scale of a few Myr (e.g. Gladman et al. 1997). The resonance ν_6 occurs when the longitude of perihelion of the body $\tilde{\omega}$ rotates with the proper frequency $\dot{\nu}_6$ (cf. Section 8.4), which is nearly equal to the precession rate of Saturn's longitude of perihelion $\dot{\tilde{\omega}}_s$. Thus, the resonance condition may be expressed $\dot{\tilde{\omega}} \approx \dot{\tilde{\omega}}_s$ (Froeschlé and Scholl 1987). The injection in this resonance is characterized by wide variations in the eccentricity of the body, so it can pass from an orbit within the main belt to a Mars- or Earth-crossing orbit.

Menichella et al. (1996) estimate that a few hundred kilometer-sized bodies per Myr can be scattered from the main belt to NEA-type orbits, via injection in one of the above resonances, which can be adequate to justify the existence of a steady-state NEA population of about 2000 bodies with diameters greater than one km, as it is currently estimated from different surveys. Even the large NEAs with diameters $D \gtrsim 5$ km may be satisfactorily explained as driven from the main asteroid belt by the previous mechanism, overcoming some previous objections pointing to the insufficient production rate of big fragments there (Migliorini et al. 1998). Furthermore, Fernández et al. (2002a) found that the population of NEAs in cometary orbits can be adequately replenished with NEAs of smaller aphelion distances diffusing outwards. That most NEAs, including those on cometary orbits, can be satisfactorily explained as coming from the main asteroid belt does not rule out that some peculiar objects, as the cases described above, might indeed be inactive comets.

9.4. Splitting events

Comets are frequently observed to show sudden brightness increases associated to outburst and splitting phenomena of their nuclei. In essence, an outburst implies an above average loss of mass, while a splitting event implies, in addition to the above-average mass loss, the fragmentation of the nucleus into two or more clearly distinguished pieces (Fig. 9.5). Comet nuclei have very low density and internal strength, as shown by the tidal breakup of D/1993F2 (Shoemaker-Levy 9) during a close approach to Jupiter prior to the collision event of this comet with the planet (Scotti and Melosh 1993, Asphaug and Benz 1994, Solem 1994). The nucleus may consist of an assemblage of weakly-bonded building blocks which, according to some authors, are held together only by self-gravitation, the so-called 'rubble-pile' model (Weissman 1986), whereby outbursts and splittings may be a consequence of such a fragile material. Cracks in the dusty surface produced by thermal or tidal stresses may leave exposed pockets of fresh ices to the solar radiation greatly enhancing sublimation, or inducing phase transition from amorphous to crystalline water ice that is highly exothermic.

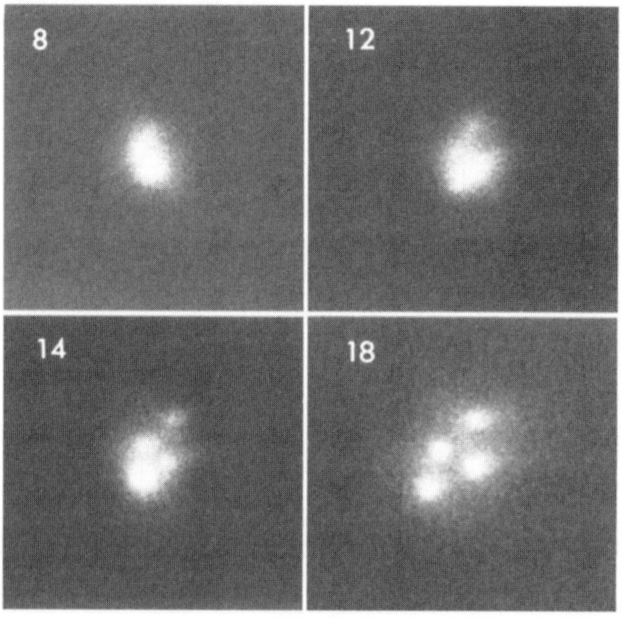

Figure 9.5. The breakup process of comet C/1975 V1 (West) in four main pieces between 8-18 March 1976 as it was registered by Scott Murrell and Claude Knuckles from New Mexico State University at Las Cruces.

Table 9.2 shows the list of comets that were observed to split, or presumed to have split because the pair of daughter comets share similar orbits. Most of the splittings do not seem to have been caused by tidal forces. Among the known pairs, only the breakup of the parent body of 42P/Neujmin 3 and 53P/van Biesbroeck can be traced back to tidal forces from Jupiter during a close encounter around 1850 (Carusi et al. 1986). We can see neither an obvious correlation with perihelion distance nor with the dynamical age of the comet.

Table 9.2: Split comets[(*)]

Nontidaly split comets	Dynamical class	q (AU)
1846 II 3D/Biela	JFC	0.86
1860 D1 Liais	LPC	1.20
1888 D1 Sawerthal	OLD	0.70
1889 O1 Davidson	OLD	1.04
1896 R2 D/Giacobini	JFC	1.45
1899 E1 Swift	NEW or YOUNG	0.33
1906 E1 Kopff	NEW	3.34
1914 S1 Campbell	OLD	0.71
1915 C1 Mellish	NEW or YOUNG	1.01
1915 W1 69P/Taylor	JFC	1.56
1942 X1 Whipple-Fedtke-Tevzadze	OLD	1.35
1947 X1 Southern Comet	OLD	0.11
1955 O1 Honda	NEW or YOUNG	0.88
1956 F1 Wirtanen	NEW	4.45
1968 U1 Wild	LPC	2.61
1969 O1 Kohoutek	YOUNG	1.72
1969 T1 Tago-Sato-Kosaka	YOUNG	0.47
1975 V1 West	OLD	0.20
1982 C1 79P/du Toit-Hartley	JFC	1.19
1985 V1 108P/Ciffréo	JFC	1.70
1986 P1 Wilson	NEW or YOUNG	1.20
1992 L1 101P/Chernykh	JFC	2.36
1994 G1 Takamizawa-Levy	YOUNG	1.36
1994 P1 141P/Machholz 2	JFC	0.75
1994g 51P/Harrington	JFC	1.57
1994w 73P/Schwassmann-Wachmann 3	JFC	0.93
1996 J1 Evans-Drinkwater	NEW or YOUNG	1.30
1996 S2 128P/Shoemaker-Holt 1	JFC	3.05
1996 57P/ du Toit-Neujmin-Delporte	JFC	1.72
1999 S4 LINEAR	LPC	0.77
2001 A2 LINEAR	LPC	0.78
2004 V5 LINEAR-Hill	JFC	4.41

(Continue Table 9.2)

Known comet pairs		
1951 J1 42P/Neujmin 3	JFC	2.04
1954 R1 53P/van Biesbroeck		2.41
1988 F1 Levy	OLD	1.17
1988 J1 Shoemaker-Holt		
1988 A1 Liller	OLD	0.84
1996 Q1 Tabur		
2002 A1 LINEAR	LPC	4.71
2002 A2 LINEAR		
Tidally split comets		
1882 R1 Great September Comet	at Sun	
1889 N1 16P/Brooks 2	at Jupiter	
1963 R1 Pereyra	at Sun	
1965 S1 Ikeya-Seki	at Sun	
1993 F2 D/Shoemaker-Levy 9	at Jupiter	

(*) Sources: Fernández and Jockers (1983), Sekanina (1997, 2004), Weaver et al. (2001), Sekanina et al. (2002)
Note: LPC refers to those long-period comets whose dynamical ages are uncertain

A comet of mass M_N and negligible internal strength will be torn apart by a massive body of mass M_P ($>> M_N$), radius R_P and density ρ_P, when the tide-raising force on the comet exceeds its self-gravity. To compute the distance of closest approach, r_{tide}, for tidal breakup, let us imagine that the comet is a "rubble-pile" consisting of two spheres of mass m and density ρ_N. The two spheres will be torn apart if the tidal force exceeds their mutual gravity, namely if

$$\frac{GM_P m}{2r^2} - \frac{GM_P m}{2(r+d)^2} \gtrsim \frac{Gm^2}{4d^2},$$

where r is the distance of closest approach of the comet to the central body and d is the distance between the centers of mass of the two spheres. By equaling the tidal force to self-gravitation, we get $r = r_{tide}$. Introducing r_{tide} above together with $M_P = 4/3\pi R_P^3 \rho_P$ and $m = 4/3\pi(d/2)^3\rho_N$, we can obtain after some operations the expression

$$r_{tide} \simeq 2.5 \left(\frac{\rho_P}{\rho_N}\right)^{1/3} R_P. \quad (9.9)$$

This is very close to the classical expression derived by the French mathematician Édouard Albert Roche in 1847 for a homogeneous liquid satellite orbiting a rigid planet: $r_{Roche} = 2.44(\rho_P/\rho)^{(1/3)} R_P$. Aggarwal and Oberbeck (1974) extended this study to the breakup of elastic solid bodies with a certain material strength, finding a smaller distance for tidal breakup: $r_{tide} = 1.38(\rho_P/\rho_N)^{(1/3)} R_P$, which is valid under the assumption that the comet is passing by (not in a collision trajectory). If we adopt $\rho_N = 0.5$ g cm^{-3}, $\rho_P = 1.4$ g cm^{-3}, which is approximately the mean bulk density of the Sun or Jupiter, we obtain $r_{tide} \simeq 1.95 R_P$. Therefore, we should expect that tidal forces from the Sun (or Jupiter) will disrupt a comet if it approaches to distances smaller than about one solar (or Jupiter) radius from its surface.

There are 602 LP comets discovered since 1850 included in Marsden and Williams's (2003) catalogue, leaving aside all the sungrazers discovered by the SOLDWIND, SMM and SOHO missions. The reason why we have considered the comet sample since the somewhat arbitrary year of 1850 is because the record of comet splittings started around mid-nineteenth century. Among the 602 LP comets, 23 experienced nontidal splittings, including comet pairs, which would give a splitting rate of $23/602 \simeq 0.04$. For the cataloged sample of 267 JF comets (updated to the end of 2004), we have likewise that 12 experienced nontidal splittings, which gives a ratio $12/267 \simeq 0.045$. A ratio $\sim 0.04 - 0.045$ is not much different from that found by Chen and Jewitt (1994) of ~ 0.06 from a survey of secondaries near their primary nuclei. Since some splittings in the observed comet sample may have passed unnoticed, our derived ratios should be taken as lower limits. Taking at face value, our ratios imply that one comet in 20-25 suffers a major splitting able to be recorded. If all the comets that reach the inner planetary region can undergo splittings through their lifetimes (as is suggested by the different dynamical ages of split comets), then the previous result may be interpreted as that one major splitting occurs in 20-25 revolutions for a given comet. Therefore, aged comets with hundreds to thousands of revolutions in the inner planetary region may have experienced on average tens to hundreds of major splittings, which gives this process an important role in limiting their physical lifetimes.

In general, the main body remains after a splitting event and the very tiny fragments released disappear after a few days or weeks. In many cases the fragments remain embedded in the active coma of the parent nucleus, so they can be detected only with highly sensitive CCD images (Fig. 9.6). Sekanina (1997) argues that nuclei of tidally split comets truly break up, while nuclei of nontidally split comets tend to peel off instead, and this would explain that multiple split events can occur on the same comet without destroying the parent nucleus.

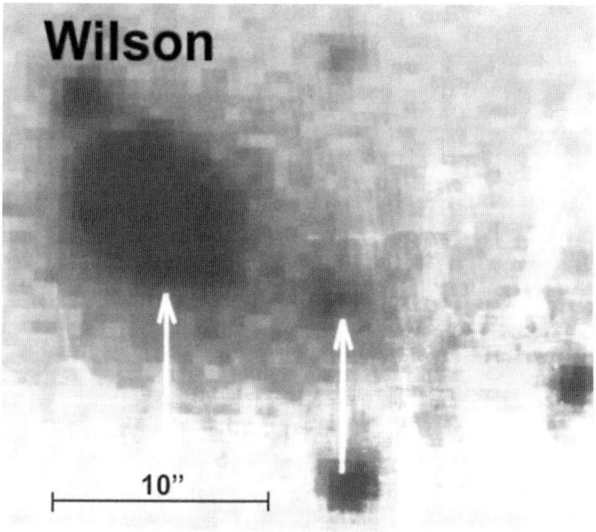

Figure 9.6. The split nucleus of comet C/1986 P1 (Wilson). The arrows indicate the locations of the primary and secondary components (Chen and Jewitt 1994).

In some cases a splitting can fragment the parent comet into two or more daughter comets that can survive for a large number of revolutions. As shown in Table 9.2, there are several cases of pairs of comets that share very similar orbits, or that their integration backwards in time leads to a common origin at some time in the past. On the other hand, there are also examples of comets that have not survived the fragmentation process and were observed to dissipate shortly after their fragmentation event (Fig. 9.7). They are presumably very small and fragile pieces from previous breakups of parent comets.

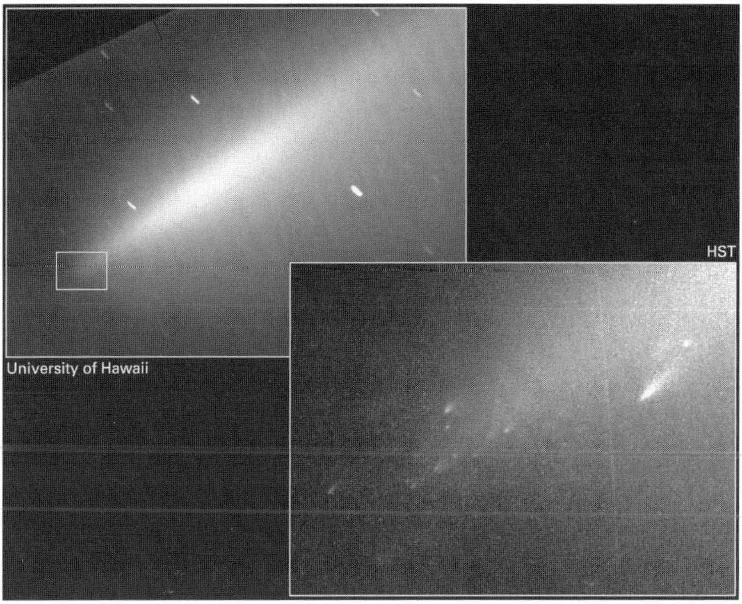

Figure 9.7. Fragmentation of comet C/1999 S4 (LINEAR) when it was at a heliocentric distance of 0.8 AU (images taken with the 2.2-m telescope of the University of Hawaii and the Wide Field and Planetary Camera 2 of the Hubble Space Telescope.

Most of the splitting and outburst events have been observed close to the Sun. Nevertheless, there are many cases throughout the inner planetary region and even beyond Jupiter, as can be seen by the frequent outbursts of comet 29P/Schwassmann-Wachmann 1. Sekanina (2002) has argued that the propensity of sungrazing comets of the Kreutz family to appear as tight pairs and clusters can be explained if the fragmentation process of their parents occurs along the orbit, even near aphelion. This raises the question of whether a comet can split "spontaneously" without the action of an external agent (solar radiation, tides raised during a close approach to the Sun or a planet, or collision with an interplanetary boulder). Sekanina (1997) argues that even when tidal forces do not lead directly to splitting, they will cause cracks in the comet material that may favor splits afterwards under the action of other agents like thermal stresses, rapid rotation or sublimation. The presence of significant amounts of CO or CO_2, much more volatile than H_2O, and/or the exothermic conversion of amorphous H_2O ice to crystalline ice could explain the persistent activity of some comets, and be in fact a cause of splittings at large heliocentric distances (see, e.g. Rickman 1994).

9.5. Possible scenarios for the physical evolution and decay of a comet

The previous discussion on the sublimation of volatile material, dust mantle buildup, outbursts and splittings, allows us to devise a possible scenario on how comets evolve physically after a large number of passages by the Sun's neighborhood. Figure 9.8 depicts possible evolutionary paths since the insertion of the comet in the inner planetary region until its demise (provided that it is not ejected by the planets in the meantime). When the comet is "young" (say, within some to a few tens passages within the inner planetary region), it is very active and the released gas arises from free sublimation, possibly from a few active areas as in the case of 1P/Halley (the fraction of active area f should be non-negligible, i. e. $\gtrsim 0.01$). If the dust mantle covers entirely the nucleus, the comet may still remain with a residual gaseous activity if the Sun's radiation sublimates the ice underneath and the sublimating gases can escape by diffusion through the porous mantle. The computation of f will give values $<< 0.01$, which should not be interpreted in terms of a fraction of free-sublimating area, but as the fraction of gas production by diffusion to that corresponding to free-

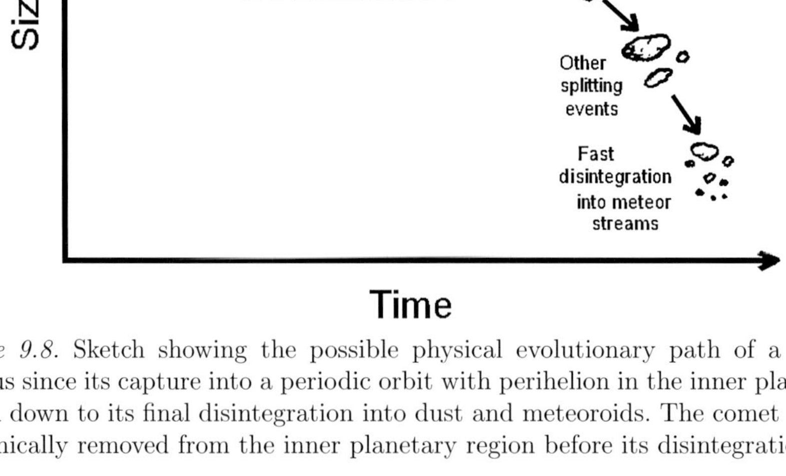

Figure 9.8. Sketch showing the possible physical evolutionary path of a comet nucleus since its capture into a periodic orbit with perihelion in the inner planetary region down to its final disintegration into dust and meteoroids. The comet can be dynamically removed from the inner planetary region before its disintegration.

sublimation through an equivalent surface area. In more aged comets, the dust mantle may grow to the point of providing an efficient insulation against the Sun's radiation; the comet becomes dormant or extinct and, if discovered, it may be misclassified as an asteroid. An outburst or splitting, which was shown above to be a rather common phenomenon, will leave exposed areas of fresh ices leading to a reactivation of the comet. In later stages, the comet may continue its disintegration into interplanetary dust and meteoroids.

Alternatively, a comet may pass from the active phase straight to the disintegration stages without periods of dormancy. We do not know yet which initial conditions are required for the comet nucleus to follow a certain physical evolutionary path, though the size may play a fundamental role, since large bodies may favor the buildup of efficient insulating dust mantles. As shown in Fig. 9.3, comets with $R \gtrsim 3$ km have fractions of active area $f \lesssim 0.01$, which lends support to the idea that such large comet nuclei may pass through inactive stages. It is also possible that smaller comets also pass through stages of dormancy, though we do not know yet whether smaller comets must fulfill other conditions in terms of physical structure or fraction of volatile content.

The debris left by a comet along its orbit as a result of the sublimation of its volatiles and/or its disintegration is called a *meteoroid stream*. It is composed of dust particles and meteoroids. When the Earth meets a meteoroid stream, a meteor shower is produced, which is observationally characterized as an unusually high frequency of meteors that seem to come from a certain direction of the sky called the *radiant*. Showers are named after the location of the sky in which their radiants are located. In 1866 Giovanni Schiaparelli noted that the orbits of the Perseid meteors were nearly identical to that of comet 109P/Swift-Tuttle, showing that there was a physical connection between comets and meteor showers. Some major showers related to some specific comets are shown in Table 9.3. In some cases the parent comets are still alive and active (e.g. comets Encke and Halley), or were observed until some time ago (e.g. comet Biela), but there are meteor showers without a cometary counterpart which suggests that their parent comets have already disintegrated.

Table 9.3: Comets and associated meteor showers(*)

Comet	Shower	Maximum date	Radiant (R.A., Dec.)
C/1861 G1 (Thatcher)	Lyrids	22 April	271.4°, +33.6°
1P/Halley	η Aquarids	3-5 May	335.6°, −1.9°
1P/Halley	Orionids	21 Oct.	94.5°, +15.8°
73P/Schwassmann-Wachmann 3	τ Herculids	3 June	228°, +39°
7P/Pons-Winnecke	Boötids	28 June	219°, +49°
2P/Encke	Daytime β Taurids	29-30 June	79°, +21°
2P/Encke	Taurids	4-7 Nov.	50.5°, +13.6°
109P/Swift-Tuttle	Perseids	12-13 Aug.	46.2°, +57.4°
21P/Giacobini-Zinner	Draconids	9 Oct.	262.1°, +54.1°
3D/Biela	Andromedids	3 Oct.	26°, +37°
55P/Tempel-Tuttle	Leonids	17 Nov.	152.3°, +22.2°
8P/Tuttle	Ursids	22 Dec.	223°, +78°

(*) Sources: Brandt and Chapman (1981), p. 157; Lodders and Fegley (1998), p. 288.

Hughes and McBride (1989) estimated a mass of 3×10^{17} g for the Perseids and 3.3×10^{16} g for the Orionids/η Aquarids, which Jewitt et al. (2003) estimate as 13% of their parent comets' masses. Therefore, several comets may be showing a substantial disintegration process (cf. Fig. 9.8), as shown by the mass deposited in their debris, though we should bear in mind that there is still a large uncertainty in the cometary masses as well as in the masses of the meteoroid streams. From the analysis of photometric dust tail data, Fulle (1990) estimated that a mass of $(5 \pm 1) \times 10^{12}$ g is ejected by comet 2P/Encke per perihelion passage in the form of dust particles and meteorids of sizes ranging between 20 μm and 20 cm. He also found an ejected mass of $(8 \pm 2) \times 10^{12}$ g for comet 6P/d'Arrest in particles between 20 μm and 10 cm. From Table 8.1 we have that the radii of these comets are: $R_N = 2.64$ km (Encke) and $R_N = 1.83$ km (d'Arrest). If we adopt a bulk density of $\rho = 0.5$ g cm^{-3}, we obtain masses $M_N = 3.85 \times 10^{16}$ g (Encke) and $M_N = 1.28 \times 10^{16}$ g (d'Arrest). If half the comet mass would be under the form of refractory grains, their lifetimes would be 3.85×10^3 revolutions and 800 revolutions, respectively. These lifetimes are consistent with those derived above for JF comets with $q < 1.5$ AU (cf. Section 9.1).

The *Infrared Astronomical Satellite* (IRAS) first detected several dust trails associated in most cases - although not in all - with known periodic comets. Sykes and Walker (1992) found that such trails were primarily composed of refractory grains of sizes ranging between 1 mm and 1 cm. The trails were found to be very compact, presumably because they were formed rather recently. With time they will scatter along the orbits of their parent comets becoming meteor streams. Sykes and Walker conclude that several times more mass is lost in these tails than the mass derived from ground-based studies of gas and dust production rates. As a consequence, the physical lifetimes of comets in the inner planetary region should be shorter than those derived from the sublimation lifetimes computed in Section 9.1. Furthermore, given the large mass of refractory material contained in the trails, JF comets may contribute a substantial fraction of the matter contained in the zodiacal dust cloud (Fernández 1988).

The mass contained in the zodiacal dust cloud is the result of a balance between losses of dust grains driven to the Sun by Poynting-Robertson drag, or ejected to interstellar space by solar radiation pressure, and gains of meteoric matter, essentially supplied by comets and asteroids. To keep the zodiacal dust complex in steady-state, an input rate of $\sim$ 10 ton s^{-1} of meteoric matter is required (e.g. Grün 1999). From *Infrared Space Observatory* observations of 2P/Encke Lisse et al. (2004) derive a mass loss rate of $70 - 280$ kg s^{-1}. If this mass loss rate is typical of other JF comets, these bodies rather than asteroids may indeed supply most of the zodiacal dust material.

9.6. Small comets

Another problem, partly related to the progressive disintegration of comets, is the population of small comets in the inner planetary region. In principle, small comets may have different origins, as for instance: (1) they were "initially" small, this understood in the sense that they were already of small size when they reached the inner planetary region for the first time, though this does not necessarily mean to be primordially small, since they could be the debris of the collisional evolution of the planetesimal population in the protoplanetary disk or in the trans-neptunian belt; (2) they are the daughter products of the fragmentation process of larger parent comets after successive passages through the inner planetary region; and/or (3) they are worn-down comets that lost most of their mass in previous passages. Irrespective of their origin, the question is how long can a small comet remain active in the inner planetary region. The answer is not simple because of the obvious difficulties to detect bodies that are intrinsically very faint.

From the study of comets that came close to the Earth, Kresák (1978) found a cutoff in the distribution of their absolute total magnitudes at $H_{10} \sim 10.5$, which should roughly correspond to a nucleus radius $R_N \sim 0.5$ km. He argued that if such intrinsically faint comets existed, they should have been picked up during their close approaches to the Earth. This conclusion was essentially corroborated by Sekanina and Yeomans (1984) from the study of the population of comets that approached the Earth to less than 2500 Earth radii. One should expect that small comets will dissipate much faster due to sublimation, outbursts and splittings. On the other hand, as argued by Brandt et al. (1996), small comets may also be created in large numbers through processes such as collisions or splittings, so what matters as regards to their population size is the balance between the destruction and creation rates. Therefore, these authors conclude, their number could be quite large, and the explanation why they have not been discovered is because there have not been dedicated search programs for small icy sublimating bodies. But, as we shall see next, the situation may be slowly changing.

Contrary to what was found by Kresák (1978) and Sekanina and Yeomans (1984), there is a growing evidence that the population of small comets, say those with radii $R_N \sim 0.1 - 0.5$ km might be quite substantial, although we are far from being able to quantify it. We show in Table 9.4 a list - for sure, incomplete - of comets for which there are photometric measurements strongly suggesting that they belong to the small comet category. All of them, with the exception of Sugano-Saigusa-Fujikawa, are of short period which means that they have been around for, at least, several revolutions. Most of the comets of the table were discovered when they were close to the Earth and/or to the Sun. Table 9.4 brings the geocentric and heliocentric distances of the comets at the moment of discovery (Δ_{disc}, r_{disc}) and the estimated radius. Two of the comets of the table, 45P and 51P, were taken from Table A2.1 of Appendix 2. Their quality classes are poor (3 and 4 respectively), though under most reasonable circumstances, they should fall within the category of small comets even considering the large error bars. Comet Sugano-Saigusa-Fujikawa was observed spectroscopically and in the infrared by Hanner et al. (1987) when it was at ≈ 0.06 AU from the Earth. From the thermal emission, the authors found that it would correspond to a nucleus of ~ 350 m radius. Comets 76P, 87P and 147P have been observed with the *Hubble Space Telescope* by Lamy and colleagues (Lamy et al. 2005) who determined their nuclear radii by the method of coma substraction explained in Section 7.2.3. Comets P/1999 RO28 and P/2001 WF2 were discovered by the sky survey program LONEOS when they were at close distance from the Earth. Although their estimated radii are still too uncertain to be included in Table A2.1 (as comets 76P, 87P and 147P), the preliminary values that can be drawn from the photometric measurements strongly suggest that these comets are indeed quite small.

Table 9.4: Small comets candidates

Comet	Discovery year	Δ_{disc} (AU)	r_{disc} (AU)	R_N (km)
45P/Honda-Mrkos-Pajdusakova	1948	0.45	0.58	0.33
51P/Harrington	1953	0.83	1.75	0.23
76P/West-Kohoutek-Ikemura	1975	1.44	1.40	0.33
87P/Bus	1981	1.30	2.28	0.28
C/1983 J1 (Sugano-Saigusa-Fujikawa)	1983	1.03	0.50	0.37
147P/Kushida-Muramatsu	1993	1.77	2.74	0.21
P/1999 RO28 (LONEOS)	1999	0.27	1.27	0.10
P/2001 WF2 (LONEOS)	2001	0.43	1.39	0.40

As said before, the list of potential small comets of Table 9.4 is probably incomplete, since there are several other candidates for which we can only give an upper limit to their brightness. We can mention among them 18D/Perrine-Mrkos that disappeared after being reported to have an absolute magnitude of 19.5 ($R_N \sim 0.4$ km), P/2000 G1 (LINEAR) and P/2002 T1 (LINEAR), discovered both at a geocentric distance of only 0.2 AU, whose radii are $\lesssim$0.6 km, as can be derived from the upper limit that can be set to their absolute nuclear magnitudes. Summing up, there is solid observational evidence supporting the existence of small comets with radii within the range 0.1 - 0.5 km in Earth-crossing or approaching orbits. Several search programs of near-Earth objects, like LINEAR or LONEOS, are proving to be potentially very useful for detection of small comets coming close to the Earth. However, it is still not possible to quantify such a population. We cannot say anything either about the presence of "mini-comets" with radii $R_N \lesssim 0.1$ km, for which there is no observational evidence. This is in sharp contrast with NEAs for which the discovery rate in the range $R \sim 0.05 - 0.5$ km is rapidly growing.

9.7. Collision with the Sun

The recent discovery of a large number of sungrazing comets by the coronograph on board of the ESA/NASA *Solar and Heliospheric Observatory* (SOHO) spacecraft has renewed the interest in this class of comets. Before the SOHO discoveries, there were 23 sungrazers discovered visually, and by the SOLWIND and Solar Maximum Mission satellites. The SOHO discoveries has raised the number to nearly 500 in a short time span of six years (1996-2002) (Fig. 9.9). About 84% of the sungrazers have orbital elements similar to those of the family of comets known as Kreutz, being probably the fragments of a single large parent comet that was tidally disrupted by the Sun. The rest of the discovered SOHO comets are clearly non-Kreutz members, and

Figure 9.9. The sungrazer SOHO-6 plunging into the Sun (left) as recorded by the LASCO camera on board the SOHO spacecraft on 23 December 1996. We can also see the inner streamer belt of the solar corona along the Sun's equator, where the low latitude solar wind originates, stretching beyond the disk that eclipses the Sun. The field of view of the coronograph is about 12 Sun's radii (SOHO).

they seem to belong to at least three other groupings called the Meyer, Marsden and Kracht groups. All the members of the Kreutz family have perihelion distances between 1 and 2 solar radii (from the center of the Sun), while the other groups have somewhat larger perihelion distances between 7 and 12 solar radii. For most of the Kreutz members - except the largest ones - such a close approach to the Sun leads fatally to its tidal disruption and total dissipation of the debris after the passage, which agrees with what was shown above in eq. (9.9). On the other hand, comets of the other groups that do not approach so closely are observed to survive the perihelion passage.

The injection of a large comet from the Oort cloud into a sungrazing orbit appears as an event of extremely low probability. Furthermore, some of the Kreutz members have orbital periods $P \lesssim 10^3$ years suggesting that the progenitor was already on an dynamically evolved orbit. It is hard to explain how the progenitor could have withstood several sungrazing passages without being tidally disrupted already in its first passage. The study of the dynamical evolution of LP comets with small perihelion distances (say $q \lesssim 2$ AU) and inclinations close to 90° has shed light into this puzzle. Thus, it has been shown that long-term secular perturbations by the planets involving the so called Kozai mechanism, where the dynamics is characterized by the libration of the comet's argument of perihelion ω around 90° or 270°, cause correlated changes in the perihelion distance, eccentricity and inclination of such comets. As a result, their perihelion distances can drop to very small values during part of the dynamical evolution. It is then possible that the progenitor of the Kreutz family was not originally a sungrazer, but evolved to such a state after several passages which led to its tidal disruption by the Sun.

Inspired in the previous dynamical scenario, Bailey et al. (1992) considered a simple model consisting of the Sun, a planet on a circular orbit (Jupiter), and the comet on an orbit of initial semimajor axis a_o, eccentricity e_o and inclination i_o. From the secular perturbation theory, which ignores the effect of close planetary encounters and mean-motion resonances, these authors found the integrals of motion: $a = a_o$ (constant), and $\theta = (1 - e^2) \cos^2 i$ = constant, which can be used to derive the minimum perihelion distance q_m that a comet can reach during its dynamical evolution, namely

$$\theta = (1-e_o^2) \cos^2 i_o = (1-e_m^2) \cos^2 i_m \simeq \frac{2q_m}{a_o}, \tag{9.10}$$

where we bear in mind that $e_m \sim 1$ (the considered comets are always on very eccentric orbits), and that the minimum q is obtained for $|\cos i_m| = 1$. Therefore, we have

$$q_m \simeq q_o \cos^2 i_o, \qquad (9.11)$$

which shows that if the initial inclination is close to 90°, the body can reach indeed very small perihelion distances. For instance, the Halley-type comet P/Hartley-IRAS with $q_o = 1.28$ AU and $i_o = 95.°72$, can reach a minimum $q_m = 0.013$ AU. As shown in Fig. 9.10, when the comet reaches sungrazing states, the inclination goes either to values close to 180° or to 0°, as expected from the condition $\theta =$ constant. It is also possible to see the coupling between the libration of ω around 90° and the peaks in i (either upward or downward) and q (downward) due to the Kozai mechanism.

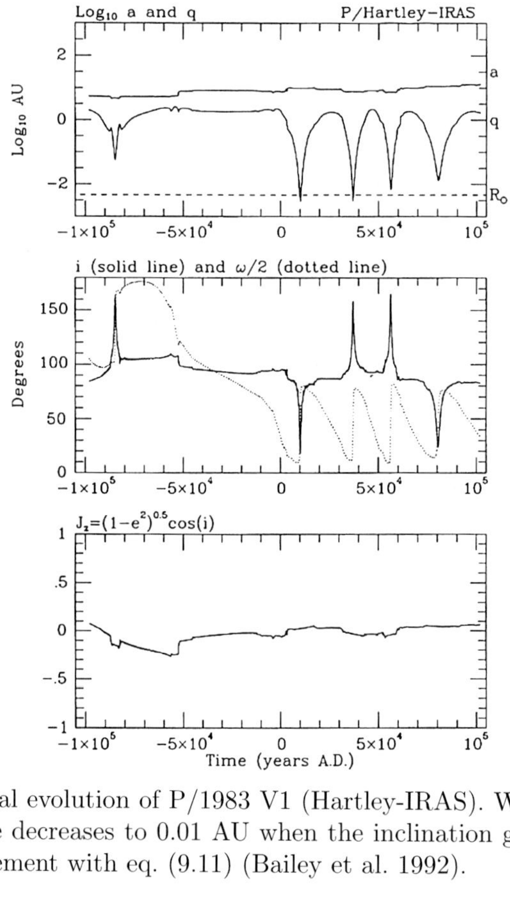

Figure 9.10. Orbital evolution of P/1983 V1 (Hartley-IRAS). We can see that the perihelion distance decreases to 0.01 AU when the inclination goes either close to 0° or 180° in agreement with eq. (9.11) (Bailey et al. 1992).

For an isotropic flux of Oort cloud comets injected in the region of the terrestrial planets (say with $q < 2$ AU), while the fraction attaining sungrazing orbits with $q < 0.01$ AU would be only 0.5%, Bailey et al. (1992) estimate that the fraction of potential sungrazers raises to about 15% when the previous dynamical mechanism is taken into account. Therefore, collision with the Sun becomes an important end state, where the term "collision" is applied either to direct collision or a grazing passage ($\lesssim 2$ $R_\odot$ from the center of the Sun) leading to the tidal disintegration of the comet.

9.8. Collisions with the planets

During their passages through the planetary region comets can collide with any of the planets. Let us start analyzing the case of long-period comets: since they come to the planetary region from random directions, any point of the sphere centered on the Sun of radius equal to that of the planet's orbit a_p (assumed to be circular), will have the same probability of being traversed by the comet. Therefore, if the planet has a collisional cross section πR_G^2, where R_G is the gravitational radius of collision, the probability of collision per perihelion passage is

$$p_{LP} = \frac{2\pi R_G^2}{4\pi a_p^2} = \frac{R_G^2}{2a_p^2}, \qquad (9.12)$$

where the factor two that multiplies the collisional cross section arises from the fact that the comet traverses the sphere twice, at the entry and at the exit. The radius R_G is related to the planet's radius R_p by: $R_G^2 = R_p^2(1 + v_{esc}^2/u^2)$, v_{esc} being the escape velocity of the planet, and u the relative velocity of the comet with respect to the planet at infinity (or, in practice, outside the planet's sphere of influence). For a body on an orbit with semimajor axis a, perihelion distance q, and inclination i, the encounter velocity with a planet, assumed to move on a circular orbit of radios a_p and velocity v_p, can be obtained from eqs. (4.44) and (7.25) as

$$U^2 = 3 - \frac{1}{A} - 2\sqrt{2Q(1 - Q/2A)}\cos i, \qquad (9.13)$$

where we have normalized the orbital parameters to those of the planet, namely $U = u/v_p$, $A = a/a_p$, and $Q = q/a_p$. For the case of a LP comet we have $Q << A$, so eq. (9.13) becomes approximately

$$U^2 = 3 - 2\sqrt{2Q}\cos i. \tag{9.14}$$

For LP comets moving on orbits with random orientations, we can adopt $\overline{\cos i} = 0$, so $\overline{U^2} = 3$.

Applying eq. (9.12) to the cases of Jupiter and the Earth with the average encounter velocity at infinity derived before, we obtain $p_{LP} = 3.3 \times 10^{-8}$ and 9.5×10^{-10} respectively. If a LP comet has on average ten passages through the planetary region, the probability of collision raises by a factor of ten, so one LP comet every 3×10^6 comets will collide with Jupiter, and one every 10^8 with the Earth.

Equation (9.12) is not longer valid for JF and HT comets or asteroids, since they do not have random orientations. For a body of inclination i, the probability of collision per orbital revolution with a planet (again assumed to be in a circular orbit of radius a_p and orbital velocity v_p) is (Öpik 1951)

$$p_{JF} = \frac{\sigma^2 U}{\pi \sin i |U_x|}, \tag{9.15}$$

where $\sigma = R_G/a_p$, and U_x is the radial component of $\vec{U}$, which is given by

$$U_x^2 = 2 - 1/A - A(1 - e^2), \tag{9.16}$$

We note that eq. (9.15) will not be valid for $i \approx 0°$. Yet Öpik (1951) notes that secular variations will act in such a way that the average $\sin i$ will be always nonnegligible. The estimated mean collision probability for a JF comet with the Earth is $p_{JF} = 1.3 \times 10^{-9}$ per year, while for Jupiter is 9×10^{-7} (Olsson-Steel 1987, Shoemaker et al. 1994). If the average lifetime of a JF comet is 10^4 yr (cf. Section 9.1), the probability of collision with the Earth per comet is thus of the order of $10^{-5} f_q$, where f_q is the fraction of time that it acquires an Earth-crossing orbit. Collisions with Jupiter are much more frequent and, in fact, dominate the collision rate with planets. The probability of collision with Jupiter is about 9×10^{-3} per JF comet. Therefore, if the total population of JF comets is $\sim 10^4$ (cf. Section 7.8) and the average lifetime $\sim 10^4$ yr, then the frequency of collisions with Jupiter is $9 \times 10^{-3} \times 10^4/10^4\text{yr} \simeq 9 \times 10^{-3}$ per year, or one collision every $\sim 10^2$ yr, which shows that Shoemaker-Levy-9-type events should be expected over time scales not much longer than the human life span.

The probability that a comet crossing the orbits of several planets will end up colliding with any of them will thus be obtained as

$$p_{plan} = \sum_{i=1}^{N} p_k \frac{\Delta t}{T}. \qquad (9.17)$$

where p_k is the collision probability with planet k, Δt is the time that the comet spends in an orbit crossing that of planet k, T is the total dynamical survival time, and N is the number of planets with which the comet can collide during its dynamical evolution. Of course, eq. (9.17) also applies to LP comets. If their perihelion distances keep more or less constant during their dynamical evolution, then $\Delta t/T = 1$ for all the planets whose orbits are crossed by the LP comets.

9.9. Collisions with asteroids and interplanetary boulders

Catastrophic collisions with interplanetary boulders might be of a certain significance as an end-state of comets. Furthermore, non-catastrophic collisions may reactivate dust-mantled comets in a dormant phase that may well look like asteroids. If low-active or inactive comets had remained undetected, the sudden activity triggered by such collisions might favor their discovery. Cometary phenomena like outbursts and splittings have also been attributed to collisions with interplanetary boulders (Harwit 1968).

The number of collisions of a comet of collisional cross-section $\sigma \simeq \pi R_N^2$ (where we neglect the gravitational focusing) with projectiles of mass $> m$ during dt is

$$dn_c = \sigma n(m, r, z) u dt, \qquad (9.18)$$

where u is the encounter velocity that can be computed from eq. (9.13), and $n(m, r, z)$ is the spatial distribution of asteroids and interplanetary boulders in the inner planetary region of mass greater than m at a heliocentric distance r and at a distance z to the ecliptic plane. It can be modeled by the expression

$$n(m, r, z) = \Gamma(m)\nu(r, z) \qquad (9.19)$$

where $\Gamma(m)$ is the cumulative number of bodies of mass greater than m per unit volume. It is well represented by the power-law

$$\Gamma(m) = Am^{-\beta}, \qquad (9.20)$$

where A is a normalization factor. For the population of asteroids in the inner planetary region (main belt plus NEAs) with mass greater than 2.8×10^{15} g a value $A = 8.5 \times 10^{-23}$ cm^{-3} is obtained (Fernández 1981c). An exponent $\beta \simeq +0.833$ is found for a population of bodies evolving collisionally (Dohnanyi 1972). $\nu(r, z)$ is a dimensionless function characterizing the spatial distribution of asteroids and interplanetary boulders that can be derived empirically from the observed population of asteroids. Some empirical curves are shown in Fig. 9.11 as a function of r and for three different values of z.

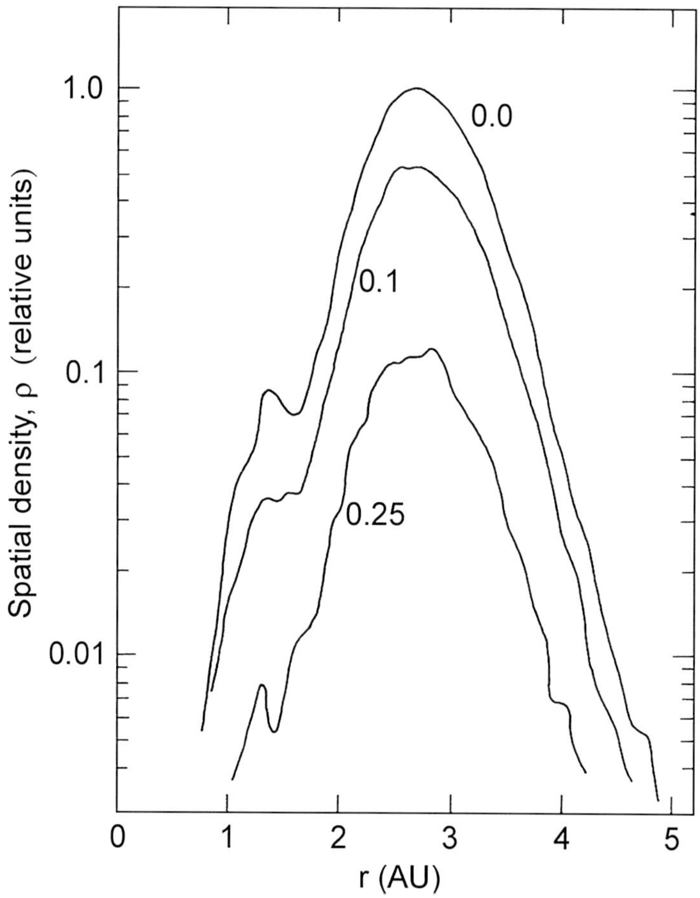

Figure 9.11. Spatial density of asteroids as a function of the heliocentric distance and for three different distances to the ecliptic plane: 0.0, 0.1, and 0.25 in units $r = 1$ (Fernández 1981c).

By integrating eq. (9.18) along the orbit of period P, we can compute the number of collisions per orbital revolution capable of disrupting the comet, namely

$$n_d = \sigma \int_0^P n(m_d, r, z) u dt, \qquad (9.21)$$

where $m_d = 2SM_N/u^2$ is the minimum mass capable of disrupting the comet of mass M_N, and S is the fragmentation energy (in erg g^{-1}), i.e. the minimum impact energy per unit mass of target required to disrupt the body. The collisional lifetime of the comet, expressed in number of revolutions, is thus

$$N_{ast} = \frac{1}{n_d} \qquad (9.22)$$

Fernández (1990) derived collisional lifetimes for a standard comet nucleus of 1 km radius, assuming for the nucleus material a fragmentation energy of 10^6 erg g^{-1} and a density of 0.5 g cm^{-3}. The collisional lifetime is shown in Fig. 9.12 and is defined as the average time - or number of revolutions - a comet can perform before suffering a catastrophic collision with an interplanetary boulder.

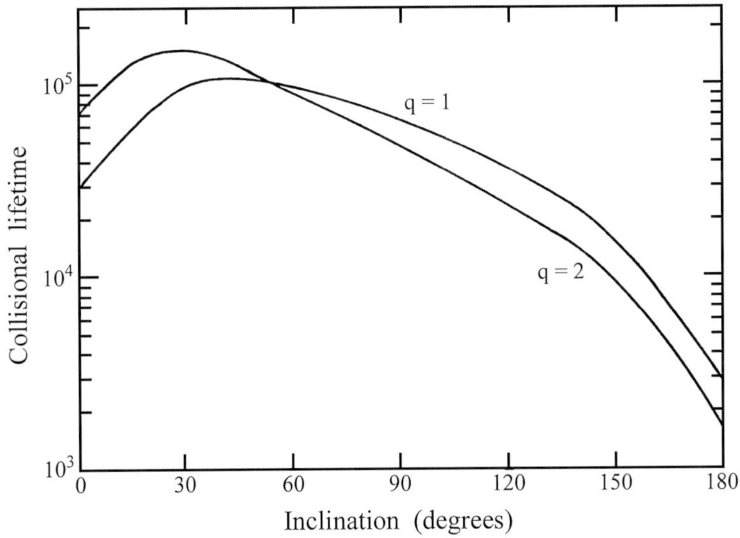

Figure 9.12. Collision lifetime (in number of revolutions) of an one-kilometer radius LP comet as a function of its orbital inclination. The results are for a fragmentation energy $S = 10^6$ erg g^{-1}. Collisional lifetimes for other fragmentation energies can be easily scaled just by multiplying by $(S/10^6)^{-1}$ (Fernández 1990).

The fragmentation energy S is a crucial parameter and unfortunately one that is very uncertain. For instance, the fragmentation energy, or "crushing strength", of a meteor can be derived from the altitude in the atmosphere at which it fragmets from aerodynamic pressure loading. If $\rho_{atm} = \rho_{atm}(h)$ is the atmospheric density, which is a function of the altitude h, the aerodynamic pressure, P_d, on a meteor with flight velocity V is given by

$$P_d = \frac{1}{2}\rho_{atm} C_D V^2. \tag{9.23}$$

where $C_D \sim 1$ is the drag coefficient. From the observation that the Draconids, associated with periodic comet 21P/Giacobini-Zinner, fragment at very high altitudes, Ceplecha and McCrosky (1976) could derive crushing strengths as low as 10^3 dyn cm^{-2}. Wetherill and ReVelle (1982) derived for the Prairie Network fireballs (of probable cometary origin) crushing strengths in the range $10^5 - 10^6$ dynes cm^{-2}, so the upper limit would be consistent with the nominal value of fragmentation energy adopted for Fig. 9.12.

From impact experiments in the laboratory, Kawakami et al. (1983) found that the specific energy required for the complete destruction of an ice target is 5×10^5 erg g^{-1}, which is consistent with the previous values derived from meteor disintegration. Similar fragmentation energies were found for porous ice targets by Ryan et al. (1999). Yet, the tidal splitting of D/1993 F2 (Shoemaker-Levy) suggests much lower material strengths, perhaps better described as agglomerations of smaller components hold together essentially by self-gravity (e.g. Scotti and Melosh 1993, Asphaug and Benz 1994).

Actually, we are interested in the minimum specific energy required for *disrupting* the body, which is understood as the process of fragmenting and dispersing the body (Davis et al. 1979). We can derive it by assuming a strengthless material, in which case the kinetic energy of the projectile has to be greater than the gravitational potential energy of the body of mass M_N, radius R_N and density ρ_N, namely

$$E_{kin} = \frac{1}{2}mu^2 = \frac{3}{5}\frac{GM_N^2}{R_N},$$

which leads to

$$S_{min} = \frac{E_{kin}}{M_N} \simeq 1.67 \times 10^{-7} R_N^2 \rho_N \quad \text{erg g}^{-1}, \tag{9.24}$$

where R_N and ρ_N are in cgs units. For a standard nucleus of $R_N = 1$ km and $\rho_N = 0.5$ g cm^{-3} we obtain $S_{min} = 835$ erg g^{-1}. For bodies with a certain degree of internal cohesion, S may raise somewhat above S_{min}, so we can conclude that the range $S \sim 10^4 - 10^6$ erg g^{-1} covers the most likely range of disruption energies of comets.

Once we compute the collisional lifetime for an one-km radius comet nucleus, it is very easy to scale up or down for other sizes and fragmentation energies from

$$N_{ast} \propto R_N^{-2} \times \Gamma(m_d) = R_N^{-2+3\beta} S^{-\beta}. \qquad (9.25)$$

As shown in Fig. 9.12, the collisional lifetimes are strongly dependent on the comet's inclination. They are shorter for comets on retrograde orbits, being of not more than a few 10^3 revolutions for $i = 180°$. For an average random LP comet of $q \sim 1$ AU we can set its typical collisional lifetime at something in between $\sim 10^3$ and $\sim 5 \times 10^4$ revolutions, for disruption energies between $10^4 - 10^6$ erg g^{-1}, bearing in mind that a wide range of values around N_{ast} are posible due to different orbital inclinations as well as to uncertainties in comet's size, internal strength, and the number and size distribution of interplanetary boulders. With these reservations, we can estimate a probability of catastrophic collision with an interplanetary boulder of $p_{ast} \sim 1/N_{ast} \sim 10^{-3}$ to 2×10^{-5} per perihelion passage. As before, if we assume that a LP comet can perform an average number of ten revolutions, we obtain a probability of collisional disruption of about 10^{-2} to 2×10^{-4} per LP comet reaching the region of the terrestrial planets (say with $q < 2$ AU). Catastrophic collisions with interplanetary boulders may play an important role as an end-state. It can be two-three orders of magnitude more likely than collision with a planet.

LEFTOVERS OF THE SOLAR SYSTEM FORMATION

Comets and asteroids are thought to be the residues of the formation of the planets, as such their study may provide important clues on the physical and chemical conditions prevailing in the early protoplanetary disk. In particular, the icy constitution of comets suggests that their birthplace was located in the region of the Jovian planets or beyond where the temperatures were low enough for water to condense. The interest in the formation of giant gaseous planets has greatly increased in the last few years by the discovery of exoplanets, which are in general of Jupiter-size or greater (for a review of the characteristics of exoplanets see, e.g. Marcy et al. 2000). Models of the formation of the solar system rested until a few decades ago on theoretical developments on the formation of a disk of gas and dust surrounding the protosun (the Laplace theory of the *solar nebula*). Today, much progress has been made in the observational front that has allowed to study very young stars surrounded by circumstellar disks, that are presumed to be the observational counterparts of the early solar nebula. The discovery of exoplanets allows us to compare our theoretical expectations of formation of planetary systems with observations (Levison et al. 1998). Furthermore, if comets were a byproduct of the formation of our solar system, they may as well be present in large numbers around other planetary systems, and pervade the interstellar space after being ejected by planetary perturbations. We will review in this chapter what is our current understanding of the processes leading to the formation of our planetary system (and others in general) and what place comets occupy in this scenario.

10.1. Early phases of star formation

Interstellar molecular clouds provide the placental material out of which new stars form. Most of the molecular mass is in the form of giant molecular clouds (GMCs) with typical masses 10^{5-6} $M_{\odot}$, diameters ~ 50 pc, average densities $\sim 10^2$ H_2 molecules cm^{-3}, and an average temperature of 10 K (Blitz 1993). The distribution of molecular gas in the Galaxy can be traced through molecules with strong transition lines, like CO at 2.6 mm wavelength. Heyer and Terebey (1998) have shown that the CO emission in the Galaxy is almost exclusively confined to the spiral arms. This close association with the spiral arms suggests that molecular clouds form from the compression of less dense atomic hydrogen clouds, and their lifetimes should be less than an arm crossing time of $\sim 10^7$ yr. GMCs may contain several places of star formation where the density is several orders of magnitude higher than the average. Such condensed regions can attain values of about 10^{4-5} H_2 cm^{-3}. The formation of massive stars can act itself as a mechanism to compress the surrounding gas, giving rise to a next round of star formation. Infrared, X-ray, and radio continuum maps reveal the existence of dense clusters of young stars in many nearby GMCs (Fig. 10.1). It has been possible to determine the mass spectrum of prestellar fragments in several star-forming regions, like ρ Ophiucus and

Figure 10.1. The Trapezium cluster, a neighborhood stellar nursery in the Orion nebula. Image taken with the Hubble near-infrared camera (NASA; K. Luhman (Harvard Smithsonian Center for Astrophysics); and G. Schneider, E. Young, G. Rieke, A. Cotera, H. Chen, M. Rieke, R. Thompson (Steward Observatory)).

Serpens. The prestellar mass spectra resemble the shape of the stellar initial mass function, which is known to approach $\Delta N/\Delta M_* \propto M_*^{-2.7}$ for 1 $M_\odot \lesssim M_* \lesssim 10$ $M_\odot$, and $\Delta N/\Delta M_* \propto M_*^{-1.2}$ for 0.1 $M_\odot \lesssim M_* \lesssim$ 1 $M_\odot$.

Star formation starts with the fragmentation of a dense region within the molecular cloud into a number of gravitationally bound cores. The cores usually appear assembled in massive clumps out of which stellar clusters form. A minimal requirement for the collapse of a core is that its self-gravity must overcome its thermal pressure. This criterion, developed by James Jeans, gives the minimum unstable mass for a given temperature T and number density of molecules n. The *Jeans mass*, M_J, is thus obtained by equalling the self-gravity to the gas pressure, namely

$$M_J \approx \left[2.5\left(\frac{3}{4\pi}\right)^{1/3}\right]^{3/2}\left(\frac{kT}{G\bar{\mu}m_H}\right)^{3/2}\frac{1}{\sqrt{\rho}} \approx 10\frac{T^{3/2}}{\sqrt{n}}\ \mathrm{M}_\odot \qquad (10.1)$$

where G is the gravitational constant, $\bar{\mu}$ is the average molecular weight ($\bar{\mu} \approx 2$ for a cloud where molecular hydrogen is the dominant species), m_H is the mass of the hydrogen atom, ρ is the mass density of the cloud and $n = \rho/\bar{\mu}m_H$. In eq. (10.1) T must be expressed in Kelvin degrees and n in cm^{-3}. For typical values of a dense core: $n = 10^5$ cm^{-3}, $T = 10$ K, we obtain $M_J = 1$ $M_\odot$, i.e. of the order of stellar masses. Pioneering surveys of dense cores in dark clouds were carried out in transition lines of NH_3 in radio.

Once a core becomes gravitationally unstable and collapses, it transits a roughly isothermal phase which tends to produce a strong central condensation of matter with a radial density gradient approaching $\rho \propto r^{-2}$ (André et al. 2000). When the opacity of the material becomes sufficiently high to impede the free release of the thermal radiation generated by the gravitational collapse, the isothermal phase gives way to a slow contraction in quasi-hydrostatic equilibrium in which the object warms up. When the central object has accumulated $\simeq 90\%$ of its final stellar mass, it becomes a pre-main-sequence star.

The IRAS satellite showed that some of the cores had embedded sources, while others were starless. The starless dense cores were identified as potential precursors of protostars, associated with the isothermal phase of collapse. Cores with embedded sources emit almost all of their radiation in the far infrared and sub-millimeter wavelength range. They are classified as class I protostars (i.e. the earliest phase). A prestellar core may last for a few times 10^6 yr. Comparison of the masses derived from observations in the 1.3 mm continuum (from about 0.05 $M_\odot$ to 3 $M_\odot$) with Jeans masses suggests that most of the starless cores are close to gravitational virial equilibrium (André et al. 2000).

10.2. Circumstellar disks

A rotating collapsing molecular core soon develops an accretion disk around it. Circumstellar accretion disks transfer matter from the surrounding molecular cloud to the central forming star. The rapid accretion of material onto the central condensation also builds up the disk mass where planets may form. Between 25 and 50% of pre-main-sequence stars in nearby dark clouds have detectable circumstellar disks (Beckwith and Sargent 1993). Observational estimates of disk masses around young pre-main-sequence stars typically give values that are less than 10% the mass of the central star, typically in the range 0.001 - 0.1 $M_\odot$. The *Hubble Space Telescope* has captured very exciting images of disks around young stars, that might resemble the early solar nebula (Fig. 10.2). The outer radius of the disk can reach a few hundreds AU (Beckwith et al. 2000). The transfer of angular momentum from the central star outward may occur via turbulent or magnetic processes, with a simultaneous transfer of disk material onto the protostar. This

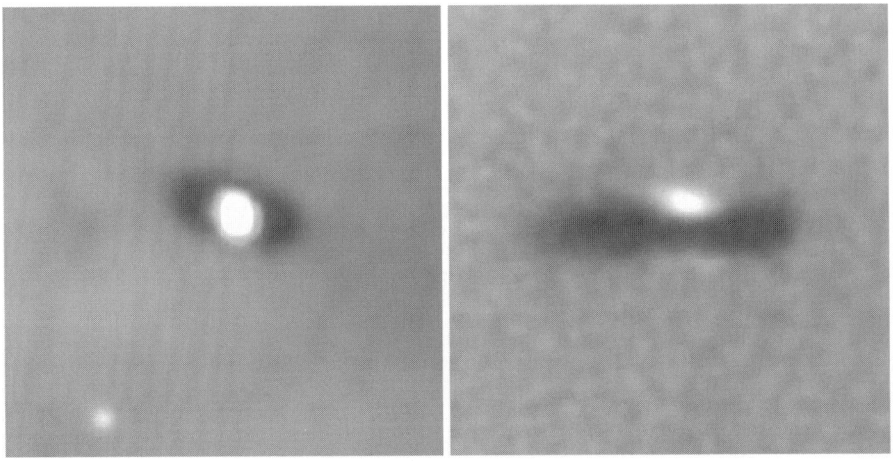

Figure 10.2. HST images of two dark silhouette disks in the Orion nebula. The left panel shows a disk inclined $\sim 45°$ with respect to the line of sight. The estimated mass is 2.4×10^{-5} $M_\odot$ and the diameter 475 AU. The right panel shows a near edge-on disk illuminated at its central regions by light coming from the central star scattered by the dust envelope (McCaughrean and O'Dell 1996).

phase may be identified with the class II or *T Tauri* phase. Observations of the excess blue emission arising from the accretion of disk material onto the protostar, suggest median mass accretion rates of $\sim 10^{-8}$ $M_{\odot}$ yr^{-1} (Hollenbach et al. 2000). Observations also show that the main accretion phase is always accompanied by the ejection of part of the accreted material in the form of prominent bipolar jets.

The dust in the inner disks around low-mass stars apparently disappears in $\sim 10^7$ yr as suggested by the observation that almost all pre-main-sequence stars in young clusters like Ophiucus ($\sim 10^6$ yr old) have disks, whereas in older regions the fraction of stars with disks drops (Strom 1995). This has also been corroborated by the 1.3 mm continuum observations of post-T Tauri stars (Gahm et al. 1994). Haisch et al. (2001) have carried out an extensive infrared survey of open clusters with ages 2.5 - 30 Myr in the L-band (3.4 μm), which directly measures the infrared excess in the Planckian curve of energy distribution, caused by micron-sized, hot ($\sim$ 900 K) dust grains in the inner regions of circumstellar dust disks. They found that most of the stars of the youngest clusters ($> 80\%$) have circumstellar disks, but this fraction rapidly decreases with the cluster age, which places an overall disk lifetime of $\sim$ 6 Myr. This short lifetime agrees with that found by Briceño et al. (2001) from a survey in the line H_{α} of hydrogen (= 6563 Å) and in the near infrared JHK colors of regions of different ages within the OB1 association. The H_{α} emission line is related to accretion of material from the circumstellar disk onto the star whereas, as mentioned, the IR emission is related to the presence of dust. The authors found that both, H_{α} and infrared emission, disappear in older stars, from which they conclude that the time scale for disk dissipation is a few Myr.

The reason why the lifetime of circumstellar disks is so short is because stars generally form in massive star-forming regions like Orion. In such regions, the forming stars and the circumstellar disks are subject to the intense UV radiation field from nearby O and B stars that heat the disk surface layers to $\sim$ 1000 K (Throop et al. 2001). Gas heated above the local escape velocity is lost from the disk, and with it the small dust grains entrained in the escaping gas flow. The very short physical lifetimes of circumstellar disks places tight constraints on the formation of giant gaseous planets (Zuckermann et al. 1995). On the other hand, Thi et al. (2001) have observed substantial quantities of H_2 gas in the debris disks of three stars aged between 8 - 20 Myr. The H_2 detection was accomplished by the measurements of the mid-infrared lines corresponding to the pure J=2→0 (28 μm) and J=3→1 (17 μm) rotational transitions with a spectrometer on board of the *Infrared Space Observatory* (ISO). The measurements show that the hydrogen to dust ratio follows cosmic abundances (about 100 to 1) so this result, if confirmed, challenges other results pointing to a rapid dissipation of the gaseous components. In any case, the amount of material left in these disks is too small to form Jupiter-sized planets, so the formation time scale of $\lesssim 10^7$ yr for these planets seems to hold.

In summary, if the disk is to lead to the formation of giant gaseous planets, they must form within $\sim 10^7$ yr. Planet formation can itself be a process leading to the dispersal of the disk. Besides planet formation, other processes may be at work to dissolve disks. Among them we can mention: (a) accretion of disk material onto the protostar, (b) stripping of the outer portions of the disk due to close stellar encounters, (c) removal of disk material swept by strong winds from the central protostar, and/or (d) photoevaporation due to the UV radiation from the central star and/or close stellar companions. For a more detailed account of these different processes see, e.g., Hollenbach et al. (2000).

10.3. The protoplanetary disk of the early Sun

The theoretical models of the formation of the solar system developed during the last half a century, combined with the wealth of observational data of the earliest phases of star formation, allows us to build a comprehensive and consistent picture of the different phases that transform a cloud of gas and dust into a planetary system. We can outline these phases as follows: After the collapse of the solar nebula, the protosun in its way to the main sequence was surrounded by a protoplanetary disk of gas and dust (Fig. 10.3). A minimum-mass model would require ~ 0.01 M$_\odot$ for the disk; that amount results from just simply adding to the planetary masses the required complement of hydrogen, helium and volatiles that were lost from the protoplanetary disk. We note that the minimum mass estimated for the protoplanetary disk of the early Sun is about the average mass found for observed disks around young stars.

The standard model of the protoplanetary disk is based on a set of very simple laws for the variation of the temperature, volume density and surface density with the heliocentric distance: $T = T_o(r/r_o)^{-n}$, $\rho = \rho_o(r/r_o)^{-m}$ and $\Sigma = \Sigma_o(r/r_o)^{-l}$. If we ideally spread the planet

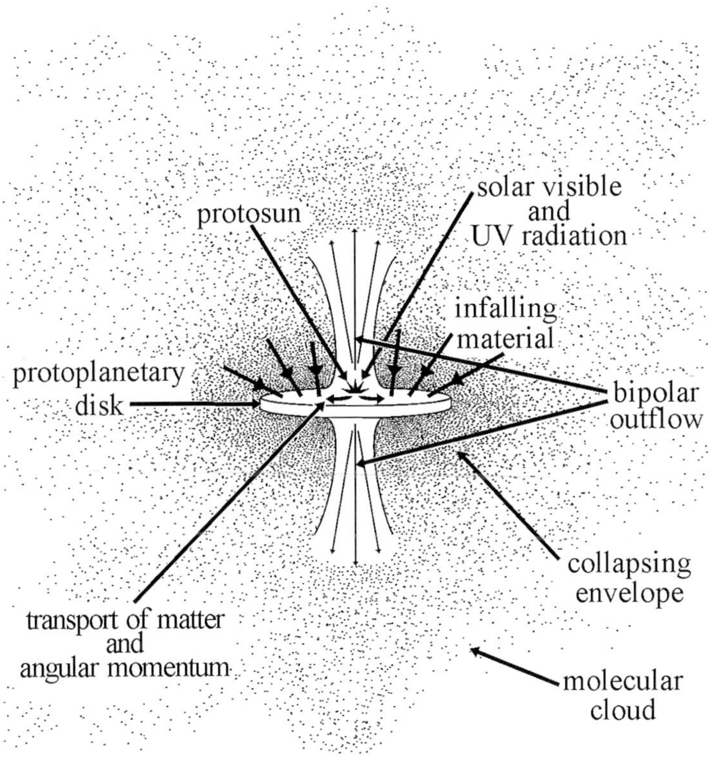

Figure 10.3. A model for the forming Sun and surrounding protoplanetary disk when it was still embedded in the placental molecular cloud.

masses in the protoplanetary disk, and allow for the hydrogen + helium and ices missing in the region of the terrestrial planets and H+He missing in the region of Uranus and Neptune, we derive an index $l \simeq 3/2$ for the surface density (Weidenschilling 1977). For a nebula in radiative equilibrium $n = 1/2$, but this will certainly be a lower limit because of the opacity of the nebular material. As an educated guess we can adopt $n = 1$ (Goldreich and Ward 1973), which will yield a range of temperatures within the planetary region consistent with the condensation of different substances of cosmogonic importance at different heliocentric distances. Furthermore, from the equation of ideal gases $p = \rho kT/\bar{\mu}m_H$, we can derive for the pressure: $p = p_o(r/r_o)^{-q}$, where $q = m + n$.

The primordial interstellar grains, or newly condensed grains, of the protoplanetary disk grew by mutual collisions and/or condensation onto the grains of molecular species still in the gaseous phase, forming icy grain mantles. In the gravitational potential dominated by the Sun, the vertically, isothermal, pressure-supported disk will fulfill the condition

$$\frac{dp}{dz} = \rho F_z, \tag{10.2}$$

where F_z is the vertical component of the Sun's gravitational force, namely

$$F_z = -\frac{GM_\odot z}{(r^2 + z^2)^{3/2}} \simeq -\Omega^2 z,$$

where z is the vertical distance to the midplane, and $\Omega = (GM_\odot/r^3)^{1/2}$ is the Keplerian mean motion at the Sun's distance r.

We have in addition

$$\frac{dp}{dz} = \frac{kT}{\bar{\mu}m_H}\frac{d\rho}{dz}, \tag{10.3}$$

where we assume that T does not vary with z. The mean square velocity of the gas molecules is $\overline{v^2} = 3kT/\bar{\mu}m_H$. By combining eqs.(10.2) and (10.3) we obtain

$$\frac{d\rho}{\rho} = -\frac{3\Omega^2}{\overline{v^2}} z dz,$$

which upon integration yields

$$\rho(z) = \rho_o \exp\left(-\frac{3}{2}\frac{\Omega^2 z^2}{\overline{v^2}}\right). \tag{10.4}$$

The half-thickness of the gaseous component of the protoplanetary disk, or scale height H_g, can be calculated by assuming a constant density $\rho(z) = \rho(0) = \rho_o$. If Σ_g is the surface gas density at a distance r, we have

$$\Sigma_g = \int_{-\infty}^{+\infty} \rho(z)dz = 2\rho_o \left(\frac{2\pi\overline{v^2}}{3\Omega^2}\right)^{1/2}. \tag{10.5}$$

The surface density is related to the scale height by the simple expression

$$\Sigma_g = 2\rho_o H_g.$$

By introducing H_g in eq. (10.5) we get

$$H_g = \left(\frac{2\pi\overline{v^2}}{3\Omega^2}\right)^{1/2}. \tag{10.6}$$

Since $\Omega^2 \propto r^{-3}$ and $\overline{v^2} \propto T \propto r^{-n}$, we have that $H_g \propto r^{(3-n)/2} = r$ for $n = 1$. Finally, $\rho \propto \Sigma / H \propto r^{-[l+(3-n)/2]} = r^{-5/2}$.

At the beginning, the dust component was intimately mixed with the gaseous component but, due to the drag force, the grains in the disk very soon started to settle towards the midplane. The characteristic descent time of grains onto the disk midplane is given by (Goldreich and Ward 1973)

$$\tau_z \simeq \frac{1}{\Omega}\frac{\Sigma_g}{\rho_p}\frac{1}{s}, \qquad (10.7)$$

where ρ_p and s are the density and radius of the dust particle respectively. For the standard model $\Sigma_g \propto r^{-3/2}$, so τ_z turns out to be independent of r and depends only on ρ_p and s.

10.4. Grain assemblage and formation of planetesimals

Grains will collect material by coagulation as they descend towards the midplane, forming branching-chain aggregates like the Brownlee particles. These particles are highly porous aggregates of sub-micron size units of chondritic material, namely more or less primitive undifferentiated stony material (Fig. 10.4). The collisional growth of an agglomerate will typically be dominated by addition of the smallest units available in the medium (Chokshi et al. 1993). Sticking of small particles can be due to van der Waals attraction or chemical interaction (Weidenschilling 1980; Chokshi et al. 1993). The sticking probability increases significantly for irregularly shaped dust grains, especially for submicron grains. The general conclusion from several laboratory experiments is that micron- and submicron-sized grains stick with high probability as long as the relative velocity is below a few m s^{-1} (see a review by Beckwith et al. 2000). There are several observational studies that seem to confirm that grain growth is taking place in circumstellar disks. For instance, Throop et al. (2001) have observed the attenuation of the bright background of the circumstellar disk 114-426 in the Orion nebula, which is seen in silhouette against nebular light, in 1870-nm Paschen α and 656-nm H_α lines with the planetary camera of the *Hubble Space Telescope*. They found a similar extinction for both wavelengths. Since small grains redden trasmitted light while large grains do not alter its color (i.e. what is called a "gray" opacity), they conclude that grains must have grown to more than 5 μm, i.e. 25 - 50 times

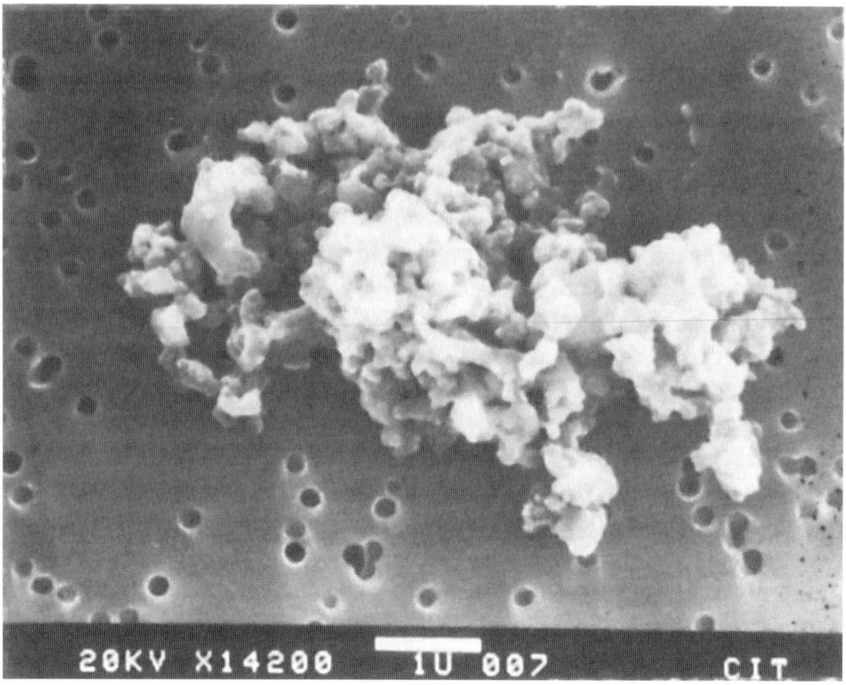

Figure 10.4. A highly porous chondritic aggregate of an interplanetary dust particle collected in the stratosphere as a micrometeorite. The scale at the bottom is 1 mμ (Fraundorf et al. 1982).

larger than typical interstellar grains, in order to render the extinction wavelength-independent.

Numerical models developed by Dominik and Tielens (1997) of collisions of aggregates of dust grains with each other or with large grains show that such collisions are quite inelastic, in which the aggregates have the capacity to dissipate large amounts of energy internally. Their model takes into account the complex physics of the contact between two dust grains, including energy dissipation due to elastic wave excitation by contact formation, and rolling and sliding motions in the contact area. The authors found that a loose aggregate can use the excess impact kinetic energy into compaction (Fig. 10.5). Only when the impact energy reaches a critical value about ten times greater than the energy required to break all the bonds between individual grains, the agglomerate will be collisionally disrupted. This condition is fulfilled for velocities of a few tens m s^{-1}, well above the expected turbulent-induced velocities of a few m s^{-1} in the nebula (Weidenschilling 1997). Bridges et al. (1996) have analyzed the growth of par-

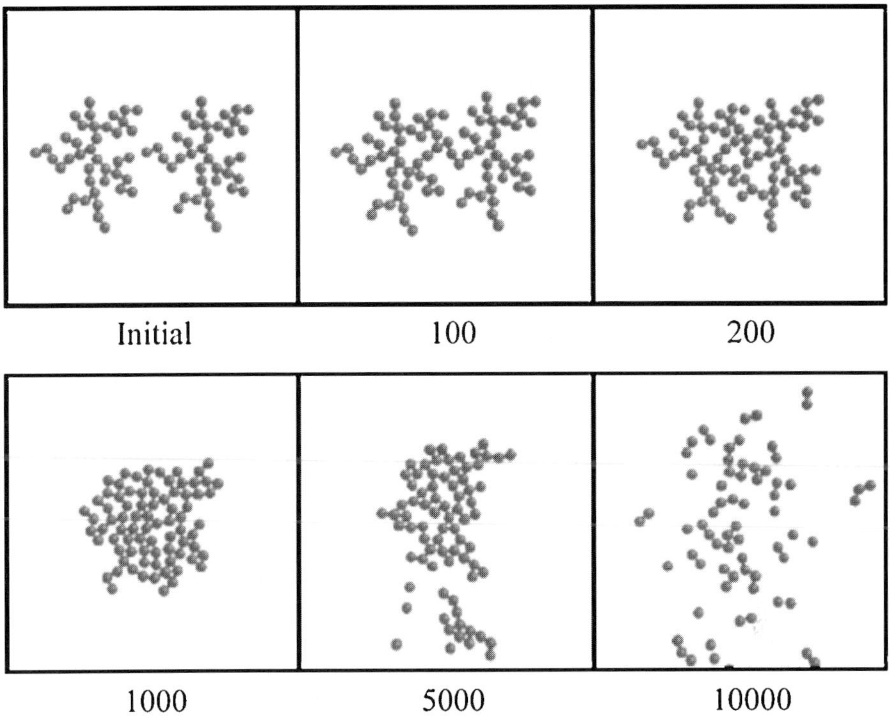

Figure 10.5. Numerical simulations describing the impact of two equal 40-aggregates at different velocities (in cm s^{-1}). The first panel shows the two aggregates prior collision. The other panels show the final results going from sticking without compaction, with compaction, and catastrophic disruption as the impact velocity increases. The grain material is ice and the individual particle radius is 1000 Å (Dominik and Tielens 1997).

ticles from centimeter-sized to meter-sized under the assumption that collisional coagulation is the dominant accretion process for all sizes. They found that several types of water-frost coated surfaces stick together when brought into contact at low enough sticking velocity (< 0.4 cm s^{-1}). Once in the nebula, cooling of grains was rapid and their surfaces became cold fingers on which water ice and other volatiles could condense. Bridges et al. concluded that the formation of surface layers of frost is a necessary step for providing a glue for the sticking of colliding particles or grain assemblages in low-temperature regions of protoplanetary accretion disks. In conclusion, we should expect that gentle collisions between aggregates or between aggregates and large grains will lead to sticking with negligible or moderate compaction.

Grains in the nebula will accrete mass at a rate of the order

$$\frac{dm}{dt} \sim \pi s^2 u \rho_d$$

$$\sim 2^{1/2}\pi^{3/2}G^{1/2}s^2\Sigma_d^{3/2}H_d^{-1/2} \propto r^{-11/4} \qquad (10.8)$$

where u is the mean relative velocity of dust particles, Σ_d their surface density, and H_d is the scale height of the dust component of the disk. At the beginnings, when the dust was intimately mixed with the gas component, we would have $H_d \sim H_g$, but afterwards when the dust started to settle we should expect $H_d < H_g$. The relative velocity u is related to the dust disk thickness by $H_d \sim u^2/2\pi G\Sigma_d$ (Weidenschilling 1995), whereby $u \sim (2\pi G\Sigma_d H_d)^{1/2}$. We can also assume that $\Sigma_d \propto r^{-3/2}$ like the gas component. Lastly $\rho_d = \Sigma_d/H_d$ is the volume density of the dust disk.

Once grains settle, or are close to the central plane, Goldreich and Ward (1973) argued that gravitational instabilities can occur in the dense dust layer leading to fragmentation into kilometer-sized bodies. The authors found that a density perturbation of size λ in a thin rotating disk can grow if the following condition is fulfilled

$$F(\lambda) = \Omega^2\lambda^2 - 4\pi^2 G\Sigma_d\lambda + 4\pi^2c^2 < 0, \qquad (10.9)$$

where c is the velocity dispersion of the dust particles. Equation (10.9) can only have solutions if $c < c_{cr}$, where the critical velocity c_{cr} is

$$c_{cr} = \frac{\pi G\Sigma_d}{\Omega}.$$

If this condition is fulfilled, there would be a range of wavelengths $\lambda_1 < \lambda < \lambda_2$ that are unstable (i.e. that lead to collapse). Short disturbances of wavelengths $\lambda < \lambda_1$ are stabilized by the random motion of the particles, while long disturbances of $\lambda > \lambda_2$ are stabilized by the shear associated to the differential rotation of the disk.

However Weidenschilling (1995) showed that collisional coagulation should be the dominant growth mechanism throughout all the size range, from micron-size to kilometer-size, since internal motions of large grains are poorly damped whereby c keeps always above c_{cr}, thus preventing the onset of instabilities. Particles must grow to at least several meters size before they can decouple from shear-induced turbulence and slow down to the low velocities required for the instability to occur. The radial velocity dispersion induced by drag will further delay the onset of instability until the particles grow to boulders of mean size in the range 10 - 100 m. Bodies of this size will not be affected either by gravitational instability because they keep high relative velocities, so they will continue growing by collisional accretion until reaching sizes 0.1-1 km, i.e. they will already be in the size range of the planetesimals proposed by Goldreich and Ward's gravitational instability mechanism.

In summary, whether the formation of planetesimals occurred via fragmentation of a thin disk in kilometer-sized chunks, or via collisional coagulation of grains or clusters of grains, the key of the whole process is in the first place the growth rate of grains that determined the speed at which they decoupled from the gas and settled into the central plane. As shown by eq. (10.8), the growth rate of grains strongly decreases with the heliocentric distance r so, for a given time, the size reached by a grain will be approximately $\propto r^{-11/4}$. Furthermore, we found that the settling time of grains $\tau_z \propto 1/s$ (cf. eq. (10.7)), then $\tau_z \propto r^{11/4}$. Weidenschilling (1980) finds that the settling time of grains at $r = 10$ AU is $\tau_z \sim 10^4$ yr, so τ_z may go up to $\sim 2 \times 10^5$ yr at 30 AU, $\sim 6 \times 10^6$ yr at 10^2 AU, and $\sim 3.8 \times 10^7$ yr at 200 AU. If the dispersal of the gas component of the primordial nebula took place on a time scale shorter than $\sim 10^7$ yr, the grains could not have settled in the central plane by gaseous drag at distances greater than about 10^2 AU. Consequently, they might have never reached the condition for further growth into comet-sized objects, either by gravitational instabilities in a thin grain disk or by collisional coagulation at heliocentric distances $\gtrsim 10^2$ AU. Perhaps this is one of the reasons for the seeming absence or scarcity of bodies in low-eccentricity and low-inclination orbits beyond $r \sim 50$ AU (see Section 8.9).

From the previous discussion we can outline a tentative scenario for planet accumulation: For distances $r \lesssim$ some tens AU grains could settle into a thin disk and grow into comet-sized planetesimals. Closer to the Sun (distances $r \lesssim 30$ AU), the formed planetesimals could continue their growth by mutual collisions to planets. For distances $r \gtrsim 10^2$ AU grains did not have enough time to settle into a thin central disk, so they continued surrounding the solar system in a broad disk whose outer boundary is difficult to assess at this point, but it could have extended up to $\sim 10^3$ AU, which corresponds to the outer radius of some of the observed protoplanetary disks.

Figure 10.6 depicts the hierarchical structure resulting from the early conditions in the protoplanetary disk, going from planets to dust as we move away from the Sun. After a time scale of $\sim 10^7$ yr, small grains

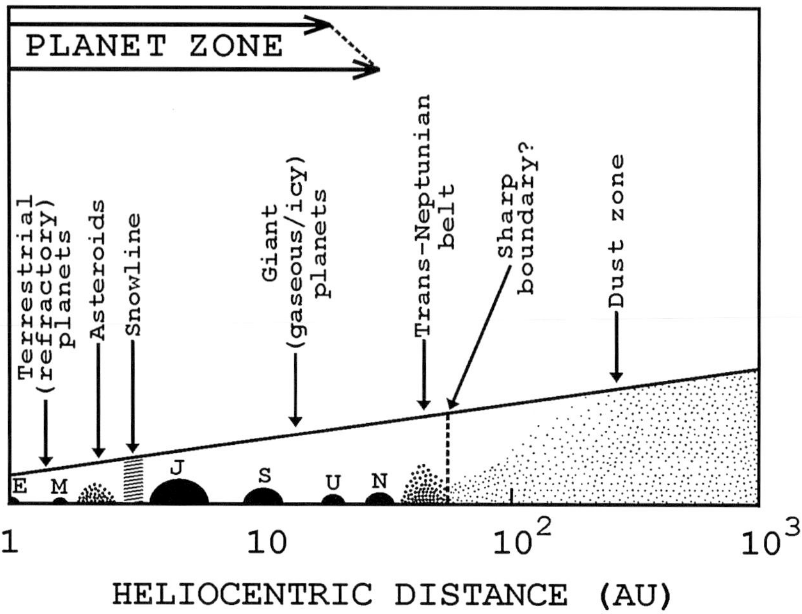

Figure 10.6. The growth of grains into planetesimals and planets strongly depends on the heliocentric distance. Beyond the snowline water could condense to form new ice grains or ice mantles around refractory cores.

might not longer be present because either they have evolved into larger grains or they left the solar system under the action of the strong solar wind. Orbital decay by Poynting-Robertson drag of centimeter-sized grains is longer than the solar system age at distances greater than $r \sim 10^2$ AU, so primordial large grains might have remained in the outer portions of the protoplanetary disk until present, provided that they could form there.

10.5. From planetesimals to planets

The accumulation process in the protoplanetary disk continued via mutual collisions among planetesimals. While the disk was "cold" ($e \approx \sin i \approx 0.01$ or smaller), the collisions were very gentle leading to accretion rather than erosion. In this way, a large number of embryo planets formed in the disk. For an embryo planet of mass M and radius R, its rate of mass accretion is

$$\frac{dM}{dt} = \pi R_G^2 u \frac{\Sigma_d}{H_d}, \tag{10.10}$$

where $R_G^2 = R^2(1 + v_{esc}^2/u^2)$ is the gravitational radius of collision, u is the relative velocity of the unperturbed trajectories of order $u \sim (e^2 + i^2)^{1/2} v_c$, where v_c is the circular Keplerian velocity at the distance r, and Σ_d/H_d is the mass density of solid matter at the distance r. Since $v_{esc}^2 = 2GM/R \propto M^{2/3}$, eq. (10.10) can be re-written in the form

$$\frac{dM}{dt} = AM^{2/3} + BM^{4/3}, \qquad (10.11)$$

where A and B are constants that depend on the bulk density of the embryo planet and u. When M is small, the rate of mass accretion is roughly proportional to $M^{2/3}$, while it is proportional to $M^{4/3}$ when M is large, i.e. the mass growth speeds up in a new regime that is called *runaway accretion*. The masses of the embryos that reach the phase of runaway accretion will detach from the rest of the planetesimal masses and become protoplanets.

The condensation of H_2O was very important, since it supplied to the disk an amount of solid material $\sim$ 10 times greater than the rocky refractory component (silicates + iron), the latter typical of the warmer inner planetary region. The region in which the temperature and pressure allowed the condensation of water is known as the *snow-line*, and it is believed to have been located at about 4 - 5 AU. We note from eq. (10.8) that an amount of solid material $\sim$ 10 times greater implies an increase in the mass accretion of dust grains by a factor of about 30. Furthermore, the sticking efficiency of icy grains or ice-coated refractory grains was greater than that of refractory grains (cf. previous section). Therefore, the accretion in the region of Jupiter and Saturn, i.e. just beyond the snowline, proceeded at a faster rate than both the inner and outer planetary region. The solid cores of Jupiter and Saturn grew on a time scale short enough to allow the capture of massive gaseous envelopes before the dispersal of the gaseous component of the protoplanetary disk. On the other hand, the number density of solid bodies in the trans-neptunian region was too low to allow the formation of large planets.

The standard model of planet formation via the accretion of solid cores faces the competing view of direct formation of Jupiter-size gaseous planets by direct collapse of portions of a massive ($\sim 0.1 - 1\ M_\odot$) gaseous protoplanetary disk (e.g. Cameron 1978, Boss 1997, Mayer et al. 2002). This "star-formation" mode has the advantage of reducing the formation time scale to $\sim 10^3$ yr, a comfortable short time as compared to the observed lifetimes of circumstellar disks, but details about the final orbits, and their eccentricities have not been still worked out, so the debate solid core-accretion versus star formation mode is still unsettled. Furthermore, torques acting between gaseous disks and protoplanets might cause their inward migration on very short time scales of a few 10^3 yr (e.g. Lin et al. 1996). However, if this process was actually very efficient, it would lead to a high demise rate of planets, as their more likely outcome would be to be engulfed by the central star, since there is not a quite satisfactory mechanism at hand to stop the migration of the planet just a few tenths AU away from the central star. There are many interesting aspects related to the way giant gaseous planets form and to what extent they migrate due to torques with the surrounding gaseous disk. The interest in these topics has been incentivated by the discovery of extrasolar planets. However, we will not delve into them and just assume for the rest of the chapter that the Jovian planets formed, presumably by core-accretion (but direct collapse is not ruled out), and will focus on the fate of the solid residual material left after their formation.

Fernández and Ip (1984, 1996) developed numerical models that considered the accretion and scattering of bodies in the Uranus-Neptune zone. They found an unexpected result: the orbit of Neptune, and to a lesser extent those of Uranus and Saturn, experienced an outward drift due to exchange of angular momentum with the interacting planetesimals. The angular momentum gained by the orbital expansion of these planets was compensated by a small drift of the massive Jupiter inwards. The theory behind this model is very simple: let us consider one of the Jovian planets, Neptune for instance, that moves on an orbit assumed to be circular of radius a_N. The angular momentum of Neptune is $H_N = M_N h_N$, where M_N is Neptune's mass (assumed to be constant and equal to the current one) and $h_N = (\mu a_N)^{1/2}$ is its specific angular momentum, where $\mu = GM_\odot$. If Neptune has a gravitational interaction with a planetesimal of mass m, its orbital radius will suffer a change Δa_N due to the exchange of angular momentum with the interacting planetesimal given by

$$\frac{\Delta a_N}{a_N} \sim -2\frac{m}{M_N}\frac{\Delta h}{h_N}, \qquad (10.12)$$

where Δh is the change in the specific angular momentum of the interacting planetesimals. In eq. (10.12) we assume that Neptune keeps the circular orbit, so the change in angular momentum only increases or decreases its orbital radius. We also assume for simplicity the planar case.

Let us consider first that all the residual planetesimals of Neptune's accretion zone were ejected to interstellar space. In the process of transfer from a near-circular orbit in Neptune's zone to a parabolic orbit, a planetesimal will gain an amount of specific angular momentum

$$\Delta h = (\sqrt{2} - 1)a_N v_N, \qquad (10.13)$$

where $v_N = (\mu/a_N)^{1/2}$ is the heliocentric circular velocity at Neptune's distance. If m_r is the total mass of planetesimals removed from Neptune's accretion zone and ejected to interstellar space, eq. (10.12) would yield in this case

$$\frac{\Delta a_N}{a_N} \sim -2(\sqrt{2} - 1)\frac{m_r}{M_N}. \qquad (10.14)$$

For instance, if Neptune ejected a total mass $m_r = 0.5M_N$, its orbital radius would have shrunk to $a'_N = a_N + \Delta a_N \sim 0.6a_N$.

If we now assume that all the residual planetesimals of Neptune's accretion zone are scattered inwards to Jupiter's influence zone, where this planet takes dynamical control and finally ejects them, the average specific angular momentum lost by each one of the interacting planetesimals (and gained by Neptune) will be approximately given by

$$\Delta h = -\left[1 - \left(\frac{2a_J}{a_N + a_J}\right)^{1/2}\right] a_N v_N, \qquad (10.15)$$

where a_J is the radius of Jupiter's orbit. In this case Neptune will increase its orbital radius by

$$\frac{\Delta a_N}{a_N} \sim +2\left[1 - \left(\frac{2a_J}{a_N + a_J}\right)^{1/2}\right] \frac{m_r}{M_N}. \qquad (10.16)$$

Again, if we assume that $m_r = 0.5M_N$, the orbital radius of Neptune's orbit will expand to $a''_N = a_N + \Delta a \sim 1.46a_N$.

In a more realistic situation we have that some of the planetesimals are hyperbolically ejected or transferred to the Oort cloud, while others are scattered inward and fall under the gravitational control of the inner Jovian planets. However, these two processes are not balanced; as shown by Fernández and Ip (1984, 1996), for Saturn, Uranus and Neptune, inward-transfer predominates over ejection, so the average

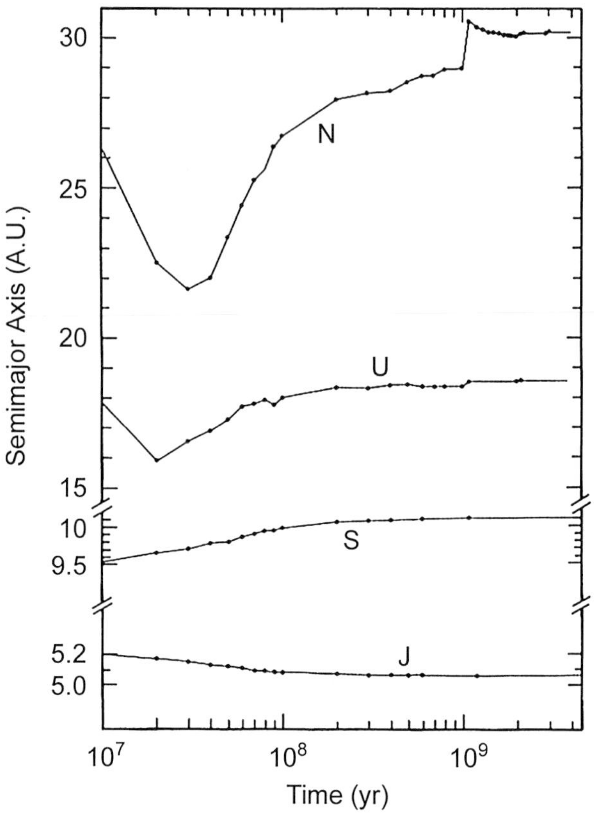

Figure 10.7. A numerical simulation that shows the time evolution of the semi-major axes of the four accreting Jovian planets as a result of exchange of angular momentum with interacting planetesimals (Fernández and Ip 1984).

angular momentum carried by the interacting planetesimals, $\overline{\Delta h}$, turns out to be negative, thus producing and outward displacement of these planets. Results of some numerical simulations are shown in Fig. 10.7 where we can see that proto-Saturn, proto-Uranus and proto-Neptune had to start their orbital evolution closer to the Sun to end up at their current distances. By contrast, the massive Jupiter as the main ejector of bodies loses some angular momentum and shrinks its orbit.

The numerical models developed by Fernández and Ip (1984, 1996) are suggestive in that initial masses 2-3 times the combined masses of Uranus and Neptune (i.e. $\sim 60-100M_{\oplus}$) were required in order to form these planets. The solid material not incorporated into the planets was lost to the inner planetary region, or to the interstellar space, or placed into the Oort cloud. A sketch of the different phases in the formation of the solar system and scattering of residual planetesimals is shown in Fig. 10.8. If the embryo planets that formed Uranus and Neptune were able to accrete some hydrogen and helium from the nebula, then their accretion time scales were probably very short, perhaps of the order of $\sim 10^7$ yr, much shorter than suggested before (e.g. Safronov 1969). A short time scale for the scattering of the residual planetesimals is also inferred, so we set the time of the massive scattering of residual planetesimals in the first $\sim 10^8$ yr of the solar system age.

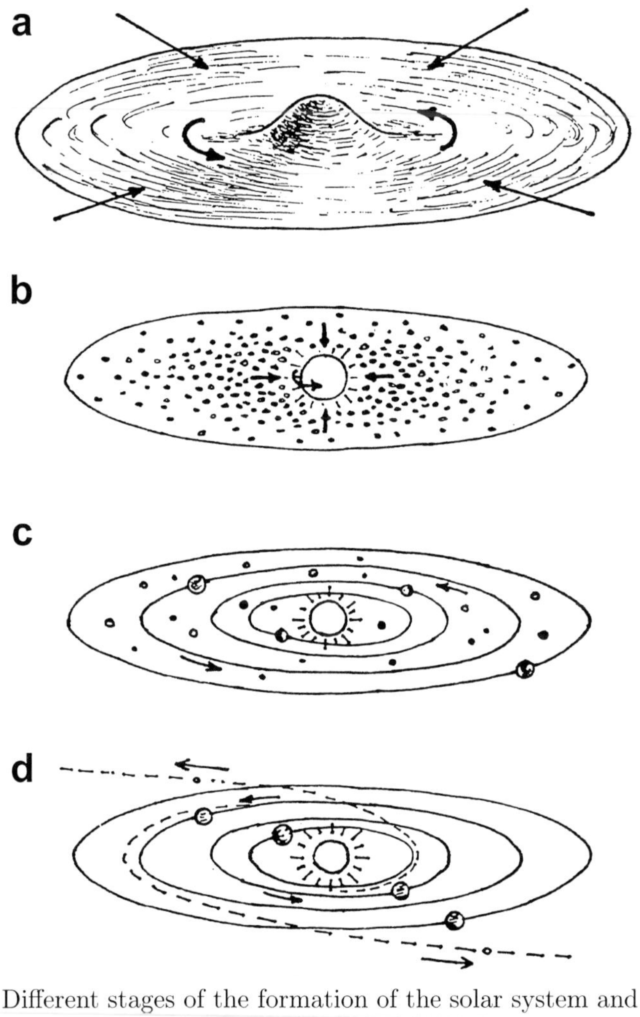

Figure 10.8. Different stages of the formation of the solar system and scattering of residual planetesimals: (a) gravitational collapse of the gas and dust cloud forming the protosun and the protoplanetary disk; (b) formation of kilometer-sized planetesimals in the protoplanetary disk; (c) accumulation of planetesimals in a few massive embryo planets; (d) gravitational scattering of the residual planetesimals.

10.6. Outward transport of planetesimals by resonant coupling with the outward-drifting proto-Neptune: The buildup of the trans-neptunian belt

As shown in Section 8.4, two bodies locked in a mean motion resonance of order $j : (j+k)$ exchange angular momentum. As analyzed there, gains of angular momentum by the resonant bodies alternate with losses, so they remain at the same average distances of the central body. We can now ask the question: what happens if the orbital radius a_1 of the inner (massive) planet of mass m_1 is increased by a small amount Δa_1? Will the external small body of mass $m_2 (<< m_1)$ remain locked in this resonance, or will the resonance break down? To analyze this problem, let us consider as in Section 8.4 that the inner body moves on a circular orbit, and the external body moves on an orbit of eccentricity $e_2 (\neq 0)$, semimajor axis a_2, and longitude of the perihelion $\tilde{\omega}_2$. The critical angle is (cf. eq. (8.2)) $\sigma = (j+k)\lambda_2 - j\lambda_1 - k\tilde{\omega}_2$, where λ_1 and λ_2 are the mean longitudes of m_1 and m_2. The angle σ will librate around the equilibrium point (e.g. the aphelion) with a period T_σ given by (Dermott et al. 1988)

$$T_\sigma \simeq T_2 \left[3(j+k)^2 \frac{m_1}{M_\odot} f(\alpha) e_2^k \right]^{-1/2}, \qquad (10.17)$$

where $\alpha = a_1/a_2$, and T_2 is the orbital period of m_2. $f(\alpha)$ is a positive function derived from the Fourier expansion of the disturbing function that can be expressed in terms of Laplace coefficients (see, e.g., Weidenschilling and Davis 1985, Patterson 1987). This function has to be evaluated for each resonance. Some numerical values of $f(\alpha)$ for particularly interesting resonances are shown in Table 10.1.

Table 10.1: Numerical values of $f(\alpha)$

Resonance	α	$f(\alpha)$
1:2	0.630	0.428
2:3	0.763	2.484
3:4	0.825	3.283

The ratio α of semimajor axes will oscillate during the libration period T_σ with an amplitude

$$\left(\frac{\Delta\alpha}{\alpha}\right)_{res} \simeq 8(j+k)\frac{m_1}{M_\odot} f(\alpha) e_2^k \frac{T_\sigma}{T_2}. \quad (10.18)$$

By combining eqs.(10.17) and (10.18) we obtain (Dermott et al. 1988)

$$\left(\frac{\Delta\alpha}{\alpha}\right)_{res} \simeq 8\left[\frac{1}{3}\frac{m_1}{M_\odot} f(\alpha) e_2^k\right]^{1/2}. \quad (10.19)$$

A perturbation on the system strong enough to produce a change in the ratio of semimajor axes α greater than that given by eq. (10.19) will break down that particular resonance. On the other hand, smooth perturbations, either on m_1 or m_2, that keep the relative change $\Delta\alpha/\alpha$ below that given by eq. (10.19), will preserve the resonance even in the case that both m_1 and m_2 migrate outwards or inwards.

Let us now consider the perturbation on the inner planet caused by a planetesimal of mass m'. In the two-body problem, the body m' will move on a hyperbolic orbit in a planetocentric frame of reference, so it will suffer a deflection γ given by

$$\tan\frac{\gamma}{2} = \frac{Gm_1}{pu^2}, \quad (10.20)$$

where p is the impact parameter and u is the encounter velocity.

If the perturbing body moves on an orbit of eccentricity e' and inclination i' (both assumed to be low), the encounter velocity u will be approximately expressed as

$$u \approx (e'^2 + i'^2)^{1/2} v_1, \quad (10.21)$$

where i' is in radians, and $v_1 = (\mu/a_1)^{1/2}$ is the (circular) velocity of m_1.

The heliocentric velocity v' of the body m' will experience a change after the encounter with the planet m_1 given by

$$\Delta v' \approx 2u \sin\frac{\gamma}{2}. \quad (10.22)$$

The change in the heliocentric velocity of the planet, Δv_1, can be derived from the condition of energy conservation, i.e.

$$m_1 v_1 \Delta v_1 = m' v' \Delta v'. \quad (10.23)$$

The change Δv_1 is related to the change in the semimajor axis, Δa_1, by

$$\Delta v_1 = \frac{\mu^{1/2}}{2a_1^{3/2}} \Delta a_1. \tag{10.24}$$

Now, let β be the angle between the vectors $\vec{v}_1$ and $\vec{u}$, and $b = u/v_1$. The velocity v' will be related to v_1 and u by

$$v' = v_1(1 + b^2 - 2b\cos\beta)^{1/2}, \tag{10.25}$$

so v' depends on the angle β that varies between 0 and 2π, i.e. it depends on the encounter geometry. In the following we shall adopt an average $< \cos\beta >= 0$.

By combining eqs.(10.20)-(10.25), we finally obtain

$$\left(\frac{\Delta a_1}{a_1}\right)_{pert} \approx 4\left(\frac{m'}{m_1}\right)(1+b^2)b\frac{\Gamma}{(1+\Gamma^2)^{1/2}}, \tag{10.26}$$

where

$$\Gamma = \frac{m_1/M_\odot}{(p/a_1)b^2}.$$

Therefore, when the perturbation in a_1 exceeds the amplitude of the resonant oscillations in α, given by eq. (10.19), the resonant coupling will break down, namely when

$$\left(\frac{\Delta a_1}{a_1}\right)_{pert} > \left(\frac{\Delta\alpha}{\alpha}\right)_{res}. \tag{10.27}$$

A massive body of mass ratio $m'/m_1 \sim 10^{-2}$, moving in an orbit of $e' \sim \sin i' \sim 0.1$, typical of a moderately excited protoplanetary disk, will produce the required strong perturbation to break the resonance if p is smaller than a few hundreds planetary radii (Fernández and Ip 1996). On the other hand, if the mass of the perturber is below $\sim 10^{-2}m_1$, the change Δa_1 in the semimajor axis of m_1 will be below the critical value to break the resonance. In this case the resonant coupling will transfer energy and angular momentum to the external mass m_2. This analysis can be applied to the case of Neptune that migrated outwards by exchange of angular momentum with the interacting planetesimals, so it may have pushed outward external planetesimals trapped in mean motion resonances like the 2:3 or 1:2.

Very few large planetoids were massive enough to break the resonant coupling, so the radial migration of Neptune might have been smooth enough over long time spans to keep the bodies trapped in their respective resonances, thus accompanying Neptune in its radial migration outward. If an encounter of Neptune with a massive planetoid occurred, the disruption of the resonant coupling would have led to the release and deposition of the trapped bodies at the heliocentric distances attained at that moment. Nevertheless, the transfer process by resonant coupling would have proceeded with other bodies.

The previous mechanism was proposed by Malhotra (1993, 1995) to explain the origin of Pluto and the Plutinos in their peculiar rather eccentric orbits locked in the 2:3 mean motion resonance with Neptune, and also of a substantial population in the 1:2 mean motion resonance. Malhotra also showed that the outer bodies, while pushed outwards by resonant coupling with the migrating planet, increased their eccentricities from very low values, typical of planetesimals in the protoplanetary disk, to values given by

$$e_{final}^2 - e_{initial}^2 \approx \frac{1}{j+1} \ln\left(\frac{a_{N,final}}{a_{N,initial}}\right), \tag{10.28}$$

where this equation has been derived for a resonance of order $k = 1$, and $a_{N,initial}$, $a_{N,final}$ are the initial and final radii of Neptune's orbit. For instance, if Neptune migrated from $a_{N,initial} \approx 25$ AU to $a_{N,final} \approx 30$ AU, the coupled planetesimals in the resonance 2:3 ($j = 2$) would have increased their eccentricities from $e_{initial} \approx 0$ to $e_{final} \approx 0.25$, typical of Pluto and as a sort of average value for the Plutinos.

The push-outward mechanism explained above has also been invoked by Levison and Morbidelli (2003) to explain the existence of the dynamically hot classical trans-neptunian belt with a sharp edge at $a \sim 50$ AU. They argue that the protoplanetary disk was truncated at ~ 30 AU, so when the accreting proto-Neptune migrated outwards, it stopped when it hit the primordial edge of the disk. According to Levison and Morbidelli, during its outward drift Neptune pushed outward by resonant coupling planetesimals from their initial location in the planetary region to the trans-neptunian belt. They favor the 1:2 mean motion resonance because it does not excite high inclinations, so the push-outward planetesimals will have kept a flat structure, as observed in the trans-neptunian belt. As described by eq. (10.28), the standard model of resonance trapping also predicts that the eccentricities of bodies pushed outward monotonically increase. However, Levison and Morbidelli found that some of the bodies have eccentricities that oscillate, due to a secular resonance embedded in Neptune 1:2 mean motion resonance, so eccentricities can be low as well, as observed in the classical trans-neptunian belt.

The most interesting feature of the mechanism proposed by Levison and Morbidelli is that it can explain why there seems to be a sharp edge in the trans-neptunian belt at $\sim$ 50 AU which, they note, it is very close to the location of the 1:2 mean motion resonance with Neptune. If the outward migration of Neptune would have been very smooth all the time, all the trapped planetesimals would have been delivered at once at $a \sim 48$ AU. However, the migration of Neptune was somewhat jumpy because of gravitational interactions with massive bodies ($m \gtrsim 10^{-2} M_{Neptune}$) that decoupled trapped planetesimals and deposited them at distances smaller than 48 AU in nonresonant orbits. At the same time that some planetesimals were decoupled, others were trapped in the 1:2 mean motion resonance, so the process could continue until Neptune reached the edge of the protoplanetary disk.

If the push-outward mechanism by resonant coupling proposed by Malhotra for the Plutinos and by Levison and Morbidelli for the classical TNOs did work, at least partially, it would have important consequences for the chemistry and mineralogy of the TNOs, and thus of comets, since their composition might reflect the properties of bodies formed in the region of the Jovian planets, rather than those expected for bodies formed in their colder current location. If there was a primordial population formed in the trans-neptunian belt that mixed up with outward-pushed planetesimals, then at least two populations with some distinct features might coexist there at present. In this regard, Gomes (2003) argues that a large number of Neptune-scattered planetesimals raised temporarily their perihelia beyond Neptune's perturbing influence because of the action of secular, Kozai or mean motion resonances. While keeping the semimajor axis constant, these resonances produce low-eccentricity and high-inclination orbits. Since the changes in eccentricity are periodic, the bodies will eventually return to Neptune-crossing orbits. Yet, Gomes argues that because of the outward migration of Neptune, a fraction of these bodies were decoupled from the resonance while they were in low-e and high-i orbits. From numerical experiments, Gomes (2003) found that around 0.2% of the primordial Neptune-scattered planetesimals could be decoupled in this way, with an average eccentricity $\bar{e} = 0.20$ and an average inclination $\bar{i} = 16.°1$. This mechanism can thus account for the observed high-inclination population in the classical trans-neptunian belt. To produce the desired population, the test bodies have to start their evolution at ~ 25 AU, again suggesting a significant mixing of populations formed at different heliocentric distances.

10.7. The scattering of residual planetesimals and the buildup of the Oort cloud

As mentioned before, the accretion of Uranus and Neptune may have been very inefficient. The reason is that once the growing proto-Uranus and proto-Neptune acquired powerful gravitational fields (say, when they grew to masses $\gtrsim$ a few $M_{\oplus}$), ejection became a much more likely outcome than collision. The efficiency of accretion as compared to ejection was a function of the relative velocities between the accreting protoplanets and the planetesimals. The average relative velocity u results from the balance between stirring by mutual gravitational perturbations and damping by inelastic collisions and, possibly, a resisting gaseous medium. Safronov (1969) found that u can be expressed in terms of the escape velocity of the largest embryo planet or the dominant massive bodies, understood as those that contain the bulk of the mass of the planetesimal population (see also Greenberg et al 1984), namely

$$u = \frac{v_{esc}}{(2\theta)^{1/2}}, \tag{10.29}$$

where $v_{esc} = (2GM/R)^{1/2}$ is the escape velocity of the largest body of mass M and radius R, and θ is the "Safronov number" for which Safronov estimates $\theta \sim 3-5$. The ejection of a planetesimal becomes a much more likely outcome than collision with the accreting embryo planet when u reaches the value

$$u = (\sqrt{2}-1)V_{esc} \tag{10.30}$$

where $V_{esc} = (2GM_{\odot}/r)^{1/2}$ is the escape velocity from the solar system at the heliocentric distance r of the accreting planet. This condition will be fulfilled for the mass

$$M \approx 0.1\frac{\theta^{3/2}}{\rho^{1/2}}\left(\frac{M_{\odot}}{r}\right)^{3/2} \tag{10.31}$$

where ρ is the mass density of the planet (assumed to be constant). For instance, if we adopt $\theta = 4$, at Neptune's distance ($r = 30$ AU) the critical mass is $M \approx 1$ $M_\oplus$ for a density $\rho = 1.5$ g cm^{-3}, while at Uranus' distance becomes $M \approx 1.8$ $M_\oplus$. Therefore, the accretion of solid bodies by protoplanets of a few $M_\oplus$ becomes problematic in the outer planetary region, unless we assume that any of the damping mechanisms mentioned above was very efficient in keeping low relatives velocities u (which is equivalent to saying that θ was higher than the value quoted above), or that the protoplanets had enlarged collisional cross sections due to extended atmospheres (which is equivalent to decreasing ρ in eq. (10.31)) (for a discussion of this problem see, e.g., Brunini and Fernández 1999).

For Jupiter and Saturn the situation was completely different since most of the accreted material was hydrogen and helium in the gaseous phase. By contrast, Uranus and Neptune accreted most of their material via collision with planetesimals. Therefore, these were the most difficult planets to form, given the unfavorable collision/ejection ratio once proto-Uranus and proto-Neptune grew above a few $M_\oplus$, and possibly their further accretion was at the expense of the ejection of a substantial amount of solid matter. Nevertheless, as we shall see below, they were probably able to accrete non-negligible amounts of hydrogen and helium before their dispersal (Lissauer et al. 1995) and this may be an important feature to set meaningful constraints on the formation time scales of these planets.

Planetesimals scattered by Uranus and Neptune were subject to a random-walk process in the energy space (see Section 4.5) until attaining semimajor axes of $10^4 - 10^5$ AU, where perturbations by the tidal field of the galactic disk and passing stars were strong enough to raise their perihelia outside the planetary region (e.g. Fernández 1980a, Duncan et al. 1987). Once that happened, comets only evolved under the much gentler dynamical action of external perturbers, and this situation lasted until they were ejected to interstellar space or were re-injected into the planetary region. Therefore, comets stored in the Oort cloud have their perihelia removed from the planetary region, which warrants long dynamical lifetimes.

The necessary condition for a body to be stored in the Oort cloud is that its perihelion distance q can experience a strong change of the order $\Delta q \sim q$ by external perturbers before being ejected by planetary perturbations. As shown in Section 6.1, changes $\Delta q \sim q$ are reached for a semimajor axis $a_{oort} \sim 10^4$ AU, or an energy $x_{oort} \sim -10^{-4}$

AU^{-1}. Since the typical energy change per perihelion passage in the Uranus-Neptune region: $|\epsilon_t| << |x_{oort}|$, most of the comets scattered by Uranus and Neptune will fall in the energy range $x_{oort} < x < 0$ before being ejected (Fernández 1978). By contrast, the typical energy change of comets scattered by Jupiter and Saturn: $|\epsilon_t| >> |x_{oort}|$, which means a very low efficiency in placing comets in the Oort cloud. In other words, in their random-walk in the energy space, most comets scattered by Jupiter and Saturn will overshoot the energy range of Oort cloud comets $x_{oort} < x < 0$ to interstellar space (cf. Fig. 4.1).

Had the early solar system been within a galactic field similar to the current one, icy planetesimals scattered by the Jovian planets would have given rise to a loosely bound Oort cloud with an inner radius $\sim 10^4$ AU. As said, Neptune, and to a lesser degree Uranus, would have been the major contributors of icy planetesimals to such a cloud. The difficulty with this picture is that such a loosely bound Oort cloud could very hardly have withstood the disrupting effect of Giant Molecular Clouds over the solar system age (Bailey 1983a). To overcome this difficulty it is necessary to postulate the existence of an inner core of the Oort cloud as the immediate source of replenishment of the external, loosely bound Oort cloud (Hills 1981, Bailey 1983a).

If the Oort cloud needs an inner core to make up for comet losses over the solar system age, the problem then shifts to explain the origin of such a core. A comet formation *in situ* (e.g. Biermann and Michel 1978) meets the difficulty of the extremely low density of the medium at large heliocentric distances. As explained before, grains are not expected to agglomerate in comet-sized bodies at distances greater than about 10^2 AU. Or maybe the inner core is not necessary as a replenishment source, since this role is played by the trans-neptunian belt and the scattered disk. As explained in Section 8.11, these reservoirs may still be feeding the Oort cloud with bodies diffusing outwards over long time scales. Another possibility is that the early galactic environment was different from the current one, thus changing the efficiency of comet capture in the Oort cloud and its lower boundary. This point will be analyzed in the next section.

10.8. The early galactic environment of the solar system

Was the primitive neighborhood of the solar system similar or quite different from the one we observe now? Since stars tend to form in clusters within molecular clouds, it is then likely that the early solar system was not isolated as it is now, but it was a member of a star cluster embedded in the placental molecular cloud. This is what is commonly observed in star-forming regions (cf. Fig. 10.1), such as the Orion molecular cloud complex (Lada et al. 1993). That situation could have lasted for $\sim$ 30 Myr, the time required for a GMC to dissipate in the galactic medium (Blitz 1993). Furthermore, the Sun would have remained within the cluster for up to a few 10^8 yr, which is the typical dissociation time of open clusters (Lyngå 1982). The fact that Uranus and Neptune contain nonnegligible amounts of hydrogen and helium, perhaps amounting to $\sim 1 - 2$ $M_\oplus$ (Hubbard 1989, Hubbard et al. 1995) suggests that they grew fast enough to be able to capture gas from the nebula before its dispersal by the strong solar wind during the T Tauri phase of the early Sun. The required time scale should have been $\sim 10^7$ yr. The final accretion stages of Uranus and Neptune would have coincided with the massive scattering of the residual planetesimals which would have taken a few tens Myr, so it is quite possible that most or a large fraction of the scattered planetesimals reached the Oort cloud while the solar system was still in a dense galactic environment.

The galactic environment of the early solar system has only recently attracted attention, mainly by the observations that confirm that stars form in clusters in dense cores of molecular gas. Among the earliest works on this subject we can mention Mottmann (1977) who argued that the Sun was a member of an open cluster about 4 Gyr ago, so close encounters with other cluster stars were frequent. Under such circumstances, a close star encounter triggered a comet shower that caused the late heavy bombardment of the terrestrial planets and tilted the orbital planes of the Jovian planets by $\sim 8°$ with respect to the Sun's spin axis. Also, Reeves (1978) attributed certain isotopic anomalies of oxygen and magnesium observed in several classes of meteorites to a supernova explosion that occurred nearby, while the early solar nebula was still in its contraction phase. Such a close supernova explosion was likely to occur only if the early Sun belonged to an OB association where the most massive members, after a fast evolution, ended up as supernovae. Later Hills (1982) discussed the formation of a comet cloud around the protosun during the early collapsing phases of the nebula within a very dense star cluster. Gaidos (1995) has further considered the dynamical consequences of a dense galactic environment on the formation of the solar system and the comet cloud. He evaluated the efficiency of each one of the Jovian planets in placing planetesimals in an Oort-type reservoir. He concluded that Saturn might have been able to form a transient comet cloud at ~ 3000 AU, but it was rapidly eroded by the strong tidal field and frequent stellar passages prevailing at that early epoch. A variant was presented by Eggers et al. (1997) who, as Gaidos and Hills, assumed that the solar system formed within a star cluster, but instead of acquiring an Oort comet cloud by trapping its own scattered bodies, it built up the comet cloud by trapping intracluster bodies scattered by the Sun itself and the other cluster stars. Of course this model rests on the assumption that all stars in the cluster built their own planetary systems and ejected a large number of residual planetesimals.

Fernández (1997) and Fernández and Brunini (2000) have recently re-analyzed how a dense galactic environment might have affected the buildup of the Oort cloud. Their calculations show that if the Sun was a member of an open cluster, the strong perturbations of other cluster stars would have decoupled comets from the planetary region already for semimajor axes of a few 10^3 AU. Figure 10.9 shows how comets trapped in the Oort reservoir after 10^8 yr would appear at a certain arbitrary time of their orbital periods, assuming that all of them lie on the ecliptic plane. As expected, the radius of the formed inner comet core is finely tuned to the strength of the field of external perturbers, which is assumed to be constituted by stars formed in the same placental gas; the stronger the field of external perturbers, the

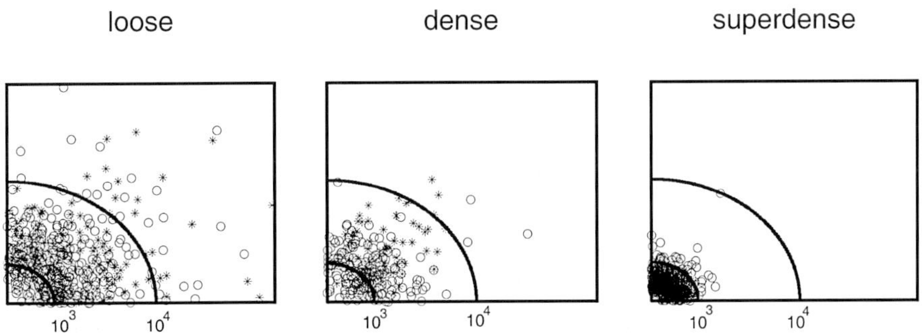

Figure 10.9. Projections onto the ecliptic plane showing how comets trapped in the Oort reservoir would appear distributed in the circumsolar space at a given time. The Sun is assumed to be within one of the open cluster models indicated above each panel (see text for details about the adopted star densities). The symbols are for bodies from the Jupiter-Saturn zone (stars) and for bodies from the Uranus-Neptune zone (open circles). The distances are expressed in AU (Fernández and Brunini 2000).

more compact the core of trapped comets is. A loose star cluster (10 stars pc^{-3}) will form a rather loose comet core of radius $\sim 10^4$ AU, a dense star cluster (25 stars pc^{-3}) will form a more compact core of a few 10^3 AU, while a superdense star cluster (100 stars pc^{-3}) will form a very compact core of radius $\sim 10^3$ AU. The relative velocity between the Sun and other cluster stars is assumed to be 1 km s^{-1} which is typical of open clusters in virial equilibrium. The star cluster was assumed to lose stars with time until its complete dissolution at 10^8 yr.

Therefore, an early very dense galactic environment might have had dramatic consequences in the buildup of the Oort cloud. The possible existence of an inner core of the Oort cloud of radius of a few 10^3 AU might be a consequence of that early environment where the Sun possibly formed. This should not be taken as an ad-hoc assumption, but based on strong observational grounds that show that most stars tend to form in clusters within molecular clouds. The recent discovery of 2003 VB_{12} (Sedna), moving in a kind of orbit expected for objects of the inner core (preliminary orbital elements $q = 76 \pm 4$ AU, $a = 480 \pm 40$ AU, $i = 11.°93$), might represent the first observed member of such a putative core. With a size very close to Pluto's (diameter $\sim 1300-1800$ km), it suggests that a total mass $\sim 5\ M_{\oplus}$ might be enclosed in the inner core, including ~ 500 Sedna-sized objects (Brown et al. 2004). An inner core of radius $\sim 10^3$ AU would imply to broaden the energy range of the Oort reservoir to $\Delta x_{oort} = 1/a_{oort} - 1/a_{\infty} \sim 10^{-3}$ AU^{-1}, which is of the order of the typical energy change of bodies scattered by Jupiter (cf. Section 4.1). Therefore, Jupiter and Saturn might have been able to place a large number of bodies from their own accretion zones in a more tightly bound Oort cloud, since planetesimals random-walking in the energy space will be about as likely to fall in the energy range of the broadened Oort reservoir as to overshoot it to interstellar space. This scenario would imply a significant mixing of bodies from different parts of the planetary region: from the Jupiter (or even the asteroidal) zone to the trans-neptunian belt, thus providing a heterogeneous population of Oort cloud comets, as the volatile content probably varies with the distance to the Sun.

10.9. The physical structure of icy planetesimals

What the models of accretion processes in the protoplanetary disk tell us about the physical structure of the formed aggregates, and how comets - if they are indeed the fossil records of that early epoch - match such models? We have already presented some ideas about sticking mechanisms and grain assemblage in Section 10.4. We will continue now delving into this problem. Donn (1963) developed a model of formation of comet nuclei by accumulation of grains in which the drag by the nebular gas facilitated the sticking upon collision by decreasing the impact velocities to meters s^{-1}. The accreted icy mass would have an initial density of about 0.3 g cm^{-3}. As discussed by Donn (1963, 1990), low impact velocities were also required to preserve the icy component of the grains. We can analyze this problem through a very simple example. Let us consider two grains of masses m_1 and m_2 colliding with an impact velocity u. Their kinetic energy is

$$E_{kin} = \frac{1}{2} m_r u^2, \qquad (10.32)$$

where $m_r = m_1 m_2/(m_1+m_2)$ is the reduced mass. If upon impact most of the kinetic energy goes into heating the grains, the increase in their temperature ΔT will be roughly obtained from

$$\frac{1}{2} m_r u^2 \sim C m_T \Delta T, \qquad (10.33)$$

where $m_T = m_1 + m_2$, and C is the average specific heat of the grains over the range of temperatures relevant to this problem (tens of Kelvin degrees). We then have

$$\Delta T \sim \frac{u^2}{8C}, \qquad (10.34)$$

where we assume that the two grains have similar masses. The specific heat of water ice is: $C_w(T) = 7.49 \times 10^4 T + 9 \times 10^5$ erg g^{-1} K^{-1}, while for silicates and organic refractories at low temperature is: $C_r = 5 \times 10^4 T$ erg g^{-1} K^{-1} (Tancredi et al. 1994). In the following we will adopt an average specific heat $C = 3.6 \times 10^6$ erg g^{-1} K^{-1} for a typical temperature $T = 50$ K. If we consider an impact velocity $u = 10^4$ cm s^{-1} we get $\Delta T \sim 3.5$ K, so the increase of temperature is very low to sublimate the ices constituting or coating the grains. Therefore, the icy component of grains will be preserved during the accumulation process.

Icy grains, or ice-coated refractory grains, with porous and dendritic-like protrusions over their surfaces, are more likely to stick together because of the interpenetration of the criss-crossed structures of the two porous layers, which is what Supulver et al. (1997) call a "velcro" model.

Grain aggregation will form fractal structures (e.g. Daniels and Hughes 1981, Meakin and Donn 1988, Donn 1990) whose number of grains is related to the size d by

$$N(d) \sim d^D, \tag{10.35}$$

where D is the fractal dimension. Simulations of accretion of dust aggregates yield values $D \sim 2$ (Meakin and Donn 1988). In this case, the density of a fractal aggregate will decrease with size according to

$$\rho(d) \propto d^{D-3} \sim d^{-1}. \tag{10.36}$$

Therefore, the fractal aggregate should be very porous with extremely low density. Yet, large grain aggregates become compacted by impacts, so larger bodies get larger mean densities. By considering compaction effects, Donn (1990) concluded that cometary nuclei with radii of a few km will be irregularly shaped and fragile, with densities of a few tenths g cm^{-3}.

We can then foresee the following stages in the accretion of icy planetesimals: (1) formation of clusters of icy or ice-coated grains with a fractal structure which may hold until sizes $\sim 10^{-2}$ cm (Weidenschilling 1997); (2) further growth by accretion of grains or other clusters with compaction of the fluffy aggregates until reaching comet sizes, keeping during the growth densities of some tenths g cm^{-3} (Donn 1990, Weidenschilling 1997); and (3) catastrophic collisions with the subsequent re-accumulation of most or part of the material, forming rubble piles. If the bodies are too small, a catastrophic collision will inevitably disperse the fragments because their gravity fields are too weak to keep the fragments gravitationally bound. Yet, for bodies with radii larger than $R \sim 10$ km the specific gravitational potential energy $e_G \sim 3GM/5R$ is similar or larger than the specific fragmentation energy S, so reaccumulation becomes an important process following catastrophic collisions (Fig. 10.10).

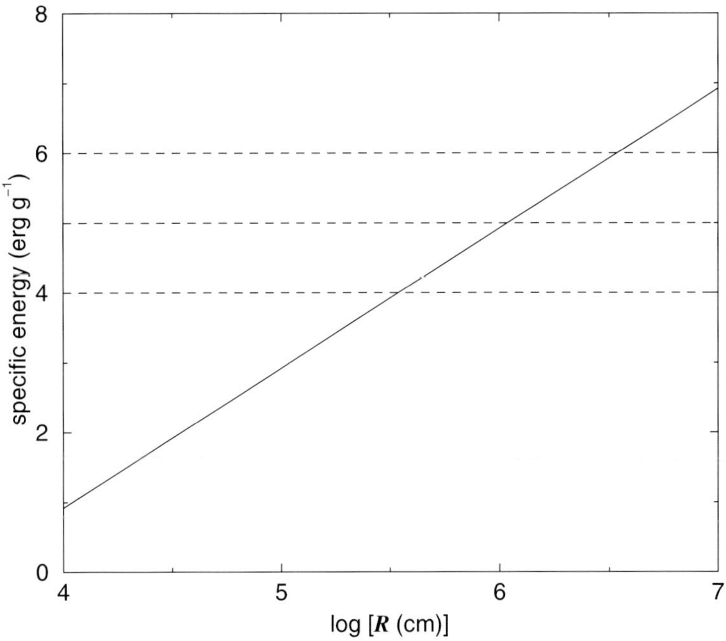

Figure 10.10. The specific gravitational potential energy as a function of size. The dashed horizontal lines are for three fragmentation energies considered to be representative of the cometary material (cf. Section 9.9).

Bearing in mind that only a fraction $f_{KE} \sim 0.01 - 0.1$ of the collisional kinetic energy is partitioned into ejecta speeds (Giblin et al. 2004), the condition for disruption becomes

$$e_G \sim S f_{EK},$$

which leads to

$$R_{dis} \sim \left(\frac{0.4 S f_{KE}}{\pi G \rho} \right)^{1/2}, \qquad (10.37)$$

where ρ is the density of the target. If we adopt $S = 2 \times 10^5$ erg g^{-1} that has been derived for porous ice targets (Giblin et al. 2004) and $\rho = 0.5$ g cm^{-3}, we obtain $R_{dis} \simeq 0.9 - 2.8$ km for $f_{KE} = 0.01 - 0.1$, respectively. Therefore, for bodies with radii $R < R_{dis}$, typical of most observed comets, catastrophic collisions would lead to the dispersion of the fragments without significant reaccumulation. Conversely, for large comets the most likely outcome would be reaccumulation into rubble piles after catastrophic collisions.

10.10. Chemistry of icy planetesimals

Comets are assemblages of minerals and ices whose chemical composition should reflect the conditions of pressure and temperature of the medium in which they formed. In our standard model for the protoplanetary disk we assumed $T(r) \propto T_o r^{-1}$ (r in AU) (cf. Section 10.3), so if $T_o = 700$ K at $r = 1$ AU, $T = 135$ K at Jupiter's distance, $T \sim 28$ K in the Uranus-Neptune zone, and $T \sim 17$ K in the trans-neptunian belt. This range of temperatures would allow the condensation of H_2O at Jupiter's zone (the "snowline"), and other more volatile substances like NH_3, CO_2, CO, NH_2 and N_2 in the outer planets zone and the trans-neptunian belt. Yet, in no place of the solar nebula temperatures were low enough to allow the condensation of H_2 and He.

Cometary nuclei probably contain interstellar material that has been partially processed in the nebula and partially diluted with nebular condensates (Irvine et al. 2000). The sublimation of much of the grain's icy component during entry into the solar nebula, by solar heating when the grains were still high above the midplane, and by friction during their infall, should have depleted the grains that fell inward of 10 - 20 AU from the center of the nebula of the volatiles inherited from the molecular cloud (Lunine et al. 2000). However, some observations suggests that part of the interstellar material was preserved in the interior of comets. From high-spatial resolution, millimeter-wave observations of Hale-Bopp, Blake et al. (1999) found that the HDO/H_2O ratio in the inner coma (and thus presumably similar to the ratio of the nucleus material) is compatible with proportions $\geq$ 15-40% of largely unprocessed interstellar material.

It is often assumed that the phase of H_2O ice, formed by deposition of water vapor onto substrates at low temperature, is determined solely by the temperature, with amorphous ice forming below 130 K, crystalline cubic ice for $130 < T < 170$ K, and exagonal ice for $T > 170$ K. However, both theoretical and experimental results indicate that the kinetics of the deposition and crystallization may control the crystallinity of ices in astrophysical situations. The phase of deposited ice can be affected by the morphology (microstructure) of the underlying substrate.

As spectroscopic studies show, comets contain highly volatile ices like CO. From molecular abundances measured in interstellar clouds, Owen and Bar-Nun (1995) predict ratios CO/$H_2O \approx 0.06$ and N_2/CO$\approx$ 1 in the gas phase. However, these ratios should not necessarily correspond to those found in comets, since the incorporation of these and other volatile species should have implied certain fractionation. One possibility is that they condensed in the gaseous phase in the outer cold regions of the protoplanetary disk, and this condensed material accreted into the planetesimals. However, it is very likely that they were taken by water when it condensed (e.g. Bar-Nun and Kleinfeld 1989).

Gas clathrates constitute an extreme form of gas trapping. Their existence was proposed by Delsemme and Swings (1952) to explain some peculiarities in the vaporization rates of comets. The idea is that some substances, like water, can form crystal structures with cavities large enough to permit occupancy by noble gas atoms or molecules from other species. Yet, water may have condensed as amorphous ice, which acts as a thick woolen carpet with a very large surface area (~ 90 m^2 g^{-1}), able to trap enormous amounts of gas at low temperatures (Bar-Nun and Kleinfeld 1989). The trapping efficiency by amorphous ice of several gases relevant in the primitive nebula: CH_4, CO, N_2, and Ar, is shown in Fig. 10.11. It is very high (a factor of ten in the ratio of gas to ice) for a low temperature of 20 K, but it drops to negligible factors $3 \times 10^{-3} - 6 \times 10^{-5}$, depending upon the gas, at a temperature $T \sim 75$ K. Icy planetesimals formed in Jupiter's region would carry much less nitrogen (most of it in the form of NH_3) and heavy noble gases than icy planetesimals formed in the trans-neptunian belt. Unless the temperature is below 10 K, ice does not trap significant amounts of He and H_2, so these gases could have only been accreted by the Jovian planets in the gaseous form. As shown in Fig. 10.11, at low temperatures ($\sim 20 - 30$ K) the trapping efficiency is similar for all the considered gases: CH_4, CO, N_2, and Ar. Bodies formed in the trans-neptunian belt would have large quantities of these gases since the estimated temperatures there were about or below this range. Yet, the trapping efficiency is strongly dependent on the considered gaseous species for temperatures $T \gtrsim 30$ K. From the previous results, Bar-Nun and Kleinfeld (1989) found that comet Halley had to be formed at ~ 50 K in order to trap the observed proportion of $\sim 7\%$ CO in amorphous ice.

Low-temperature trapping of gases on ices may be a possible source of rare gas enrichment, able to explain the anomalies observed in the atmospheres of the terrestrial planets, if comets were indeed a major source (see next chapter). Neon is not trapped in ice for $T \geq 25$ K, so despite its greater cosmic abundance (as compared to the other

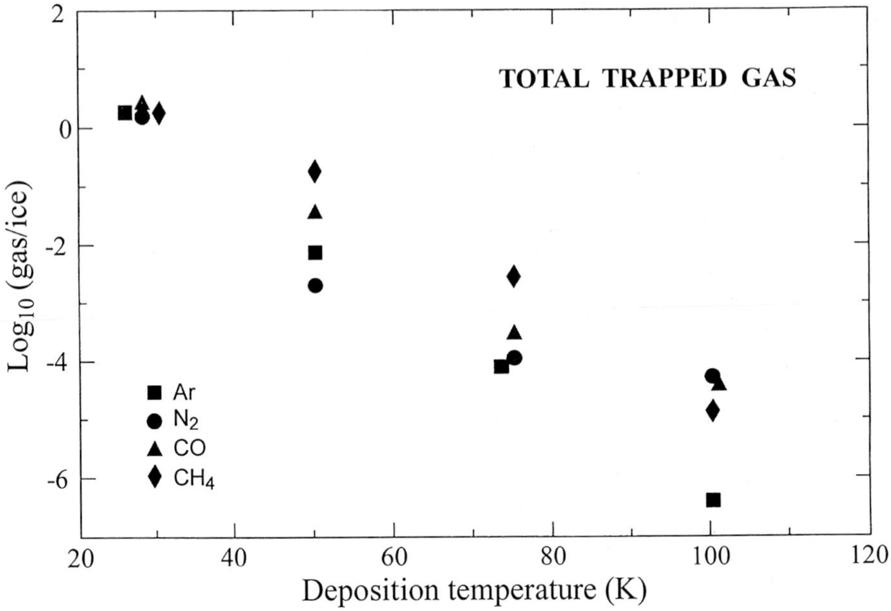

Figure 10.11. The gas to ice mass ratio of four species of gases trapped in amorphous ice. The starting mixture has equal abundances of each gas (Owen and Bar-Nun 1995).

noble gases), comets should be deficient in Ne if they formed in the planetary region. On the other hand, comets formed in the trans-neptunian belt might be neon-enriched. The detection of the noble gases He, Ne, Ar and Kr in comets has been challenging because all of them have resonance transitions in the far- and extreme-ultraviolet. Stern et al. (2000) were able to detect the 1048 Å and 1066 Å argon features in the coma of comet C/1995 O1 (Hale-Bopp) by means of an extreme-UV spectrometer on board of a sounding rocket. They found a high abundance of argon, which they express as a function of the oxygen abundance: $[Ar/O]_{HB} = (1.8{\pm}0.96)[Ar/O]_{cosmic}$, i.e. Hale-Bopp appears to be enriched in Ar as compared to cosmic abundances. The authors conclude that the ices in Hale-Bopp never reached temperatures above 35-40 K, otherwise the Ar would have been lost. As shown in Fig. 10.11, the trapping efficiency of Ar by amorphous ice drops by more than two orders of magnitude, when we pass from $T \sim 25$ K to $T \sim 50$ K. From the failure to detect the 630 Å line of Ne, Krasnopolsky et al. (1997) found that Hale-Bopp has to be depleted in neon by more than 25 times relative to cosmogonic proportions, indicating that the ice was warmed above 25 K, since, as mentioned above, Ne is not trapped in ice for $T \geq 25$ K. Therefore, the noble gas content of Hale-Bopp gives a possible narrow range of temperatures between $\sim 25 - 40$ K in which the water vapor could have condensed.

The ortho/para ratio (OPR) of cometary water is characterized by the spin temperature of this species. This parameter is believed to be of primordial character because conversions between the *ortho* (hydrogen nuclear spins parallel) and *para* (spins anti-parallel) states through collisions or radiative transitions are strictly forbidden. The real meaning of the spin temperature is not understood, however. It could be the temperature of water at the moment of its chemical formation, or it might reflect re-equilibrium with the internal temperature of the nucleus. The low spin temperatures that characterize this ratio in several comets, 25 - 35 K, are usually considered to represent the temperature of formation of the cometary ices. However, it should be cautioned that there exists relatively little data on ortho/para conversion for the different processes affecting comet water, so that it is unclear how the OPR will be affected by the condensation of water vapor onto grains, by storage in the Oort cloud or trans-neptunian belt for 4.6 Gyr, by the warming of the ices prior to and during sublimation, and by possible processing in the coma (Mumma et al. 1993). The OPR of NH_2 was measured by Kawakita et al. (2001) in comet C/1999 S4 (LINEAR). By assuming that NH_2 arises from the photodissociation of ammonia (NH_3) and that the OPR in ammonia was unchanged in the nucleus, the authors found an OPR of 1.17 ± 0.04 which corresponds to a spin temperature of 28 ± 2 K, compatible with the range of spin temperatures found for H_2O.

Summing up, several lines of chemical evidence suggest that the cometary water ice condensed in a region of the nebula with temperatures in the range 25 - 40 K. This would correspond to the outer zone of the protoplanetary disk beyond Saturn. The uncertainties are still too large, either with the temperatures at which the comet ices condensed, and with the temperature profile in the protoplanetary disk, to assess whether the observed comets formed in the zone of the outer planets or in the trans-neptunian belt, or perhaps throughout a wide range of heliocentric distances.

10.11. The collisional history of comets

Comets have been subject to mutual collisions in all the environments they are assumed to have formed. Stern and Weissman (2001) find that kilometer-sized objects formed in the region of the Jovian planets (5-30 AU) have erosional lifetimes (due to mutual collisions and collisions with small debris) much shorter than the time scales for dynamical ejection under a wide range of possible cosmogonical scenarios. Thus, it is very unlikely that comet-sized planetesimals could find their way to the Oort cloud unscathed, and possibly most of them were ground down to dust particles. If this was the case, most of the comets we observe today would be the fragments of larger parent bodies with collisional lifetimes longer than the scattering time scale to the Oort cloud.

We can roughly compute the collisional time scale in the proto-planetary disk by assuming that all the residual mass left after the formation of the Jovian planets was in comet-sized planetesimals of one-km radius, and that the total residual mass was about the same as the solid mass incorporated into the Jovian planets. The number of collisions n_c experienced by one of the planetesimals of cross-section $\sigma \simeq \pi R^2$, encounter velocity at infinity $u \simeq (e^2 + i^2)^{1/2} v_c$, during a time span τ is

$$n_c = \sigma u \frac{\Sigma_p}{\bar{m} H_p} \tau, \qquad (10.38)$$

where e and i are the typical eccentricity and inclination of a planetesimal, v_c the heliocentric circular velocity at a distance r, Σ_p is the surface density of the residual solid matter, $H_p \sim i \times r$ (i is in radians) the scale height of the residual planetesimals, and $\bar{m} = 4/3\pi R^3 \rho_p$ is the average mass.

We can adopt $\Sigma_p = \Sigma_o (r/r_o)^{-3/2}$, where the normalization constants are $\Sigma_o = 50$ g cm^{-2} at $r_o = 1.5 \times 10^{13}$ cm ($= 1$ AU). With these constants we obtain $\Sigma_p \simeq 0.3$ g cm^{-2} at Neptune's distance, which is roughly the value obtained by smearing out the masses of Uranus and Neptune in their accretion zones (Greenberg et al. 1984). Furthermore, we can adopt $\rho_p = 0.5$ g cm^{-3} for the mass density of planetesimals.

With these numerical values and setting the condition $n_c = 1$ we obtain the collision time scale

$$\tau_{coll} = \frac{ir\bar{m}}{\sigma(e^2 + i^2)^{1/2}(\mu/r)^{1/2}\Sigma_o(r/r_o)^{-3/2}} \simeq 150r^3 \text{ yr}, \qquad (10.39)$$

where r is expressed in AU.

Equation (10.39) is valid for the region of the Jovian planets where water condensed. From this equation we derive collisional time scales of $\sim 2.1 \times 10^4$ yr for the Jupiter's region, and $\sim 4 \times 10^6$ yr for the Neptune's region, i.e. well below the dynamical time scales for ejection from their respective accretion zones.

As discussed in Chapter 8, bodies in the trans-neptunian belt have also been subject to a collisional erosive process that greatly contributed to reduce the primordial mass by two orders of magnitude at present. As Stern (1996) shows by numerically modelling the accretion-erosion process in the trans-neptunian belt, most comets coming from this region may not be structurally primordial, but the products of a collisional cascade. From numerical modelling, Farinella and Davis (1996) also found that most TNOs with diameters $D \lesssim 20$ km suffered catastrophic collisions during their lifetime and that bodies with diameters smaller than a few km are very likely collisionally-derived fragments from larger parent bodies. Therefore, the original icy planetesimals (or cometesimals) that formed in the protoplanetary disk could have probably been subject to collisional disruption and reaccumulation of fragments that formed rubble-piles, the latter more likely for larger comets with radii greater than several kilometers as discussed in Section 10.9. In this scenario, comets will correspond to fragments or rubble piles from the collisional disruption of parent cometesimals.

10.12. How well preserved is the comet material?

What is the relationship between icy planetesimals or *cometesimals* and the comets we observe at present? Even when comets are assumed to be the most pristine bodies in the solar system, we should not consider them as cometesimals preserved unaltered until now. As we saw in the previous section, at early times they were subject to frequent mutual collisions when their number density in the planetary region was very high. It is quite possible that such bodies, or the fragments of larger parent bodies, have suffered physical changes during the solar system lifetime that altered the pristine material. Nevertheless, the elemental abundances of comets are the closest to the solar (and cosmic) abundances, though some fractionation occurred, the most obvious one has to do with the depletion of hydrogen and helium.

As discussed in Section 10.9, the primordial fluffy material produced by the accretion of grains or grain aggregates could have been thermally processed and substantially compacted in the interior of large bodies by self gravity. Therefore, the so-called "cometary material" (rich in ices) may have different physical structures according to the size of the parent body from which they come. Yet, most of the cometary material seems to be very fragile and of low-density (see discussion in Sections 9.4 and 9.9). As described before, grains accreting at low encounter velocities will form low-density aggregates of $\lesssim 0.5$ g cm^{-3}, analogous to wind-blown snow. For the Draconids, Donn and Rahe (1982) estimated densities of about 0.01 g cm^{-3}. This is indeed an extremely porous material, presumably more porous than the original material in the interior of the nucleus because of the sublimation of the ices. Nevertheless, just by simply adding the icy component to the refractory residual material of the Draconids, we should expect densities of the primordial material not higher than about 0.1 g cm^{-3}.

The previous discussion suggests that most of the cometary material seems not to have been "hardened" by high pressures and baking in warm interiors of large parent bodies. Let us bear in mind that the central pressure of a body of density ρ (assumed to be uniform) and radius R is

$$P_c = \frac{2\pi}{3} G \rho^2 R^2, \qquad (10.40)$$

namely, the central pressure increases with the squared radius of the body. One-kilometer radius cometesimals of density $\rho = 0.5$ g cm^{-3} would reach $P_c = 3.5 \times 10^2$ dyn cm^{-2}, i.e. too low to alter the fluffy structure of the material. Therefore, if comets come from cometesimals of similar size, we should expect that such material would have been well preserved in its pristine low-density, low-strength state. As said, some or most comets may be the products of collisional cascades from large parent bodies. Such bodies may have suffered different degrees of compaction of the material. From eq. (10.40) we see that bodies with sizes in the range 10-100 km reach central pressures in the range $3.5 \times 10^4 - 3.5 \times 10^6$ dyn cm^{-2}, i.e. large enough to crush and compact the material. In this regard we should expect to find striking variations in the physical structure, and perhaps in the geochemistry, even for comets formed in the same region of the protoplanetary nebula, ranging from hardened material to loose aggregates. As discussed in the previous section, most of the comets may have suffered catastrophic collisions with reaccumulation of part of the debris in a rubble-pile, resembling the model proposed by Weissman (1986). So there are several possible paths for the physical evolution of the comet nucleus, which would allow a wide range of mass densities and internal strengths of the nucleus material.

The surface layers of comets suffered also changes caused by their exposure to interstellar and solar UV radiation and to charged particle bombardment. Both agents induce the breakup of chemical bonds and chemical reactions leading to the darkening of the surface by the formation of long-chain hydrocarbons (the 'ultrared' matter discussed in Section 8.12). Also, rare encounters with highly luminous O and supergiant stars and supernovae explosions in the Sun's neighborhood ($r \lesssim 20$ pc) might have heated the surface of comets stored in the Oort cloud or the trans-neptunian belt to temperatures ~ 30 K, or even 50-60 K (Stern 2003). Such heating would have depleted the outer layers of comets (to depths of 5-50 m to which the thermal wave penetrates) of highly volatile species, like N_2, He, Ne, CO, CH_4 and Ar. Therefore, comets preserved intact during the last 4 Gyr would present volatile-depleted surfaces as compared to fragments chipped off larger parent bodies in the recent past. Such fragments would maintain their highly volatile species on or close to their surfaces. In short, comets are Rosetta stones preserving the footprints of the events that shaped our solar system, waiting to be deciphered.

COMETS AND LIFE

Comets have been held responsible either for bringing life to Earth or for destroying it as a consequence of catastrophic impacts. In 1871 the German physiologist Hermann von Helmholtz suggested that "comets and meteors which swarm everywhere through space...may scatter germs wherever a new world has reached the stage in which it is a suitable place for organic beings" (cited by Oró and Lazcano 1997, p. 5). Indeed, the idea that comets were suitable vectors for transporting living things among stars and planets goes back to Isaac Newton and William Herschell among others. Therefore, the interest in comets goes beyond their astrophysical and cosmogonical aspects and also concerns their relevance to the origin and development of life on Earth. Being comets rich in water, other volatiles, and many carbon-bearing molecules, they could have greatly contributed to the development of the early terrestrial oceans and atmosphere, and could have supplied the necessary organic ingredients to build more complex macromolecules that led to the first lifeforms. A more extreme assumption is that life was brought to Earth directly by comets, which either originated in its interior or it formed elsewhere and was later mixed with the material that accreted into comets. On the other hand, collisions of large comets with the Earth might have triggered mass extinctions. In this regard comets are double-edged swords: either they might have helped life to start when the Earth was lifeless, or might have killed it after it started. We will analyze below these relevant topics and finish the chapter - and the book - presenting a brief summary of the comet missions for the next few years and their potential to address these problems.

11.1. The volatile content of the Earth: endogenous or exogenous?

The Earth's outer veneer is rich in water and carbon which were essential ingredients for the development of life. We first have to ask if these materials can be explained from degassing of the rocks that formed the Earth, or we need an exogenous source. The first observation is that the biogenic elements (C, H, O, N) are extremely depleted in the

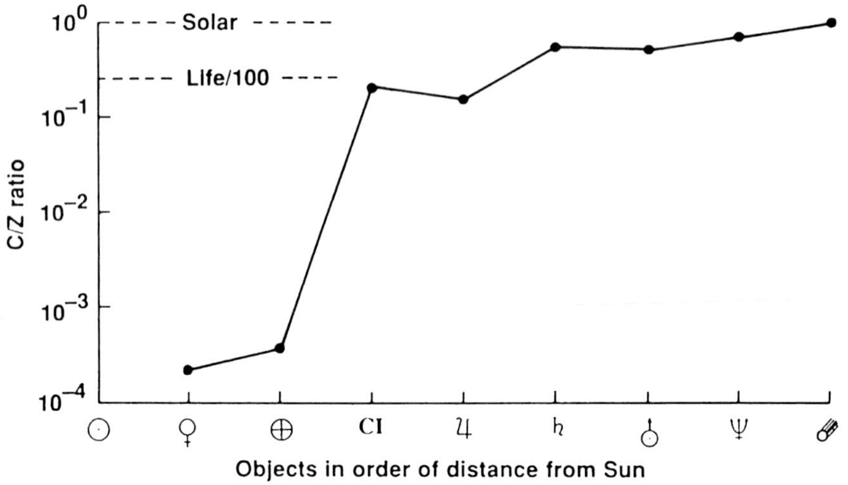

Figure 11.1. Ratio of carbon (C) atoms to all atoms heavier than He (Z) throughout the solar system (McKay 1991).

region of the terrestrial planets, as can be seen in Fig. 11.1 for the case of carbon. The reason is that the temperatures in the inner part of the protoplanetary disk were presumably too high ($\sim 10^3$ K) to allow the condensation of water and carbon-bearing molecules. Therefore, these materials remained in the gas phase and were not incorporated into the terrestrial planets (Delsemme 1997). The Earth formed by the accretion of material of its influence zone on a time scale of $\sim$ 100 Myr. Monte Carlo simulations by Wetherill (1985) showed that the Earth probably suffered the impact of a Mars-sized body in its latest stages of accretion. Such a megaimpact melted the Earth throughout, and part of the material of both the Earth and the projectile vaporized and recondensed in geocentric orbit forming the Moon. Furthermore, this megaimpact and other massive impacts would have removed a primordial terrestrial atmosphere formed by degassing. It is then very likely that the current Earth's atmosphere and the ocean water were acquired once the stage of violent bombardment subsided, presumably from planetesimals not belonging to its own accretion zone but from elsewhere.

It could happen that the Earth accreted substantial quantities of water before the denser Ni-Fe metallic component migrated to the center to form the core about 4.4 Gyr ago. But in that case the water molecules would have been destroyed in contact with the metallic iron via the reaction $H_2O + Fe \rightarrow FeO + H_2$, thus producing iron oxide and liberating molecular hydrogen that escaped dragging with it other gases present at that moment (Chyba et al. 1994).

The scenario depicted above thus suggests a primitive Earth poor in carbon and volatiles. These materials, on the other hand, were abundant beyond the threshold of 2.6 AU from the Sun. We are naturally led to the conclusion that the Earth accreted an outer veneer of volatile-rich material coming from regions beyond ~ 2.6 AU after its core formed about 4.4 Gyr ago. Comets are of course good candidates as a potential source of this material.

11.2. The post-accretion heavy bombardment

As discussed in Sections 10.5-10.7, the current structure of our solar system was the result of a very complex process of accretion of solid bodies and gas that involved the migration of the growing Jovian planets, and the redistribution of solid matter by gravitational scattering within the planetary region. The latter process is of upmost importance for the delivery of volatiles to the terrestrial planets, in particular Earth and Mars (e.g. Wetherill 1975, Chyba 1987, Ip and Fernández 1988). The accretion process of the Jovian planets became at a point highly inefficient because their powerful gravitational fields favored ejection of the interacting planetesimals as compared to collision (cf. Section 10.7). The scattering of bodies by the Jovian planets implied an important mixing of matter throughout the planetary region.

Laboratory analyses of lunar and terrestrial rocks have allowed to reconstruct the early impact cratering record, indicating that the surfaces of the Earth and the Moon, and probably those of the other terrestrial planets, were subject to a heavy bombardment that finished about 3800 Myr ago (e.g. Sleep et al. 1989). This violent period is what we call the post-accretion heavy bombardment. As shown in Fig. 11.2, the bombardment intensity (as measured by the largest impact energy of the colliding bodies) shows a sharp decrease by more than 6 orders of magnitude between the time of planet formation at 4600 Myr and 3800 Myr. In the first $10^7 - 10^8$ yr, the flux of planetesimal impactors could have been $\sim 10^9$ times more intense than the current flux (Hartmann et al. 2000). Figure 11.2 shows the estimated energies and ages (with the respective uncertainties) of the projectiles that formed the lunar basins Imbrium and Orientale, and other large craters. The Earth's cratering record is much sparser; the only significant events for which

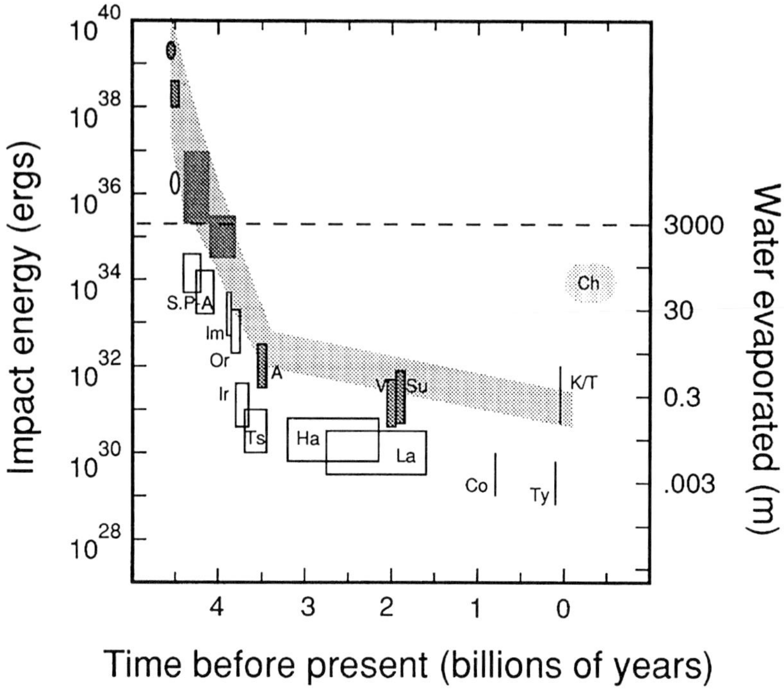

Figure 11.2. The largest impacts on Earth (filled boxes) and Moon (open boxes). Ovals are self energies of formation. The dashed line corresponds to an ocean-vaporizing impact energy. Boxes earlier than Imbrium (Im) are theoretical estimates (Zahnle and Sleep 1997).

we have reliable data are: the body at the Cretaceous-Tertiary (K/T) event, the bodies that formed the craters at Manicougan (M), Sudbury (S) and Vredevort (V), and the body (A) reponsible for the Archaean spherule beds.

The rapid decrease in the impact rate during the post-accretion heavy bombardment followed the depletion of the sources of projectiles of shorter dynamical lifetimes. At the beginnings the high collision rate was due to the sweep up of residual planetesimals in the zone of the terrestrial planets (Wetherill 1985). Once the accretion zones of the terrestrial planets were swept clean, bodies coming from the main asteroid belt, the Uranus-Neptune zone and the trans-neptunian belt were probably responsible for keeping the high flux of impactors. Such reservoirs, containing up to several tens $M_\oplus$ would have kept a high rate of collisions for ~ 800 Myr. On the other hand, the accretion zones of Jupiter and Saturn were rapidly depleted, possibly long before the terrestrial planets completed their formation, so they did not contribute to the stage of post-accretion heavy bombardment. After 3800 Myr, the impact cratering rate still shows a steady decrease but at a much lower rate, indicating the depletion of the reservoirs of projectiles within the planetary region of shorter dynamical lifetimes. The reservoirs with long dynamical lifetimes, basically the asteroid belt, Jupiter's Trojans, trans-neptunian objects and the Oort cloud, have been responsible for the post-3800 Myr impact rate. As Fig. 11.2 shows, the decrease during this later period has been very smooth.

A very interesting finding is that the end of the post-accretion heavy bombardment might roughly coincide with the start of life on Earth. There is firm paleontological evidence that communities of photosynthetic micro-organisms thrived on Earth about 3.5 Myr ago. It is very hard to find older preserved life signatures because older rocks have suffered different degrees of metamorphism. Yet Schidlowski (1988) argued that the oldest sedimentary rocks found on the Earth with an age of about 3800 Myr show a ratio between the carbon isotopes ^{12}C to ^{13}C suggesting a biogenic origin. If this interpretation is correct, it would mean that life on Earth started as soon as the heavy bombardment subsided. There is even the possibility that life had a start one or several times before that, only to succumb after a sterilizing megaimpact. Another alternative is that life started very early and survived in the underground at several kilometers depth, conquering the surface when the stage of heavy bombardment ended.

Oró (1961) already suggested that comets might have been an important source of organic matter to the primitive Earth, though his calculation was based on a simple extrapolation to the past of the observed comet flux at present. Under this assumption, the estimated amount of water supplied by comets turns out to be several orders of magnitude smaller than the Earth's water content, estimated to be $\sim 3 \times 10^{24}$ g, including the water contained in the oceans, atmosphere, crust and mantle (Morbidelli et al. 2000). After Oró's original paper, there have been several works presenting some estimates of the cometary (icy) mass accreted by the primitive Earth that considered the post-accretional massive flux of icy bodies from the zone of the Jovian planets. Some of the results are presented in Table 11.1, which were taken from Oró and Lazcano (1997) table, updated with some recent estimates by Fernández and colleagues. Except for Oró's estimate, the rest should be taken as upper limits, since they did not take into account losses of volatiles due to megaimpacts as explained before. The second column of the table brings the estimated time span of the comet bombardment.

Table 11.1: Cometary matter trapped by the Earth$^{(*)}$

Cometary matter (g)	Time (years)	Reference
$2.0 \times 10^{14-18}$	2.0×10^9	Oró (1961)
$1.0 \times 10^{25-26}$	Late-accretion	Whipple (1976)
3.5×10^{21}	Late-accretion	Sill and Wilkening (1978)
7.0×10^{23}	4.5×10^9	Chang (1979)
2.0×10^{22}	4.5×10^9	Pollack and Yung (1980)
1.0×10^{23}	2.0×10^9	Oró et al. (1980)
$1.0 \times 10^{24-25}$	1.0×10^9	Delsemme (1984, 1991)
$6.0 \times 10^{24-25}$	1.0×10^9	Ip and Fernández (1988)
$1.0 \times 10^{23-26}$	4.5×10^9	Chyba et al. (1990)
$3.0 \times 10^{24-25}$	a few 10^8	Fernández and Ip (1997)
$4.5 \times 10^{24-25}$	a few 10^7	Brunini and Fernández (1999)

(*) the quoted values prior to 1997 are from Oró and Lazcano (1997) table.

As Table 11.1 shows, about 10-20 times the amount of water contained in the oceans could have reached the Earth via collisions of icy bodies from the outer planetary region, although what fraction of this material might have effectively been retained is still quite uncertain. The answer depends on whether the Earth and the other terrestrial planets were already formed when the influx of planetesimals from the outer planetary region reached its peak. If the terrestrial planets were still in the process of accretion, most of the accreted volatile material might have been lost in subsequent megaimpacts (Fernández and Ip 1997). Therefore, the timing of the formation of the outer planets and the terrestrial planets becomes a relevant issue as regards to the volatile contribution of icy planetesimals to the latter, for a premature formation and scattering of bodies by Uranus and Neptune would have met the terrestrial planets still in a violent environment, subject to lose any accreted volatile material by later megaimpacts.

11.3. Comet contribution to the atmospheres of the terrestrial planets

The acquisition of atmospheres by the terrestrial planets was a key factor in the later development of life on Earth, and perhaps on early Mars.

The noble gases proportions offer important clues to learn about the atmospheric origin because such gases are chemically inert and, apart from helium, they do not readily escape from planetary atmospheres. Their low abundance in air, as compared to the solar abundances, implies that they could not have been accreted directly from the solar nebula. Had the Earth's atmosphere been primordial (and thus with solar abundances), the neon content would have been roughly similar to the carbon content, which is in stark contrast with the strong neon depletion. Hence it comes the idea that the Earth's atmosphere (and presumably those of the other terrestrial planets) was "secondary", which means that it was accreted from materials depleted in noble gases after the solar nebula dissipated. If the Earth ever had a primary atmosphere of nebular material of solar composition, either it was extremely thin or very efficiently lost.

It was believed before that the late gaseous veneer of the terrestrial planets was provided by carbonaceous meteorites, based on the fact that the relative abundances of the nonradiogenic isotopes of the noble gases neon (^{20}Ne), argon (^{36}Ar), and krypton (^{84}Kr) are virtually identical in the Earth's atmosphere and in that type of meteorites (Fig. 11.3). Carbonaceous meteorites come from the outer part of the asteroid belt and are rich in carbon and water. Yet, the meteoritic hypothesis for the secondary atmospheres of the terrestrial planets poses serious difficulties. For instance, it was found that the xenon (^{130}Xe) abundance in the Earth's atmosphere is only about 4% of that expected if it would have been supplied by meteorites. An underabundance of xenon was also found in Mars and Venus (see Fig. 11.3). It was suggested that the terrestrial xenon could have been adsorbed inside crustal rocks, but no buried xenon has been found despite intense searches, so it seems very unlikely that the Earth's crust can harbor a significant part of the missing atmospheric xenon (Zahnle 1993).

In Venus neon and argon are much more abundant than in Earth. Again, we find here a very different pattern from that of meteorites, so an extra source of volatiles seems to be required. Owen and Bar-Nun (1995) argue that comets were the extra source, in particular those formed in the Jupiter-Saturn region. They suggest that these comets were poor in C, N and noble gases which were not efficiently trapped in amorphous ice given the high temperatures of the medium (cf. Fig. 10.11). There was also some contribution from comets of the Uranus-Neptune region, much richer in C, N and noble gases. Therefore, the Owen-Bar Nun model explains the different propor-

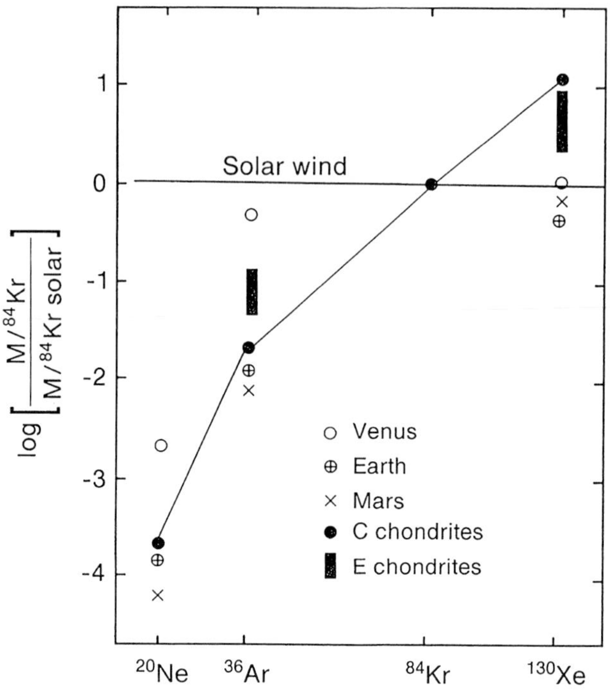

Figure 11.3. The abundance ratio of a given nonradiogenic noble gas to that of ^{84}Kr in the atmospheres of the terrestrial planets and in volatile-rich meteorites, normalized to the respective ratios for the solar wind as indicative of the solar abundances (Hunten et al. 1988).

tions of noble gases as due to the varying contributions of the different sources: meteorites, Jupiter-Saturn comets, and Uranus-Neptune (or trans-neptunian) comets. From this model, Venus's atmosphere, much richer in noble gases, could have received a greater proportion of Uranus-Neptune or trans-neptunian cometary matter, perhaps as a matter of chance (e.g. collision of a large comet from the outer solar system with Venus.)

The isotopic composition gives also important clues about the early evolution of the planetary atmospheres. For instance, the isotopic ratio ^{15}N/^{14}N is greater in the martian atmosphere, as compared to that observed in Earth. This enrichment has been interpreted as due to selective thermal escape of the lighter ^{14}N from the martian atmosphere. From this observation only, Owen and Bar-Nun (1995) estimate that the Mars's atmosphere should have been at least 10 times denser than the current one. Furthermore, there are observations that required a strong impact erosion of the martian atmosphere during the heavy bombardment: the nucleogenic isotopes of xenon and argon are enhanced with respect to the nonradiogenic ones. The episodic erosion would have allowed the preferential buildup of the radiogenic species. It is then possible that the primitive martian atmosphere was ~ 1000 times denser than the current one (Owen and Bar-Nun 1995), thus allowing the presence of liquid water on its surface.

11.4. Comet contribution to the impact rate for the last 3800 Myr

As Fig. 11.2 shows, the rate of impacts of interplanetary boulders with the Earth and the other terrestrial planets has shown on average a very smooth decline over the last 3800 Myr. Therefore, the current impact rate gives a good idea of the collisional history of the terrestrial planets during this long period, and what was the contribution of the different sources with long dynamical time scales: main-belt asteroids, Jupiter's Trojans, Oort cloud comets, and trans-neptunian objects. It is also worthwhile to stress that this period roughly coincides with that in which life took a hold on the Earth, so collisions, previously beneficial for the development of life, became a threat for the already existing lifeforms.

Since comets were the first celestial bodies whose orbits were found to cross that of the Earth, it became clear, at least since Edmond Halley computed the orbits of a few well observed comets, that some of them could be potential colliders with our planet. But only in recent times we have been able to get good statistics of comet passages and sizes and, hence, to assess the damage caused by the impact of one of these bodies. At the typical impact velocities of comets with the Earth (between $\sim 20 - 60$ km s^{-1}), an one-km-sized interplanetary body will have enough energy to cause global environmental damage because of devastating wildfires, tsunamis, changes in the atmospheric and oceanic chemistry and the injection of sun-darkening submicrometer dust particles in the stratosphere (e.g. Chapman and Morrison 1994). For a given projectile mass, the energy delivered will depend on the impact velocity that is on average greater for long-period comets (~ 55 km s^{-1}) than for JF comets or Earth-crossing asteroids (~ 20 km s^{-1}). On the other hand, for a given size comets contain less mass than asteroids because the former have less density, so this partly compensates for the different impact velocities. Hence, in the following we will standardize our results to an one-km diameter body - whether a comet or an asteroid - considering this size as the threshold for causing global damage.

The rate of passages of LP comets in Earth-crossing orbits brighter than a total absolute magnitude $H_{10} = 10.5$ ($R_N \gtrsim 0.5$ km) is ~ 7 yr^{-1} (cf. Section 2.7). If their collision probability with the Earth is $p_{LP} = 9.5 \times 10^{-10}$ per orbital revolution (cf. Section 9.8), the frequency of collisions is

$$f_{LP} \simeq 7 \times 9.5 \times 10^{-10} \simeq 6.7 \times 10^{-9} \text{ yr}^{-1}$$

i.e. one collision every $\approx 1.5 \times 10^8$ yr.

For JF comets we have about 15 comets in Earth-crossing orbits with absolute nuclear magnitude $H_N < 18.5$ ($R_N > 0.7$ km) (cf. Section 7.8). If we extrapolate this result down to $R_N = 0.5$ km by means of the size distribution of eq. (7.14), we obtain ~ 40 JF comets with radii > 0.5 km. An increase by a factor of two in the size of the population of JF comets seems to agree with the current discovery rate of near-Earth JF comets thanks fundamentally to dedicated search programs like LINEAR and LONEOS. The average collision probability per JF comet was found to be $p_{JF} = 1.3 \times 10^{-9}$ yr^{-1} (Section 9.8), so the probability of collision for the whole sample is

$$f_{JF} \simeq 40 \times 1.3 \times 10^{-9} = 5.2 \times 10^{-8} \text{ yr}^{-1}$$

i.e. a frequency of collisions of one JF comet of radius $R_N > 0.5$ km every 1.92×10^7 yr.

For Halley-type comets Levison et al. (2002) estimate a number of 84 active comets with $q < 1.3$ AU down to $D \simeq 1$ km, from which about 50 may be Earth-crossers. By comparing orbital distribution models with the observed population of HT comets and asteroids in HT orbits (assumed to be dormant HT comets), the authors estimate a ratio of four dormant HT comets for every active one. But some of the inactive bodies in HT orbits may well be *bona fide* asteroids from the main asteroid belt, so Levison et al.'s estimate should be taken as an upper limit. Therefore, by allowing that only a fraction of the inactive bodies in HT orbits are dormant HT comets, we can estimate a population of about 100 HT comets in Earth-crossing orbits. The HT population has a non-random distribution of orbital planes with a median inclination of 55° (Levison et al. 2004). Furthermore, if we adopt an average perihelion distance $\bar{q} \simeq 0.7$ AU for Earth-crossing

HT comets, and assume that $q << a$, we obtain an average encounter velocity at infinity $\bar{u}_{HT} \simeq 38.4$ km s^{-1} and a radial component $u_x \simeq 18$ km s^{-1} (cf. eqs. (9.14), (9.16)). By introducing these numerical values in eq. (9.15) we get a collision probability $p'_{HT} \simeq 1.64 \times 10^{-9}$ per orbital revolution. If we adopt an average orbital period of 70 yr for HT comets, the collision probability becomes $p_{HT} = p'_{HT}/70 \simeq 2.3 \times 10^{-11}$ yr^{-1}. Consequently, the probability of collision for the whole sample is

$$f_{HT} \simeq 100 \times 2.3 \times 10^{-11} = 2.3 \times 10^{-9} \text{ yr}^{-1}$$

The frequency of collisions of HT comets with the Earth is only about 35% of that of LP comets.

We have a combined collision frequency of one comet per 1.64×10^7 yr. If we assume that the average mass is 10^{16} g, the total mass supplied to the Earth during its lifetime is $\sim 2.8 \times 10^{18}$ g. This amount of mass is too little to account for the mass of the oceans (1.4×10^{24} g), so its interest does not rest on the matter supplied, but on the danger such collisions pose to life on Earth. A few of the colliding comets might have been 10 km-size or larger, and their collisions might have led to mass extinctions. This is the edge of the sword that really matters when collisions of comets in the last 4 Gyr are considered.

We can compare the collision rate of comets with that of Earth-crossing asteroids (ECAs). Bottke et al. (2002) estimate a population of 960 ± 120 near-Earth asteroids (NEAs) brighter than $H = 18$ ($D > 1$ km) of which 68% are Earth-crossers. If we assume that the collision probability per ECA is the same as for JF comets multiplied by the factor $\sim 7/4$ (that takes into account the on average smaller orbital period of ECAs as compared to that of JF comets), we find a collision probability for the whole sample of

$$f_{ECA} \simeq 653 \times 1.3 \times 10^{-9} \times 7/4 = 1.5 \times 10^{-6} \text{ yr}^{-1}$$

i.e. one collision every 6.7×10^5 yr. This result agrees quite well with some other estimates (e.g. Stuart 2001) that take into account corrections for detection bias in the observed sample of NEAs.

We see that comets make up only about 4% of all the collisions with the Earth for objects greater than one km in size. The rest corresponds to asteroids. Yet, the share of comets in the impact rate may be larger if we take into consideration the following two aspects: (1) From time to time comet showers, triggered by close star passages or encounters with GMCs (cf. Section 6.5) may greatly increase the comet impact rate. As discussed in Section 6.6, the lack of clusterings in the crater ages caused by comet showers set an upper limit on the highest comet flux at about 3000 times the current flux of LP comets. If we average this flux over the mean interval of $\sim$ 30 Myr between comet showers, we get a time-average comet flux $\sim$ 100 times the current flux of LP comets. This is still not a big deal as regards to the amount of water contributed to the Earth's oceans, but it could rise the comet contribution to the time-average impact rate with the Earth to up to $\sim$ 30% of the total. (2) A fraction of the ECAs might be dormant or difunct comets, which again should increase the comet contribution to the impact rate, and lower that from asteroids. Bottke et al. (2002) estimate that $\sim$ 6% of the ECAs are extinct comets, while Rickman et al. (2001) suggests that the population of extinct or dormant comets might be twice as large as the active one. It is then possible that under certain extreme combinations of effects (1) and (2), comets might become the dominant source of projectiles in the Earth's neighborhood, though in general we should expect that comets constitute a minor fraction ($\lesssim$10%) of the collisional population. Table 11.2 summarizes the main results presented in this section. The impact velocities are given by $v_i = (u^2 + v_{esc}^2)^{1/2}$, where u is the encounter velocity at infinity and $v_{esc} \simeq 11.2$ km s^{-1} is the Earth's escape velocity.

Table 11.2: Impact rate (Number of bodies with $D > 1$ km per 100 Myr)

Object	Impact velocity (km s^{-1})	Impact rate
LP comets	56	0.67
JF comets	18	5.2
HT comets	40	0.23
Shower comets	56	$\lesssim$ 70
ECAs	18	150

We can compare our previous results with those obtained from other authors. Bailey (1991) found that the comet contribution to the terrestrial cratering rate is $\lesssim$10% at all diameters. Weissman (1990) estimates that long- and short-period comets provides about 12% of the cratering flux, and an extra 17% is provided by comet showers, while the rest is due to asteroids. Shoemaker et al. (1994) estimate that active and extinct periodic comets may account for about 20% of the production of terrestrial impact craters larger than 20 km diameter. All these results do not differ significantly from one another, and neither with those presented here. Given the large uncertainties in the sizes of the different populations, as well as the physical fate of comets (disintegration versus deactivation), our results cannot be more accurate than within an order of magnitude.

Even if the asteroids dominate the impact rate with the Earth, the situation may be different among large projectiles, say greater than several km diameter (Shoemaker et al. 1990). This is clear to understand by noting that there have been cases of giant comets observed in historical times. Furthermore, giant comets observed at present, like 29P/Schwassman-Wachmann 1 or Chiron, might evolve to the inner planetary region and become potential colliders. Comet 1P/Halley is an Earth-crosser with an effective diameter $D \sim 10$ km, so it is large enough to cause major changes in our climate and disruptions of our ecosystem if it collided with the Earth. Comet 28P/Neujmin 1 that comes rather close to Earth ($q = 1.55$ AU) is another giant comet of $D \sim 19$ km. On the other hand, collisions among main-belt asteroids inject in NEO orbits fragments that very seldom exceed a few km diameter, so it is extreme unlikely to produce giant asteroid fragments as potential colliders with the Earth as happens with comets. The largest Earth-crossing asteroids are 3200 Phaeton ($D \sim 6.9$ km), 1580 Betulia ($D \simeq 7.4$ km), and 1627 Ivar ($D \simeq 8.1$ km), all of them smaller than 1P/Halley. The Asteroid/Comet impact ratios shown in Table 11.3 are derived from our previous results assuming that 15% of the projectiles of sizes smaller than a few km are comets and the rest asteroids. The extrapolations to larger diameters ($D > 5$ km) have been derived from Shoemaker at al. (1990).

Table 11.3: Asteroid/Comet impact ratio for different sizes

Diameter:	$D > 1$ km	$D > 5$ km	$D > 10$ km	$D > 15$ km
A/C ratio:	6.7	2.0	0.19	~ 0

Therefore, if impacts indeed triggered some of the biological mass extinctions that punctuated life evolution on Earth, the main suspects should be comets rather than asteroids (Shoemaker et al. 1994). This may be particularly the case of the mass extinction at the Cretaceous/Tertiary (K/T) boundary about 65 Myr ago that led to the demise of the dinosaurs and in general all the species with body masses greater than 25-30 kg, for which Alvarez et al. (1980) suggested an impact of a $\sim$ 10-km-sized asteroid as the root cause. According to Table 11.3, for this size range comets should be prime suspects rather than asteroids.

11.5. The deuterium problem

The answer to the question of how much water and organic material on the Earth comes from cometary bodies of the Uranus-Neptune accretion zone or the trans-neptunian belt is still uncertain. Yet, the study of the composition of the deuterium/hydrogen (D/H) isotopic ratio of water in comets, and how it compares with that of the Earth's oceans, may help to shed light on this issue. For the oceans the ratio is found to be $D/H \simeq 1.56 \times 10^{-4}$, while for the three comets in which it was so far measured it was found: $D/H = (316 \pm 34) \times 10^{-6}$ for 1P/Halley (Eberhardt et al. 1995), $D/H = (330 \pm 80) \times 10^{-6}$ for C/1995 O1 (Hale-Bopp) (Meier et al. 1998), and $D/H = (290 \pm 100) \times 10^{-6}$ for C/1996 B2 (Hyakutake) (Bockelée-Morvan et al. 1998), i.e. about twice the value of the Earth's oceans (Fig. 11.4).

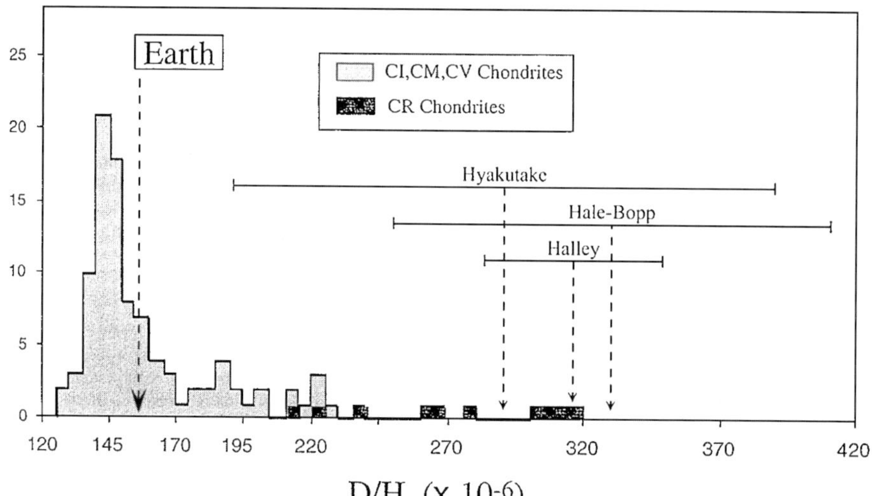

Figure 11.4. Distribution of the D/H isotopic ratio in a sample of carbonaceous meteorites (taken from Robert 2003). The value corresponding to the ocean water, and the values measured in comets 1P/Halley, C/1995 O1 (Hale-Bopp) and C/1996 B2 (Hyakutake) (with the corresponding error bars) are also indicated.

We do not know yet if the three comets mentioned above are representative of the whole comet sample. They presumably come from the Oort cloud, whose population may have originated in the Uranus-Neptune region, the trans-neptunian region, or even a small fraction in the Jupiter-Saturn region. Therefore, all the range of heliocentric distances from $\sim 4-45$ AU might be represented in Oort cloud comets, though it is very likely that predominate those formed in the outer fringes of the planetary or trans-neptunian region. If we finally find that all the comets are enriched in deuterium (with respect to the Earth's oceans), then we have to look for another source of volatiles more suitable for the Earth.

Delsemme (1999) has argued that bodies formed in the Jupiter's zone might have had a depletion of deuterium due to sublimation and recondensation of water in the warmer environment of Jupiter. According to Delsemme, the water vapor in a hot region (temperatures $700-2000$ K) was able to exchange deuterium with nebular hydrogen through the reaction

$$\mathrm{HDO} + \mathrm{H_2} \rightleftharpoons \mathrm{H_2O} + \mathrm{HD}$$

At high temperatures the reaction moves to the right, thus the water gets depleted in deuterium while the hydrogen molecules get enriched in this isotope. When the temperature dropped below the freezing point of water in the solar nebula, the condensed water was depleted in deuterium, decreasing the D/H ratio to $\sim 1.2 \times 10^{-4}$. This is in fairly good agreement with the peak in the (D/H)-distribution of carbonaceous chondrites shown in Fig. 11.4. By contrast, Delsemme argues that water ice never sublimated in the Uranus-Neptune region, thus keeping its primordial D/H $\simeq 3.2 \times 10^{-4}$. Accordingly, a mixture of $\sim 80\%$ of comets from the Jupiter's zone, plus $\sim 20\%$ of comets from the Uranus-Neptune zone might have provided the right D/H ratio to the ocean water. Morbidelli et al. (2000) also argue for an important contribution of deuterium-depleted hydrated carbonaceous bodies formed in the outer asteroid belt, perhaps attaining $\sim 90\%$ of the water currently present in our planet. The remaining $\sim 10\%$ was supplied by comets of the Neptune's region and the trans-neptunian belt.

Laufer et al. (1999) also favor a mixed origin of the water contained in the Earth's ocean, suggesting that it was delivered by both comets and rocky material from the Earth's region itself. Yet, the problem as we saw is the short collisional lifetime of bodies of the Earth's zone and their small water content. Sources of projectiles of short collisional lifetimes would require the sequestration of water in the Earth's interior and its outgassing after the end of the accretion process. This alternate scenario requires further study but, as we discussed before, the megaimpacts that suffered the Earth during its first few hundreds Myr probably led to the loss of endogenic gases and any water accummulated would have been destroyed by reaction with metallic iron before its migration to the core. Furthermore, given the high temperatures in the inner solar nebula where the terrestrial planets formed, the planetesimals formed there were very dry.

Can we find suitable reservoirs around the snowline able to last for several 10^8 yr until the megaimpacts on the terrestrial planets subsided? In principle, there are three potential reservoirs: (1) Jupiter's Trojans; (2) the Jupiter-Saturn region; and (3) the outer edge of the asteroid belt. The Trojan asteroids librate around the Lagrangean points L4 and L5 in Jupiter's orbit. Most Trojans are stable for the age of the solar system, but there is a slow diffusion due to planetary gravitational perturbations and mutual collisions. Once outside the Trojan region, the bodies evolve in orbits like those of Jupiter family comets so they become a potential source of long-lived projectiles. Residual planetesimals in the Jupiter-Saturn region have very short dynamical lifetimes, the region being depopulated on time scales $\sim 10^5$ yr (Gladman and Duncan 1990, Grazier et al. 1999), so it does not seem to be a suitable population to provide icy material to the terrestrial planets region over time scales three orders of magnitude longer. Gladman and Duncan (1990) have also explored the cleaning of the outer edge of the asteroid belt ($3.1 < a < 3.9$ AU) over 12 Myr. They found that 14 out of 80 test particles that were removed (i.e. they became either Mars or Jupiter crossers), are basically those with semimajor axes close to mean motion resonances with Jupiter. Many asteroids in this region (probably with a high carbon and water content) may have dynamical time scales (to become planet crossers) of several 10^8 yr, so they might be an alternative source of water and organic matter, in agreement with Delsemme's and Morbidelli et al.'s studies.

We have still to build a fully consistent picture of the volatile sources of the terrestrial planets. As pointed out by Morbidelli et al. (2000), there is a seemingly inconsistency between the mostly "meteoritic" origin of the ocean water and the mostly "cometary" source for the atmospheres of the terrestrial planets. It could be possible that most of the water was provided by carbonaceous asteroids that survived erosive impacts which, on the other hand, removed early atmospheres, whereas comets from the outer planetary region provided the later veneer of water and part of the atmospheric gases (CO, N_2 and noble gases) that gave origin to the current atmospheres of the terrestrial planets. Undoubtedly, we need to measure the D/H ratio in more comets, in particular those of the Jupiter's family and other bodies of the solar system, like asteroids from the outer belt and Trojans, to advance further in this important issue.

11.6. May comets harbor lifeforms?

The surprinsingly rapid start of life on Earth - in geologic terms - has raised the question about whether life originated from purely endogenic processes, or whether it was favored to a greater or lesser degree by exogenous agents. The most extreme of the latter type of theories is panspermia that assumes that life was created elsewhere in the universe and brought to Earth by some vector, as for instance comets.

Comets are very rich in organic matter (see Table 3.2). Therefore, besides water comets supplied the organic ingredients to form the broth in which more complex polymers and finally microorganisms emerged. Could microorganisms have formed within comet nuclei themselves and seeded the primitive Earth, thus answering the question about why life started so early? Could comets offer a friendly environment for the development of lifeforms? Two distinguished scientists as Fred Hoyle and Chandra Wickramasinghe have argued that bacteria, algae and viruses populate the interstellar medium and comets, based on the similarities between the spectrum of interstellar dust with those produced by micro-organisms (Hoyle and Wickramasinghe 1979). However, the interpretation of the broad spectral features in terms of algae and bacteria is not unique, and other more conventional non-biological sources are as well possible, so we should take their claim with extreme caution.

The origin and early evolution of life on Earth argues in favor of liquid water as a necessary ingredient. There are no known organisms on Earth that can thrive on pure ice or that can extract liquid water from ice using metabolic energy (see, e.g. McKay 1997). Therefore, the search for life in comets - or in other celestial bodies - should follow the search for present or past existence of liquid water. Several authors have considered the possibility that the heat released by short-lived radioactive isotopes, in particular ^{26}Al, would have maintained a liquid water core for a time long enough to allow the development of microorganisms (e.g. Wallis 1980, Prialnik et al. 1987). ^{26}Al is a very powerful heat source of very short lifetime (7.4×10^5 yr). The product of its decay is ^{26}Mg, and an excess of this isotope (that is stable) was found in Ca-Al inclusions in the Allende meteorite. ^{26}Al was probably extant in the protoplanetary disk, probably produced by nearby ordinary novae and/or massive stars present in the early dense galactic environment of the Sun (cf. Section 10.8).

For a ratio ^{26}Al/^{27}Al$\simeq 5\times10^{-5}$ for the cometary dust, Podolak and Prialnik (1997) estimate an average heating rate of $\bar{Q}\sim 2\times10^{-3}$ erg g^{-1} s^{-1} from ^{26}Al decay. Let us consider a spherical cometary nucleus of radius R_N, density ρ, and thermal conductivity K. The energy produced within a volume of radius $r(\leq R_N)$ will be transferred outwards by thermal conduction, so if we neglect other energy sinks, the energy balance equation gives

$$\frac{4}{3}\pi r^3\rho\bar{Q}\simeq 4\pi r^2 K\frac{dT}{dr},$$

which upon integration between the limits $0<r<R_N$ leads to

$$T(r)\simeq T_o+\frac{\bar{Q}\rho}{6K}(R_N^2-r^2). \tag{11.1}$$

For loose snow Wallis (1980) adopts: $K=1.67\times10^4$ erg cm^{-1} s^{-1} K^{-1} and $\rho=0.25$ g cm^{-3}, whereas for solid ice + dust (at ~200 K): $K=2.93\times10^5$ erg cm^{-1} s^{-1} K^{-1} and $\rho=1$ g cm^{-3}. By introducing the appropriate numerical values in eq. (11.1) for loose snow, we can see that interior temperatures in the comet nucleus raise above the melting point of ice in a 10 km-nucleus. We must also check that the interior pressure be above the triple point ($\gtrsim 6\times10^3$ dyn cm^{-2}). From eq. (10.40) we find that a 10-km radius nucleus has a central pressure of $\sim10^4$ dyn cm^{-2}, which is just above that required for keeping liquid water. Wallis (1980) thus concluded that comet nuclei greater than ~ 10 km radius could have had melted cores that lasted for about one Myr, i.e. until the ^{26}Al source was exhausted (Fig. 11.5). These melted cores might have been surrounded by a 1-km thick ice shell as the water vapor generated in the central core leaked through cracks and interstices in the overlying layers and then recondensed into ice. He argues that the central fluid core would have provided a very suitable and protected environment for colonies of bacteria that could have survived the refreezing of the fluid core for several Gyr, so such frozen bacteria might still be apt for triggering life if they meet a suitable environment. Wallis then considers comets as a vector for panspermia.

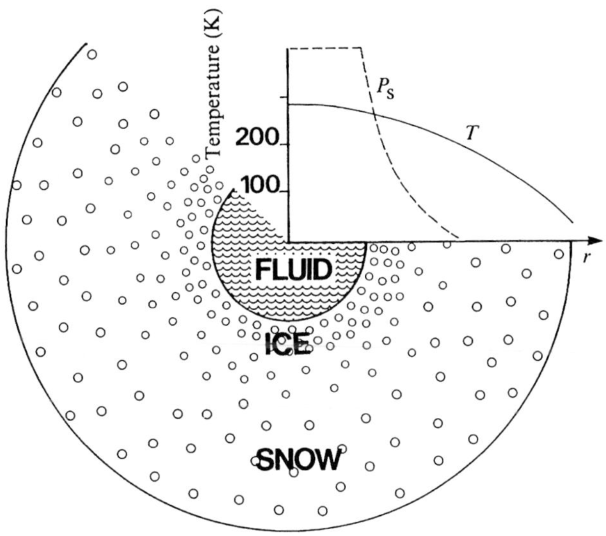

Figure 11.5. Hypothetical liquid core of a comet nucleus. The calculated vapor pressure P_s and temperature T are plotted as a function of the distance r to the center. In the central region P_s is high enough to allow the condensation of the H_2O vapor (Wallis 1980).

More recently, Podolak and Prialnik (1997) have re-evaluated the melting of ice in the cometary interior and its maintainance, reaching a more pessimistic conclusion. They argue that efficient cooling mechanisms, such a heat conduction through the ice and heat carried by the flow of gas through the porous nucleus would have made difficult to reach the above conditions for core melting. Thus, if we allow for phase transition from amorphous to crystalline ice when the core reaches a temperature $T \sim 137$ K, this will result in the release of a latent heat of 9×10^8 erg g^{-1}. This heat is sufficient to transform the adjacent layer. This layer too releases latent heat and crystallizes, and the process continues moving outwards until the subsurface layers. The heating and crystallization process stops in the outer layers because they are efficiently cooled by radiation. Because the thermal conduction of crystalline ice is about one order of magnitude greater than that of amorphous ice, the rate of cooling is much more rapid, so the temperature through the nucleus drops quickly.

There is an additional problem so far overlooked: comets are far from being considered safe heavens that have gone through very peaceful existences. As we analyzed before (cf. Section 10.11), in the early solar system comets were subject to mutual collisions that led to successive fragmentations and re-accummulations of fragments (Stern and Weissman 2001), thus making virtually impossible that liquid cores could have been preserved undisturbed. Giant comets ($\gtrsim$ 100 km) might have been spared from catastrophic collisions, and they may be the best candidates for searching liquid environments in their interiors.

Another possibility, analyzed by McKay (1997), is that the solar nebula could have contained dormant lifeforms that were accreted together with the solid material into the planetesimals. This enters into the panspermia scheme in which life pre-dates the Earth and is ubicuitous in the universe, present in interstellar matter where the Sun and other stars form. In this scenario comets would have been the suitable vectors to carry such lifeforms from the nebular material to the Earth's surface. We do not know enough yet about the survival of dormant microorganisms under hostile environments, with high dose of radiation and in absence of liquid water, to assess the plausibility of this scenario.

11.7. Space missions to comets

Key questions about the chemical and physical nature of cometary nuclei and their relevance to the origin and development of life on Earth will be answered only through space missions. The very successful flotilla to 1P/Halley: the European *Giotto*, the Soviet *Vega 1* and *2*, and the Japanese *Suisei* and *Sakigake*, encouraged the planning of more ambitious missions. NASA's *Deep Space 1* imaged very impressively the nucleus of 19P/Borrely during a flyby in September 2001. Unfortunately, the *Comet Nucleus Tour* (CONTOUR) mission, planned to flyby comets 2P/Encke and 73P/Schwassmann-Wachmann 3, was lost after launch. Despite this setback, there are three new comet missions, each of them with its specific goals and novel instrumentation, which promise to greatly advance in our understanding of these bodies. Here is a summary of the new missions:

11.7.1. Stardust

It is a NASA flyby mission to gather samples of comet dust and return them to Earth. It was launched in February 1999 and swept through the coma of comet 81P/Wild 2 in January 2004 taking during the flyby encounter very impressive images of the nucleus (cf. Fig. 3.3). For collecting the dust particles the spacecraft carries a shield of a lightweight silica-based insulating material called aerogel (density 0.02 g cm^{-3}). When particles hit the aerogel, they drill through the material, gradually slowing down, creating furrows that scientists will use to track the paths of the dust particles. The aerogel has a density low enough to collect the particles without destroying them. Most of the particle's kinetic energy is absorbed in the aerogel, keeping the particles at temperatures low enough (below $\sim 500°$ C) to preserve the organic materials. The long journey to comet Wild 2 started in 1999 because two gravity assist encounters with Earth were used to boost the spacecraft to a more eccentric orbit of encounter with the comet (Fig. 11.6). The spacecraft with its invaluable cargo of cometary dust will return to Earth in January 2006.

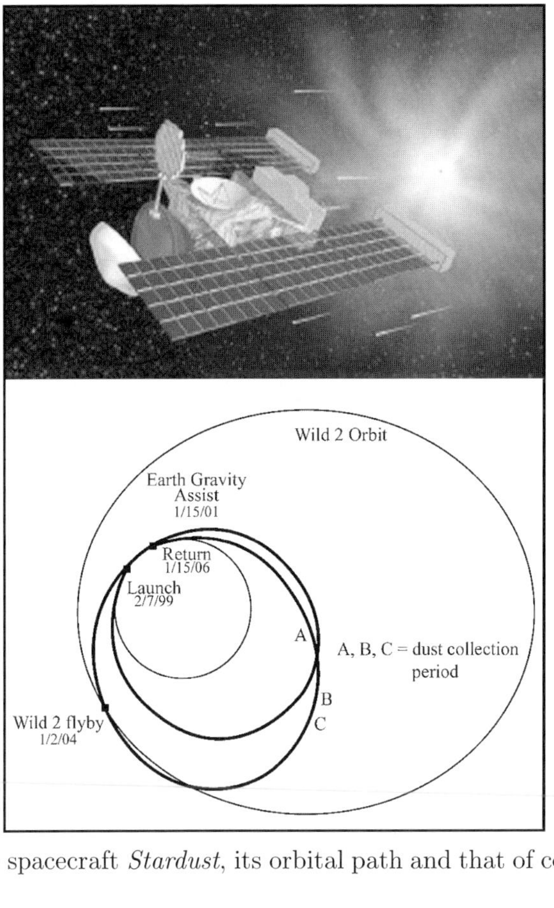

Figure 11.6. The spacecraft *Stardust*, its orbital path and that of comet 81P/Wild 2 2 (NASA).

11.7.2. Deep Impact

The spacecraft will carry a probe to collide with comet 9P/Tempel 1 in a attempt to peer beneath its surface. The spacecraft is expected to arrive in July 2005. The impactor will be equipped with a camera (Fig. 11.7). The spacecraft will travel at a relative velocity of 10 km s^{-1}.

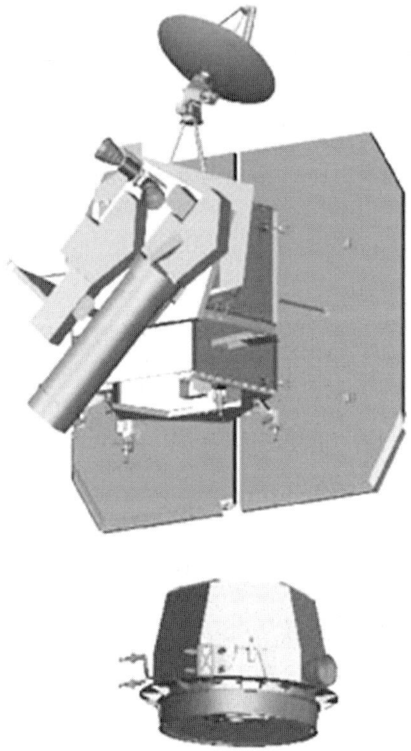

Figure 11.7. Deep Impact: Flyby spacecraft with its 7.5 m^2 solar array and the impactor (bottom) (NASA).

The impactor will be released at a distance of about 864,000 km to hit the $\sim$ 6 km diameter comet nucleus. The energy from the impact will excavate a crater about 100 m wide and 30 m deep. The study of the crater growth rate and final morphology will provide important clues about the physical structure of the outer layers of the comet nucleus. The impactor will hit on the sunlit side so another camera and an infrared spectrometer on the flyby spacecraft, along with ground-based observatories, will study the resulting icy debris and exposed pristine interior material. The impact experiment will be very useful to study the differences between the outer layers and the subsurface of the comet nucleus. If the nucleus contains a dust mantle that rests over an ice-rich subsurface, then the crater will trigger gaseous activity by exposing fresh ices to the Sun's radiation. On the other hand, if the icy material is exhausted in the subsurface, no gaseous activity will be triggered by the impact.

11.7.3. Rosetta

This ESA mission was first planned for a rendezvous with comet 46P/Wirtanen. It was rescheduled for launch in February 2004 to rendezvous comet 67P/Churyumov-Gerasimenko in November 2014. The comet has an effective radius of about 2.1 km and moves on an orbit of orbital period $T = 6.57$ yr, perihelion distance $q = 1.24$ AU, aphelion distance $Q = 5.74$ AU, and inclination $i = 7^{\circ}.12$. When the rendezvous is complete, the spacecraft will proceed with a maneuver to place it in orbit around the comet. The surface will be imaged from distances between 5 and 25 comet nucleus radii. Simultaneously, spectrometers will scan the surface in various spectral ranges down to the infrared range to determine the mineral and chemical composition of the surface material. Rosetta will carry a lander (Fig. 11.8) with a series of complex instruments for the analysis of surface samples.

Figure 11.8. An artist conception shows the *Rosetta* orbiter and the box-shaped lander as it starts its landing maneuvering on the nucleus of comet 67P/Churyumov-Gerasimenko. During the descent maneuver, the lander will unfold three legs for a gentle touch down (ESA).

The *Rosetta* orbiter will search for suitable landing sites over the pockmarked surface. Once a safe landing site is selected, the *Rosetta* lander will separate from the orbiter and slowly descend on the comet's surface. After touchdown, the lander will anchor itself to the surface by means of a harpoon, to prevent it escaping from the comet's extreme weak gravity field, and start probing it. The lander will return close-up pictures of the surface, drill into the dark organic crust and sample the primordial material. The internal structure of the nucleus will also be probed by means of radio signals transmitted from the orbiter to the lander through the nucleus and back again.

By flying alongside the comet for more than one year, the *Rosetta* spacecraft will be able to monitor the dramatic transformation that takes place in the comet as it approaches the Sun. Rosetta's exploration will end in December 2015, six months after the comet passes perihelion. To gain enough energy to place the spacecraft on a rendezvous trajectory with 67P/Churyumov-Gerasimenko, it will require three gravity assist manoevers with the Earth and one with Mars (Fig. 11.9). The spacecraft will take advantage of its passages through the asteroid belt to fly by a couple of asteroids and take close-up images of their surfaces. The tentative targets and flyby dates are: 2867 Steins in September 2008 and 21 Lutetia in July 2010.

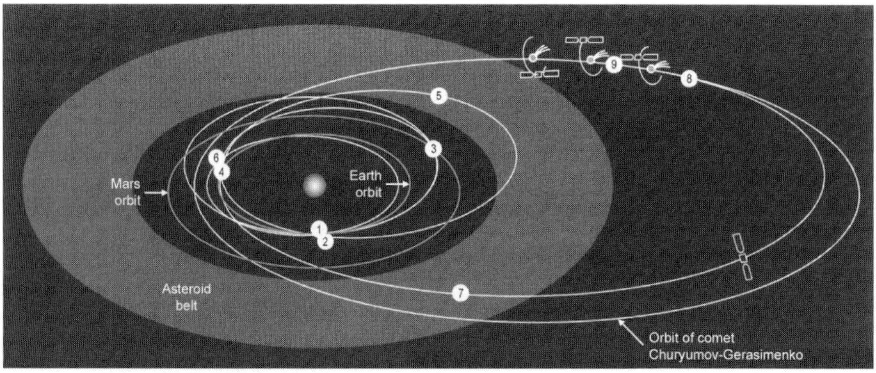

Figure 11.9. The *Rosetta* trajectory. (1) launch (March/2004); (2) first Earth gravity assist (March/2005); (3) Mars gravity assist (March/2007); (4) second Earth gravity assist (November/2007); (5) passage through the asteroid belt; (6) third Earth gravity assist (November/2009); (7) second passage through the asteroid belt; (8) rendezvous with comet 67P/Churyumov-Gerasimenko (January-May/2014); (9) lander delivery (November/2014) (adapted from ESA Bulletin, No. 117, February 2004).

No doubt, we are in a very exciting moment as regards to cometary science. The space missions, together with new sky surveys and ground-based observations with large telescopes, will provide us very valuable data about the comet population, size distribution, chemistry, physical structure, and origin. So, as a closing reflection, we can say that comets promise to continue in the forefront of public attention, now not as messengers of impending disasters, but as space targets where mankind will peer into its own origin.

Appendix 1

Table A1.1: Astronomical and physical constants

Velocity of light	$c = 2.9979 \times 10^{10}$ cm s^{-1}
Gravitational constant	$G = 6.673 \times 10^{-8}$ cm^3 g^{-1} s^{-2}
Plank constant	$h = 6.6261 \times 10^{-27}$ erg s
Boltzmann constant	$k = 1.3807 \times 10^{-16}$ erg K^{-1}
Stefan-Boltzmann constant	$\sigma = 5.6705 \times 10^{-5}$ erg cm^{-2} s^{-1} K^{-4}
Avogadro number	$N_A = 6.0221 \times 10^{23}$ mol^{-1}
Molar gas constant	$\mathcal{R} = 8.3146 \times 10^7$ erg mol^{-1} K^{-1}
Atomic mass unit (m^{12}C/12)	$amu = 1.6605 \times 10^{-24}$ g
Mass of proton	$m_p = 1.6726 \times 10^{-24}$ g
Mass of electron	$m_e = 9.1094 \times 10^{-28}$ g
Electron charge	$e = 4.8032 \times 10^{-10}$ esu
Energy in electron volt	1 eV $= 1.6022 \times 10^{-12}$ erg
Solar mass	$M_\odot = 1.9891 \times 10^{33}$ g
Solar radius	$R_\odot = 6.9595 \times 10^{10}$ cm
Solar effective temperature	$T_\odot = 5778$ K
Solar absolute luminosity	$L_\odot = 3.8268 \times 10^{33}$ erg s^{-1}
Solar constant (at 1 AU)	$S = 1.3676 \times 10^6$ erg cm^{-2} s^{-1}
Earth mass	$M_\oplus = 5.9736 \times 10^{27}$ g
Mean Earth radius	$M_\oplus = 6.371 \times 10^8$ cm
Julian day	1 d = 24 h =86400 s
Julian year	365.25 d = 31557600 s
Gregorian calendar year	365.2425 d
Mean sidereal day	86164.09054 s
Sidereal year	365.25636 d
Astronomical unit	1 AU $= 1.49598 \times 10^{13}$ cm
Light-year	9.4605×10^{17} cm
Parsec	1 pc $= 3.0857 \times 10^{18}$ cm
Radian	1 rad $= 180^\circ/\pi = 206264.8''$

Table A1.2: Conversion of cgs units to SI units and other units

1 cm = 10^{-2} m = 10^{8} Å (Ångström) = 10^{4} μm
1 g = 10^{-3} kg
1 dyne = 10^{-5} N (Newton) (kg m s^{-2})
1 erg = 10^{-7} J (Joule) (kg m^2 s^{-2}) = 2.39×10^{-8} cal
1 erg s^{-1} = 10^{-7} W (Watt)
1 dyn cm^{-2} = 0.1 Pascal (N m^{-2}) = 10^{-3} millibar = 0.1 Pa = 9.869×10^{-7} atm
1 Gauss = 10^{-4} Tesla
1 esu (electrostatic unit) = $\frac{10}{c}$ C (Coulomb) (c speed of light in cm s^{-1})
1 Hz (Hertz) = 1 s^{-1}

Appendix 2

Table A2.1: Absolute nuclear magnitudes and radii of JF comets

Comet	q (AU)	$H_N^{(1)}$	$\mathrm{QC}^{(*),(1)}$	$R_N^{(\dagger),(1)}$ (km)	$R_N(La)^{(2)}$ (km)
2P/Encke	0.338	16.0	3	2.10	
4P/Faye	1.657	16.3	1	1.83	1.77
6P/d'Arrest	1.353	16.5	1	1.66	
7P/Pons-Winnecke	1.258	16.3	3	1.83	
9P/Tempel 1	1.500	15.6	1	2.52	3.13
10P/Tempel 2	1.482	14.9	2	3.48	4.63
14P/Wolf	2.413	16.2	4	1.91	
15P/Finlay	1.034	17.2	4	1.21	
16P/Brooks 2	1.835	16.6	4	1.59	
17P/Holmes	2.165	16.6	1	1.59	1.71
19P/Borrelly	1.358	15.9	1	2.19	2.42
21P/Giacobini-Zinner	1.034	17.6	3	1.00	
22P/Kopff	1.584	16.3	1	1.83	1.67
24P/Schaumasse	1.205	17.8	4	0.91	
26P/Grigg-Skjellerup	0.997	17.2	1	1.21	
28P/Neujmin 1	1.552	12.7	1	9.58	
30P/Reinmuth 1	1.878	17.6	3	1.00	
31P/Schwassmann-Wachmann 2	3.409	15.2	2	3.03	
32P/Comas-Solá	1.846	15.6	3	2.52	
33P/Daniel	2.157	17.8	4	0.91	
36P/Whipple	3.088	16.0	1	2.10	
37P/Forbes	1.446	17.6	2	1.00	0.81
40P/Väisälä 1	1.783	16.5	3	1.66	
41P/Tuttle-Giacobini-Kresák	1.052	18.4	4	0.69	
42P/Neujmin 3	2.001	18.4	4	0.69	
43P/Wolf-Harrington	1.582	16.0	2	2.10	
44P/Reinmuth 2	1.890	16.7	3	1.52	1.61
45P/Honda-Mrkos-Pajdušáková	0.528	20.0	3	0.33	0.34
46P/Wirtanen	1.059	18.8	1	0.58	0.62
47P/Ashbrook-Jackson	2.305	15.5	1	2.64	2.80
48P/Johnson	2.308	15.9	2	2.19	
49P/Arend-Rigaux	1.369	14.8	1	3.64	
50P/Arend	1.917	17.7	4	0.96	0.95
51P/Harrington	1.568	20.8	4	0.23	

(cont.)

52P/Harrington-Abell	1.756	17.4	4	1.10	
53P/Van Biesbroeck	2.415	15.0	1	3.31	
56P/Slaughter-Burnham	2.543	16.8	1	1.45	
58P/Jackson-Neujmin	1.381	18.7	4	0.60	
59P/Kearns-Kwee	2.339	17.6	3	1.00	0.79
60P/Tsuchinshan 2	1.770	18.4	4	0.69	
61P/Shajn-Schaldach	2.330	18.0	2	0.83	0.64
63P/Wild 1	1.961	16.8	2	1.45	1.45
64P/Swift-Gehrels	1.339	16.3	4	1.83	
65P/Gunn	2.446	14.3	2	4.57	
67P/Churyumov-Gerasimenko	1.292	16.0	1	2.10	
68P/Klemola	1.755	15.6	3	2.52	
69P/Taylor	1.948	16.0	4	2.10	
70P/Kojima	2.003	17.1	3	1.26	1.86
71P/Clark	1.559	18.0	2	0.83	0.68
73P/Schwassmann-Wachmann 3	0.937	17.7	3	0.96	
74P/Smirnova-Chernykh	3.546	15.1	3	3.16	2.23
75P/Kohoutek	1.775	16.3	3	1.83	
77P/Longmore	2.310	15.8	4	2.30	
78P/Gehrels 2	2.000	16.4	3	1.74	
79P/du Toit-Hartley	1.199	17.2	4	1.21	
81P/Wild 2	1.590	16.2	2	1.91	
82P/Gehrels 3	3.627	18.1	3	0.79	0.73
84P/Giclas	1.846	17.5	3	1.05	0.90
86P/Wild 3	2.310	19.0	2	0.53	0.43
87P/Bus	2.181	19.0	4	0.53	0.28
88P/Howell	1.406	17.7	4	0.95	
89P/Russell 2	2.290	17.3	4	1.15	
90P/Gehrels 1	2.966	15.5	3	2.64	
91P/Russell 3	2.510	17.1	4	1.26	
92P/Sanguin	1.807	17.2	2	1.21	
94P/Russell 4	2.231	16.1	3	2.00	
97P/Metcalf-Brewington	2.611	16.8	4	1.45	
98P/Takamizawa	1.585	15.3	4	2.89	
99P/Kowal	4.673	14.2	3	4.80	
101P/Chernykh	2.356	15.9	3	2.19	
103P/Hartley 2	1.032	17.2	3	1.21	1.20
104P/Kowal 2	1.397	16.8	4	1.45	

(cont.)

105P/Singer-Brewster	2.032	18.0	4	0.83	
106P/Schuster	1.550	18.0	3	0.83	0.94
108P/Ciffréo	1.713	18.0	3	0.83	
110P/Hartley 3	2.478	16.1	2	2.00	2.15
111P/Helin-Roman-Crockett	3.473	17.3	4	1.15	
112P/Urata-Niijima	1.458	18.2	3	0.76	0.90
113P/Spitaler	2.127	17.3	3	1.15	
114P/Wiseman-Skiff	1.569	17.9	3	0.87	0.78
116P/Wild 4	2.170	15.0	4	3.31	
117P/Helin-Roman-Alu	3.715	14.8	3	3.64	
118P/Shoemaker-Levy	2.011	16.2	3	1.91	
119P/Parker-Hartley	3.045	16.3	3	1.83	
120P/Mueller 1	2.739	18.0	4	0.83	
121P/Shoemaker-Holt 2	2.664	16.1	4	2.00	
123P/West-Hartley	2.129	16.1	3	2.00	
124P/Mrkos	1.467	16.4	2	1.74	
125P/Spacewatch	1.529	18.0	3	0.83	
128P/Shoemaker-Holt 1	3.047	16.1	3	2.00	
129P/Shoemaker-Levy 3	2.807	16.5	4	1.66	
130P/McNaught-Hughes	2.116	16.6	3	1.59	
131P/Mueller 2	2.412	18.1	4	0.79	
134P/Kowal-Vávrová	2.575	16.8	3	1.45	
135P/Shoemaker-Levy 8	2.721	16.9	3	1.38	
137P/Shoemaker-Levy 2	1.869	15.4	2	2.75	
143P/Kowal-Mrkos	2.547	14.3	3	4.57	
144P/Kushida	1.431	17.3	3	1.15	
152P/Helin-Lawrence	3.110	16.0	3	2.10	
154P/Brewington	1.590	16.5	4	1.66	
P/1995 A1 (Jedicke)	4.083	15.2	3	3.03	
P/1996 A1 (Jedicke)	4.055	14.1	2	5.03	
P/1997 C1 (Gehrels)	3.565	15.6	3	2.52	
P/1997 G1 (Montani)	4.214	15.5	4	2.64	
P/2002 BV (Yeung)	2.244	15.0	4	3.31	

(*) QC : Quality Class which is defined according to the estimated uncertainty of H_N. We have: QC 1 for an estimated uncertainty less than ± 0.3 mag; QC 2 between ± 0.3 and ± 0.6 mag; QC 3 between ± 0.6 and ± 1 mag; QC 4 between about ± 1 and ± 1.5 mag, although some QC-4 magnitudes can be considered as - meaningful - upper limits to the true nuclear magnitudes (Tancredi et al. 2000).

(†) The radius is computed from eq. (2.8) by assuming a standard geometric albedo $p_v = 0.04$.

Sources: (1) Tancredi et al. (2005), (2) Lamy et al. (2005)

References

1. A'Hearn M.F., Millis R.L., and Birch P.V. 1979. Gas and dust in some recent periodic comets. *Astron. J.* 84, 570-579.
2. A'Hearn M.F., Millis R.L., Schleicher D.G., Osip D.J., and Birch P.V. 1995. The ensemble properties of comets: Results from narrowband photometry of 85 comets, 1976-1992. *Icarus* 118, 223-270.
3. Aggarwal H.R., and Oberbeck V.R. 1974. Roche limit of a solid body. *Astrophys. J.* 191, 577-588.
4. Allen R.L., Bernstein G.M., and Malhotra R. 2002. Observational limits on a distant cold Kuiper belt. *Astron. J.* 124, 2949-2954.
5. Alvarez L.W., Alvarez W., Asaro F., and Michel H.V. 1980. Extraterrestrial cause for the Cretaceous-Tertiary extinction. *Science* 208, 1095-1108.
6. Alvarez W., and Muller R.A. 1984. Evidence from crater ages for periodic impacts on the Earth. *Nature* 308, 718-720.
7. Anderson J.D., Lau E.L., Scherer K., Rosenbaum D.C., and Teplitz V.L. 1998. Kuiper belt constraint from Pioneer 10. *Icarus* 131, 167-170.
8. Anderson J.D., and Standish Jr. E.M. 1986. Dynamical evidence for planet X. In *The Galaxy and the Solar System* (R. Smoluchowski, J.N. Bahcall, and M.S. Matthews, eds.), pp. 286-296, Univ. Arizona Press, Tucson.
9. André P., Ward-Thompson D., and Barsony M. 2000. From prestellar cores to protostars: The initial conditions of star formation. In *Protostars and Planets IV* (V. Mannings, A.P. Boss, and S.S. Russell, eds.), pp. 59-96, Univ. Arizona Press, Tucson.
10. Antonov V.A., and Latyshev I.N. 1972. Determination of the form of the Oort cometary cloud as the Hill surface in the galactic field. In *The Motion, Evolution of Orbits, and Origin of Comets* (G.A. Chebotatev, E.-I. Kazimirchak-Polonskaya, and B.G. Marsden, eds.), IAU Symp. 45, pp. 341-345, Reidel, Dordrecht.
11. Arnold J.R. 1965. The origin of meteorites as small bodies. II. The model. *Astrophys. J.* 141, 1536-1547.
12. Asphaug E., and Benz W. 1994. Density of comet Shoemaker-Levy 9 deduced by modelling breakup of the parent 'rubble pile'. *Nature* 370, 120-124.
13. Bahcall J.N. 1984. Self-consistent determination of the total amount of matter near the Sun. *Astrophys. J.* 276, 169-181.
14. Bailey M.E. 1983a. The structure and evolution of the solar system comet cloud. *Mon. Not. Roy. Astr. Soc.* 204, 603-633.
15. Bailey M.E. 1983b. Theories of cometary origin and the brightness of the infrared sky. *Mon. Not. Roy. Astr. Soc.* 205, 47P-52P.
16. Bailey M.E. 1990. Cometary masses. In *Baryonic Dark Matter* (D. Lynden-Bell, and G. Gilmore, eds.), pp. 7-35, Kluwer, Dordrecht.
17. Bailey M.E. 1991. Comet craters versus asteroid craters. *Adv. Space Res.* 11, (6)43-(6)60.
18. Bailey M.E. 1994. Formation of outer solar system bodies: comets and planetesimals. In *Asteroids, Comets, Meteors 1993* (A. Milani, M. Di Martino, and A. Cellino, eds.), pp. 443-459, Kluwer, Dordrecht.

19. Bailey M.E., Chambers J.E., and Hahn G. 1992. Origin of sungrazers: a frequent cometary end-state. *Astron. Astrophys.* 257, 315-322.
20. Bailey M.E., Clube S.V.M., and Napier W.M. 1990. *The Origin of Comets*, Pergamon, Oxford, England.
21. Bailey M.E., Wilkinson D.A., and Wolfendale A.W. 1987. Can episodic comet showers explain the 30-Myr cyclicity in the terrestrial record?. *Mon. Not. Roy. Astr. Soc.* 227, 863-885.
22. Bar-Nun A., Kleinfeld I. 1989. On the temperature and gas composition in the region of comet formation. *Icarus* 80, 243-253.
23. Bauer J.M., Meech K.J., Fernández Y.R., Pittichova J., Hainaut O., Boehnhardt H., Delsanti A.C. 2003. Physical survey of 24 Centaurs with visible photometry. *Icarus* 166, 195-211.
24. Beckwith S.V.W. and Sargent A.I. 1993. The occurrence and properties of disks around young stars. In *Protostars and Planets III* (E.H. Levy, and J.I. Lunine, eds.), pp. 521-542, Univ. Arizona Press, Tucson.
25. Beckwith S.V.W., Henning T., and Nakagawa Y. 2000. Dust properties and assembly of large particles in protoplanetary disks. In *Protostars and Planets IV* (V. Mannings, A.P. Boss, and S.S. Russell, eds.), pp. 533-558, Univ. Arizona Press, Tucson.
26. Belbruno E., and Marsden B.G. 1997. Resonance hopping in comets. *Astron. J.* 113, 1433-1444.
27. Bernstein G.M., Trilling D.E., Allen R.L., Brown M.E., Holman M., and Malhotra R. 2004. The size distribution of trans-neptunian bodies. *Astron. J.* 128, 1364-1390.
28. Bertaux J.L., Blamont J.E., and Festou M. 1973. Interpretation of hydrogen Lyman-alpha observations of comets Bennett and Encke. *Astron. Astrophys.* 25, 415-430.
29. Biermann L. 1978. Dense interstellar clouds and comets. In *Astronomical Papers Dedicated to Bengt Stromgren* (A. Reiz and T. Anderson, eds.), pp. 327-335, Copenhagen Observatory.
30. Biermann L., Huebner W.F., and Lüst R. 1983. Aphelion clustering of 'new' comets: Star tracks through Oort's cloud. *Proc. Natl. Acad. Sci. USA* 80, 5151-5155.
31. Biermann L., and Michel K.W. 1978. On the origin of cometary nuclei in the presolar nebula. *Moon and Planets* 18, 447-464.
32. Biver N. 2001. Correlation between visual magnitudes and the outgassing rate of CO in comets beyond 3 AU. *Int. Comet Quart.* 23, 85-93.
33. Blake G.A., Qi C., Hogerheijde M.R., Gurwell M.A., and Muhleman D.O. 1999. Sublimation from icy jets as a probe of the interstellar volatile content of comets. *Nature* 398, 213-216.
34. Blamont J.E., and Festou M. 1974. Observation of the comet Kohoutek (1973f) in the resonance light ($A^2\Sigma^+ - X^2\Pi$) of the OH radical. *Icarus* 23, 538-544.
35. Blitz L. 1993. Giant molecular clouds. In *Protostars and Planets III* (E.H. Levy and J.I. Lunine, eds.), pp. 125-162, Univ. Arizona Press, Tucson.

36. Bockelée-Morvan D., Crovisier J., Mumma M.J., and Weaver H.A. 2005. The composition of cometary volatiles. In *Comets II* (M. Festou, H.U. Keller, and H.A. Weaver, eds.), Univ. Arizona Press, Tucson, in press.
37. Bockelée-Morvan D., Gautier D., Lis D.C., Young K., Keene J., Phillips T., Owen T., Crovisier J., Goldsmith F., Bergin E.A., Despois D., and Wootten A. 1998. Deuterated water in Comet C/1996 B2 (Hyakutake) and its implications for the origin of comets. *Icarus* 193, 147-162.
38. Bolatto A.D., Fernández J.A., and Carballo G.F. 1995. Asymmetric nongravitational forces on long-period comets. *Planet. Space Sci.* 43, 709-716.
39. Bondi H., and Hoyle F. 1944. On the mechanism of accretion by stars. *Mon. Not. Roy. Astr. Soc.* 104, 273-282.
40. Boss A.P. 1997. Giant planet formation by gravitational instability. *Science* 276, 1836-1839.
41. Bottke Jr. W.F., Jedicke R., Morbidelli A., Petit J.-M., and Gladman B. 2000. Understanding the distribution of near-Earth asteroids. *Science* 288, 2190-2194.
42. Bottke Jr. W.F., Morbidelli A., Jedicke R., Petit J.-M., Levison H.F., Michel P., and Metcalfe T.S. 2002. Debiased orbital and absolute magnitude distribution of the near-Earth objects. *Icarus* 156, 399-433.
43. Brandt J.C., A'Hearn M.F., Randall C.E., Schleicher D.G., Shoemaker E.M., and Stewart A.I.F. 1996. On the existence of small comets and their interactions with planets. *Earth, Moon, and Planets* 72, 243-249.
44. Brandt J.C., and Chapman R.D. 1981. *Introduction to Comets*, Cambridge University Press, Cambridge.
45. Briceño C., Vivas A.K., Calvet N., Hartmann L., Pacheco R., Herrera D., Romero L., Berlind P., Sánchez G., Snyder J.A., and Andrews P. 2001. The CIDA-QUEST large-scale survey of Orion OB1: Evidence for rapid disk dissipation in a dispersed stellar population. *Science* 291, 93-96.
46. Bridges F.G., Supulver K.D., Kimberly D., Lin D.N.C., Knight R., and Zafra M. 1996. Energy loss and sticking mechanisms in particle aggregation in planetesimal formation. *Icarus* 123, 422-435.
47. Brin G.D., and Mendis D.A. 1979. Dust release and mantle development in comets. *Astrophys. J.* 229, 402-408.
48. Brooke T.Y., and Knacke R.F. 1986. The nucleus of comet P/Arend-Rigaux. *Icarus* 67, 80-87.
49. Brown M.E., Blake G.A., and Kessler J.E. 2000. Near-infrared spectroscopy of the bright Kuiper belt object 2000 EB173. *Astrophys. J.* 543, L163-L165.
50. Brown M.E., and Trujillo C.A. 2004. Direct measurement of the size of the large Kuiper belt object (50000) Quaoar. *Astron. J.* 127, 2413-2417.
51. Brown M.E., Trujillo C., and Rabinowitz D. 2004. Discovery of a candidate inner Oort cloud planetoid. *Astrophys. J.* 617, 645-649.
52. Brown R.H., Cruikshank D.P., and Pendleton Y. 1999. Water ice on Kuiper belt object 1996 TO66. *Astrophys. J.* 519, L101-L104.
53. Brownlee, D.E., Horz F., Newburn R.L., Zolensky M., Duxbury T.C., Sandford S., Sekanina Z., Tsou P., Hanner M., Clark B.C., Green S.F., and Kissel

J. 2004. Surface of young Jupiter family comet 81P/Wild 2: View from the Stardust spacecraft. *Science* 304, 1764-1769.

54. Brunini A. 2004. Private communication.
55. Brunini A., and Fernández J.A. 1999. Numerical simulations of the accretion of Uranus and Neptune. *Planet. Space Sci.* 47, 591-605.
56. Brunini A., and Melita M.D. 2002. The existence of a planet beyond 50 AU and the orbital distribution of the classical Edgeworth-Kuiper-belt objects. *Icarus* 160, 32-43.
57. Byl J. 1983. Galactic perturbations on near-parabolic cometary orbits. *Moon and Planets* 29, 121-137.
58. Cameron A.G.W. 1978. Physics of the primitive solar nebula and of giant gaseous protoplanets. In *Protostars and Planets* (T. Gehrels, ed.), pp. 453-487, Univ. Arizona Press, Tucson.
59. Campins H., A'Hearn M.F., and McFadden L.-A. 1987. The bare nucleus of comet Neujmin 1. *Astrophys. J.* 316, 847-857.
60. Carruthers G.R., Opal C.B., Page T.L., Meier R.R., and Prinz D.K. 1974. Lyman-α imagery of comet Kohoutek. *Icarus* 23, 526-537.
61. Carusi A., Kresák L., Kresáková M., and Valsecchi G.B. 1991. Observations of periodic comet d'Arrest in 1678 and implications for its evolutionary history. *Astron. Astrophys.* 252, 377-384.
62. Carusi A., Kresák L., Perozzi E., and Valsecchi G.B. 1986. *Long-Term Evolution of Short-Period Comets*, Adam Hilger, Bristol.
63. Carusi A., Kresák L., Perozzi E., and Valsecchi G.B. 1987. High-order librations of Halley-type comets. *Astron. Astrophs.* 187, 899-905.
64. Ceplecha Z., and McCrosky R.E. 1976. Fireball end heights: A diagnostic for the structure of meteoric material. *J. Geophys. Res.* 81, 6257-6275.
65. Chamberlin A.B., McFadden L.-A., Schulz R., Schleicher D.G., and Bus S.J. 1996. 4015 Wilson-Harrington , 2201 Oljato, and 3200 Phaethon: Search for CN emission. *Icarus* 119, 173-181.
66. Chambers, J.E. 1997. Why Halley-types resonate but long-period comets don't: A dynamical distinction between short- and long-period comets. *Icarus* 125, 32-38.
67. Chang S. 1979. Comets: Cosmic connections with carbonaceous meteorites, interstellar molecules and the origin of life. In *Space Missions to Comets* (M. Neugebauer, D.K. Yeomans, J.C. Brandt, and R.W. Hobbs, eds.), pp. 59-111, NASA CP 2089, Washington, DC.
68. Chapman C.R., and Morrison D. 1994. Impacts on the Earth by asteroids and comets: assessing the hazard. *Nature* 367, 33-40.

Chebotarev G.A. 1966. Cometary motion in the outer solar system. *Soviet Astr. - AJ* 10, 341-344.

69. Chen J., and Jewitt D. 1994. On the rate at which comets split. *Icarus* 108, 265-271.
70. Chiang E.I., and Brown M.E. 1999. Keck pencil-beam survey for faint Kuiper belt objects. *Astron. J.* 118, 1411-1422.
71. Chokshi A., Tielens A.G.G.M., and Hollenbach D. 1993. Dust coagulation. *Astrophys. J.*, 407, 806-819.

72. Christy J.W., and Harrington R.S. 1978. The satellite of Pluto. *Astron. J.* 83, 1005-1008.

73. Chyba C.F. 1987. The cometary contribution to the oceans of the primitive Earth. *Nature* 330, 632-635.

74. Chyba C.F., Thomas P.J., Brookshaw L., and Sagan C. 1990. Cometary delivery of organic molecules to the early Earth. *Science* 249, 366-373.

75. Chyba C.F., Owen T.C., and Ip W.-H. 1994. Impact delivery of volatiles and organic molecules to Earth. In *Hazards due to Comets and Asteroids* (T. Gehrels, ed.), pp. 9-58, Univ. Arizona Press, Tucson.

76. Clark B.E., Veverka J., Helfenstein P., Thomas P.C., Bell J.F., Harch A., Robinson M.S., Murchie S.L., McFadden L.A., and Chapman C.R. 1999. NEAR photometry of asteroid 253 Mathilde. *Icarus* 140, 53-65.

77. Combes M., Moroz V.I., Crovisier J., Encrenaz T., Bibring J.-P., Grigoriev A.V., Sanko N.F., Coron N., Crifo J.F., Gispert R., Bockelée-Morvan D., Nikolsky Yu. V., Krasnopolsky V.A., Owen T., Emerich C., Lamarre J.M., and Rocard F. 1988. The 2.5-12 μm spectrum of comet Halley from the IKS-VEGA experiment. *Icarus* 76, 404-436.

78. Combi M.R., and Delsemme A.H. 1980. Neutral cometary atmospheres. I. An average random walk model for photodissociation in comets. *Astrophys. J.* 237, 633-640.

79. Cremonese G., and Fulle M. 1989. Photometrical analysis of the neck-line structure of comet Halley. *Icarus* 80, 267-279.

80. Cremonese G., and Fulle M. 1990. The dust tail of comet Wilson 1987VII. *Astron. J.* 100, 1285-1292.

81. Crézé M., Chereul E., Bienaymé O., and Pichon C. 1998. The distribution of stars in phase space mapped by Hipparcos. I. The potential well and local dynamical mass. *Astron. Astrophys.* 329, 920-936.

82. Crovisier J., Colom P., Gérard E., Bockelée-Morvan D., and Bourgois G. 2002. Observations at Nançay of the OH 18-cm lines in comets: The data base. Observations made from 1982 to 1999. *Astron. Astrophys.* 393, 1053-1064.

83. Cruikshank D.P., Roush T.L., Bartholomew M.J., Geballe T.R., Pendleton Y.J., White S.M., Bell J.F. III, Davies J.K., Owen T.C., de Bergh C., Tholen D.J., Bernstein M.P., Brown R.H., Tryka K.A., and Dalle Ore C.M. 1998. The composition of Centaur 5145 Pholus. *Icarus* 135, 389-407.

84. Daniels P.A., and Hughes D.W. 1981. The accretion of cosmic dust - a computer simulation. *Mon. Not. Roy. Astr. Soc.* 195, 1001-1009.

85. Davies J., Spencer J., Sykes M., Tholen D., and Green S. 1993. *IAU Circ. No. 5698.*

86. Davis D.R., Chapman C.R., Greenberg R., Weidenschilling S.J., and Harris A.W. 1979. Collisional evolution of asteroids: populations, rotations, and velocities. In *Asteroids* (T. Gehrels, ed.), pp. 528-557, Univ. Arizona Press, Tucson.

87. Davis M., Hut P., and Muller R.A. 1984. Extinction of species by periodic comet showers. *Nature* 308, 715-717.

88. Delahodde C.E., Meech K.J., Hainaut O.R., and Dotto E. 2001. Detailed phase function of comet 28P/Neujmin 1. *Astron. Astrophys.* 376, 672-685.

89. Delsemme A.H. 1973. Origin of the short-period comets. *Astron. Astrophys.* 29, 377-381.

90. Delsemme A.H. 1977. The pristine nature of comets. In *Comets, Asteroids, Meteorites* (A.H. Delsemme, ed.), pp. 3-13, Univ. of Toledo, Ohio.

91. Delsemme A.H. 1981. Chemical composition of cometary nuclei. In *Comets* (L.L. Wilkening, ed.), pp. 85-130, Univ. Arizona Press, Tucson.

92. Delsemme A.H. 1984. The cometary connection with periodic chemistry. *Origins Life* 14, 51-60.

93. Delsemme A.H. 1987. Galactic tides affect the Oort cloud: An observational confirmation. *Astron. Astrophys.* 187, 913-918.

94. Delsemme A.H. 1991. Nature and history of the organic componds in comets: An astrophysical view. In *Comets in the Post-Halley Era* (R.L. Newburn Jr., M. Neugebauer, and J. Rahe, eds.), pp. 377-427, Kluwer, Dordrecht.

95. Delsemme A. 1997. The origin of the atmosphere and of the oceans. In *Comets and the Origin and Evolution of Life* (P.J.Thomas, C.F. Chyba, and C.P. McKay, eds.), pp. 29-67, Springer, New York.

96. Delsemme A.H. 1999. The deuterium enrichment observed in recent comets is consistent with the cometary origin of seawater. *Planet. Space Sci.* 47, 125-131.

97. Delsemme A.H., and Miller D.C. 1971. Physico-chemical phenomena in comets-III. The continuum of comet Burnham (1960II), *Planet. Space Sci.* 19, 1229-1257.

98. Delsemme A.H., and Rud D.A. 1973. Albedos and cross-sections for the nuclei of Comets 1969 IX, 1970 II and 1971 I. *Astron. Astrophys.* 28, 1-6.

99. Delsemme A.H., and Swings P. 1952. Hydrates de gaz dans les noyaux cométaires et les grains interstellaires. *Ann. Astrophys.* 15, 1-6.

100. Dermott S.F., Malhotra R., and Murray C. 1988. Dynamics of the Uranian and Saturnian satellite systems: A chaotic route to melting Miranda? *Icarus* 76, 295-334.

101. Dohnanyi J.S. 1969. Collisional models of asteroids and their debris. *J. Geophys. Res.* 74, 2531-2554.

102. Dohnanyi J.S. 1972. Interplanetary objects in review: statistics of their masses and dynamics. *Icarus* 17, 1-48.

103. Dominik C., and Tielens A.G.G.M. 1997. The physics of dust coagulation and the structure of dust aggregates in space. *Astrophys. J.* 480, 647-673.

104. Donn B. 1963. The origin and structure of icy cometary nuclei. *Icarus* 2, 396-402.

105. Donn B. 1990. The formation and structure of fluffy cometary nuclei from random accummulation of grains. *Astron. Astrophys.* 235, 441-446.

106. Donn B., and Rahe J. 1982. Structure and origin of cometary nuclei. In *Comets* (L.L. Wilkening, ed.), pp. 203-226, Univ. Arizona Press, Tucson.

107. Drouart A., Dubrulle B., Gautier D., and Robert F. 1999. Structure and transport in the solar nebula from constraints on deuterium enrichment and giant planets formation. *Icarus* 140, 129-155.

108. Duncan M.J., and Levison H.F. 1997. A disk of scattered icy objects and the origin of Jupiter-family comets. *Science* 276, 1670-1672.

109. Duncan M.J., Levison H.F., and Budd S.M. 1995. The dynamical structure of the Kuiper belt. *Astron. J.*, 110, 3073-3081.

110. Duncan M., Quinn T., and Tremaine S. 1987. The formation and extent of the solar system comet cloud. *Astron. J.* 94, 1330-1338.
Duncan M., Quinn T., and Tremaine S. 1988. The origin of short-period comets. *Astrophys. J.* 328, L69-L73.

111. Eberhardt P., Reber M., Krankowsky D., and Hodges R.R. 1995. The D/H and $^{18}O/^{16}O$ ratios in water from Comet P/Halley. *Astron. Astrophys.* 302, 301-316.

112. Edgeworth K.E. 1938. *The Evolution of the Solar System.* Trustees of the National Library of Ireland. Unpublished manuscript Ms. No. 16869/47 & 48.

113. Edgeworth K.E. 1943. The evolution of our planetay system. *J. British Astron. Soc.* 53, 181-188.
Edgeworth K.E. 1949. The origin and evolution of the solar system. *Mon. Not. Roy. Astr. Soc.* 109, 600-609.

114. Eggers S., Keller H.U., Kroupa P., and Markiewicz W.J. 1997. Origin and dynamics of comets and star formation. *Planet. Space Sci.* 45, 1099-1104.

115. Elmegreen D.M. 1998. *Galaxies and Galactic Structure*, Prentice Hall, New Jersey.

116. Everhart E. 1967a. Intrinsic distributions of cometary perihelia and magnitudes. *Astron. J.* 72, 1002-1011.

117. Everhart E. 1967b. Comet discoveries and observational selection. *Astron. J.* 72, 716-726.

118. Everhart E. 1968. Change in total energy of comets passing through the Solar System. *Astron. J.* 73, 1039-1052.

119. Everhart E. 1972. The origin of short-period comets. *Astrophys. Lett.* 10, 131-135.

120. Everhart E. 1976. The evolution of comet orbits. In *The Study of Comets* (B. Donn, M. Mumma, W. Jackson, M. A'Hearn, and R. Harrington, eds.) IAU Coll. No. 25, pp. 445-464, NASA SP-393.

121. Farinella P., and Davis D.R. 1996. Short-period comets: Primordial bodies or collisional fragments? *Science* 273, 938-941.

122. Feder W. 1968. *An Introduction to Probability Theory and its Applications. Third Ed.*, Vol. I, Wiley, New York.

123. Fernández J.A. 1978. Mass removed by the outer planets in the early Solar System. *Icarus* 34, 173-181.

124. Fernández J.A. 1980a. Evolution of comet orbits under the perturbing influence of the giant planets and nearby stars. *Icarus* 42, 406-421.

125. Fernández J.A. 1980b. On the existence of a comet belt beyond Neptune. *Mon. Not. Roy. Astr. Soc.* 192, 481-491.

126. Fernández J.A. 1981a. On the observed excess of retrograde orbits among long-period comets. *Mon. Not. Roy. Astr. Soc.* 197, 265-273.

127. Fernández J.A. 1981b. New and evolved comets in the solar system. *Astron. Astrophys.* 96, 26-35.

128. Fernández J.A. 1981c. The role of collisions with interplanetary particles in the physical evolution of comets. *The Moon and the Planets* 25, 507-519.

129. Fernández J.A. 1982. Dynamical aspects of the origin of comets. *Astron. J.* 87, 1318-1332.

130. Fernández J.A. 1984. The distribution of the perihelion distances of short-period comets. *Astron. Astrophys.* 135, 129-134.

131. Fernández J.A. 1985a. The formation and dynamical survival of the comet cloud. In *Dynamics of Comets: Their Origin and Evolution* (A. Carusi, and G.B. Valsecchi, eds.), IAU Coll. No. 83, pp.45-70, Reidel, Dordrecht.

132. Fernández J.A. 1985b. Dynamical capture and physical decay of short-period comets. *Icarus* 64, 308-319.

133. Fernández J.A. 1988. End-states of short-period comets and their role in maintaining the zodiacal dust cloud. *Earth, Moon and Planets* 41, 155-161.

134. Fernández J.A. 1990. Collisions of comets with meteoroids. In *Asteroids, Comets, Meteors III* (C.-I. Lagerkvist, H. Rickman, B.A. Lindblad and M. Lindgren, eds.), pp. 309-312, Uppsala Universitet.

135. Fernández J.A. 1992. Comet showers. In *Chaos, Resonance and Collective Dynamical Phenomena in the Solar System* (S. Ferraz-Mello, ed.), IAU Symp. No. 152, pp. 239-254, Kluwer, Dordrecht.

136. Fernández J.A. 1997. The formation of the Oort cloud and the primitive galactic environment. *Icarus* 129, 106-119.

137. Fernández J.A. 1999. Cometary dynamics. In *Encyclopedia of the Solar System* (P.R. Weissman, L.-A. McFadden, and T.V. Johnson ,eds.), pp. 537-556, Academic Press, San Diego, USA.

138. Fernández J.A. 2002. Long-period comets and the Oort cloud. *Earth, Moon and Planets* 89, 325-343.

139. Fernández J.A., and Brunini A. 2000. The buildup of a tightly bound comet cloud around an early Sun immersed in a dense galactic environment: Numerical experiments. *Icarus* 145, 580-590.

Fernández J.A., and Gallardo T. 1994. The transfer of comets from parabolic orbits to short-period orbits. *Astron. Astrophys.* 281, 911-922.

Fernández J.A., and Gallardo T. 1999. From the Oort cloud to Halley-type comets. In *Evolution and Source Regions of Asteroids and Comets* (J. Svoreň, E.M. Pittich, and H. Rickman, eds.), Proc. IAU Coll. 173, pp. 327-338, Astron. Inst. Slovak Acad. Sci., Tatranská Lomnica.

140. Fernández J.A., Gallardo T., and Brunini A. 2002a. Are there many inactive Jupiter-family comets among the near-Earth asteroid population?. *Icarus* 159, 358-368.

141. Fernández J.A., Gallardo T., and Brunini A. 2004. The scattered disk population as a source of Oort cloud comets: Evaluation of its current and past role in populating the Oort cloud. *Icarus* 172, 372-381.

142. Fernández J.A., and Ip W.-H. 1981. Dynamical evolution of a cometary swarm in the outer planetary region. *Icarus* 47, 470-479.

143. Fernández J.A., and Ip W.-H. 1984. Some dynamical aspects of the accretion of Uranus and Neptune: The exchange of orbital angular momentum with planetesimals. *Icarus* 58, 109-120.

144. Fernández J.A., and Ip W.-H. 1987. Time-dependent injection of Oort cloud comets into Earth-crossing orbits. *Icarus* 71, 46-56.

145. Fernández J.A., and Ip W.-H. 1991. Statistical and evolutionary aspects of cometary orbits. In *Comets in the Post-Halley Era* (R.L. Newburn Jr., M. Neugebauer, and J. Rahe, eds.), pp. 487-535, Kluwer, Dordrecht.

146. Fernández J.A., and Ip W.-H. 1996. Orbital expansion and resonant trapping during the late accretion stages of the outer planets. *Planet. Space Sci.* 44, 431-439.

147. Fernández J.A., and Ip W.-H. 1997. Accretion of the outer planets and its influence on the surface impact process of the terrestrial planets. In *Astronomical and Biochemical Origins and the Search for Life in the Universe* (C.B. Cosmovici, S. Bowyer, and D. Werthimer, eds.), pp. 235-244, Editrice Compositori, Bologna.

148. Fernández J.A., and Jockers K. 1983. Nature and origin of comets. *Rep. Prog. Phys.* 46, 665-772.

149. Fernández J.A., Tancredi G., Rickman H., and Licandro J. 1999. The population, magnitudes, and sizes of Jupiter family comets. *Astron. Astrophys.* 352, 327-340.

150. Fernández Y.R., Lisse C.M., Ulrich K.H., Peschke S.B., Weaver H.A., A'Hearn M.F., Lamy P., Livengood T.A., and Kostiuk T. 2000. Physical properties of the nucleus of 2P/Encke. *Icarus* 147, 145-160.

151. Fernández Y.R., Jewitt D., and Sheppard S.S. 2001. Low albedos among extinct comet candidates. *Astrophys. J.* 553, L197-200.

152. Fernández Y.R., Jewitt D.C., and Sheppard S.S. 2002b. Thermal properties of Centaurs Asbolus and Chiron. *Astron. J.* 553, L197-200.

153. Fernández Y.R., Jewitt D.C., and Sheppard S.S. 2005. Albedos of asteroids in comet-like orbits. *Astron. J.*, in press.

154. Fesenkov V.G. 1922. Sur les perturbations séculaires dans le mouvement des cometes non périodiques par des étoiles voisines. *Publ. Russian Astrophys. Observatory* 1, 186-195.

155. Festou M. 1981. The density distribution of neutral compounds in cometary atmospheres. I. Models and equations. *Astron. Astrophys.* 95, 69-79.

156. Festou M. 1986. The derivation of OH gas production rates from visual magnitudes of comets. In *Asteroids, Comets, Meteors II* (C.-I. Lagerkvist, B.A. Lindblad, H. Lundstedt, and H. Rickman, eds.), pp. 299-303, Uppsala Universitet Reprocentralen.

157. Festou M., Rickman H., and Kamél L. 1990. Using comet light-curve asymmetries to predict comet returns. *Nature* 345, 235-238.

158. Fink U., and Hicks M.D. 1996. A survey of 39 comets using CCD spectroscopy. *Astrophys. J.* 459, 729-743.

159. Finson M.L., and Probstein R.F. 1968a. A theory of dust comets. I. Model and equations. *Astrophys. J.* 154, 327-352.

160. Finson M.L., and Probstein R.F. 1968b. A theory of dust comets. II. Results for comet Arend-Roland. *Astrophys. J.* 154, 353-380.

161. Fogg M.J. 1989. The relevance of the background impact flux to cyclic impact/mass extinction hypotheses. *Icarus* 79, 382-395.

162. Fouchard M. 2004. New fast models of the Galactic tide. *Mon. Not. Roy. Astr. Soc.* 349, 347-356.

163. Fraundorf P., Brownlee D.E., and Walker R.M. 1982. Laboratory studies of interplanetary dust. In *Comets* (L.L. Wilkening, ed.), pp. 383-409, Univ. Arizona Press, Tucson.

164. Froschlé Ch., and Scholl H. 1987. Orbital evolution of asteroids near the secular resonance ν_6. *Astron. Astrophys.* 179, 294-303.

165. Fulle M. 1990. Meteoroids from short period comets. *Astron. Astrophys.* 230, 220-226.

166. Gahm G.F., Zinnecker H., Pallavicini R., and Pasquini L. 1994. A search for cold dust around post-T Tauri candidates. *Astr. Astrophys.* 282, 123-126.

167. Gaidos E.J. 1995. Paleodynamics: Solar system formation and the early environment of the Sun. *Icarus* 114, 258-268.

168. Gallardo T., and Ferraz-Mello S. 1998. Dynamics in the exterior 2:3 resonance with Neptune. *Planet. Space Sci.* 46, 945-965.

169. Garcí a-Sánchez J., Preston R.A., Jones D.L., Weissman P.R., Lestrade J.-F., Latham D.W., and Stefanik R.P. 1999. Stellar encounters with the Oort cloud based on *Hipparcos* data. *Astron. J.* 117, 1042-1055.

170. Gehrz R.D., and Ney E.P. 1992. 0.7- to 23-μm photometric observations of P/Halley 1986III and six recent bright comets. *Icarus* 100, 162-186.

171. Giblin I., Davis D.R., and Ryan E.V. 2004. On the collisional disruption of porous icy targets simulating Kuiper belt objects. *Icarus* 171, 487-505.

172. Gladman B., and Duncan M. 1990. On the fates of minor bodies in the outer Solar System. *Astron. J.* 100, 1680-1693.

173. Gladman B., Kavelaars J.J., Nicholson P.D., Loredo T.J., and Burns J.A. 1998. Pencil-beam surveys for faint trans-Neptunian objects. *Astron. J.* 116, 2042-2054.

174. Gladman B., Michel P., and Froeschlé C. 2000. The near-Earth object population. *Icarus* 146, 176-189.

175. Gladman B.J., Migliorini F., Morbidelli A., Zappalà V., Michel P., Cellino A., Froeschlé C., Levison H.F., Bailey M., and Duncan M. 1997. Dynamical lifetimes of objects injected into asteroid belt resonances. *Science* 277, 197-201.

Goldreich P., and Ward W.R. 1973. The formation of planetesimals. *Astrophs. J.* 183, 1051-1061.

176. Goldreich P., Lithwick Y., and Sari R. 2002. Formation of Kuiper-belt binaries by dynamical friction and three-body encounters. *Nature* 420, 643-646.

177. Gomes R.S. 2003. The origin of the Kuiper belt high inclination population. *Icarus* 161, 404-418.

178. Gomes R.S., Gallardo T., Fernández J.A., and Brunini A. 2005. On the origin of the high-perihelion scattered disk: The role of the Kozai mechanism and mean motion resonances. *Cel. Mech. Dyn. Astr.*, in press.

179. Grazier K.R., Newman W.I., Kaula W.M., and Hyman J.M. 1999. Dynamical evolution of planetesimals in the outer solar system. I. The Jupiter/Saturn zone. *Icarus* 140, 341-352.

180. Green D.W., and Morris C.S. 1987. The visual brightness behavior of P/Halley during 1981-1987. *Astron. Astrophys.* 187, 560-568.

181. Green D.W.E. 1996. On the history of total-visual-magnitude estimation methods. *Int. Comet Quart.* 18, 186-204.

182. Greenberg R., Weidenschilling S.J., Chapman C.R., and Davis D.R. 1984. From icy planetesimals to outer planets and comets. *Icarus* 59, 87-113.

183. Grieve R.A.F., and Shoemaker E.M. 1994. The record of past impacts on Earth. In *Hazards due to Comets and Asteroids* (T. Gehrels, ed.), pp. 417-462, Univ. Arizona Press, Tucson.

184. Groussin O., Lamy P., and Jorda L. 2004a. Properties of the nuclei of Centaurs Chiron and Chariklo. *Astron. Astrophys.* 413, 1163-1175.

185. Groussin O., Lamy P., Jorda L., and Toth I. 2004b. The nuclei of comets 126P/IRAS and 103P/Hartley 2. *Astron. Astrophys.* 419, 375-383.

186. Grün E. 1999. Interplanetary dust and the zodiacal cloud. In *Encyclopedia of the Solar System* (P.R. Weissman, L.-A. McFadden, and T.V. Johnson, eds.), pp. 673-696, Academic Press, San Diego, USA.

187. Grün E., and Jessberger E.K. 1990. Dust. In *Physics and Chemistry of Comets* (W.F. Huebner, ed.), pp. 113-176, Springer, Heidelberg.

188. Gutiérrez P.J., Rodrigo R., Ortiz J.L., and Davidsson B.J.R. 2003. An investigation of errors in estimates of the cometary nuclei active area fractions. *Astron. Astrophys.* 401, 755-761.

189. Haisch Jr. K.E., Lada E.A., and Lada C.J. 2001. Disk frequencies and lifetimes in young clusters. *Astrophys. J.* 553, L153-L156.

190. Hamid S.E., Marsden B.G., and Whipple F.L. 1968. Influence of a comet belt beyond Neptune on the motions of periodic comets. *Astron. J.* 73, 727-729.

191. Hanner M.S. 1985. A comparison of the dust properties in recent periodic comets. *Adv. Space Res.* 4 No.9, 189-196.

192. Hanner M.S., and Campins H. 1986. Thermal emission from the dust coma of comet Bowell and a model for the grains. *Icarus* 67, 51-62.

193. Hanner M.S., Newburn R.L., Spinrad H., and Veeder G.J. 1987. Comet Sugano-Saigusa-Fujikawa (1983V) - A small, puzzling comet. *Astron. J.* 94, 1081-1087.

194. Hartmann W.K., Ryder G., Dones L., and Grinspoon D. 2000. The time-dependent intense bombardment of the primordial Earth/Moon system. In *Origin of the Earth and Moon* (R.M. Canup, and K. Righter, eds.), pp. 493-512, Univ. Arizona Press, Tucson.

195. Hartmann W.K., Tholen D.J., and Cruikshank D.P. 1987. The relationship of active comets, "extinct" comets, and dark asteroids. *Icarus* 69, 33-50.

196. Hartmann W.K., Tholen D.J., Meech K.J., and Cruikshank D.P. 1990. 2060 Chiron - Colorimetry and cometary behavior. *Icarus* 83, 1-15.

197. Harwit M. 1968. "Spontaneously" split comets. *Astrophys. J.* 151, 789-790.

198. Haser L. 1957. Distribution d'intensité dans la tête d'une comète. *Bull. Acad. Roy. Belgique* 43, 740-750.

199. Havnes O. 1970. The effect of repeated close approaches to Jupiter on short-period comets. *Icarus* 12, 331-337.

200. Heisler J. 1990. Monte Carlo simulations of the Oort comet cloud. *Icarus* 88, 104-121.

201. Heisler J., and Tremaine S. 1986. The influence of the galactic tidal field on the Oort comet cloud. *Icarus* 65, 13-26.

202. Heisler J., and Tremaine S. 1989. How dating uncertainties affect the detection of periodicity in extinctions and craters. *Icarus* 77, 213-219.

203. Heisler J., Tremaine S., and Alcock C. 1987. The frequency and intensity of comet showers from the Oort cloud. *Icarus* 70, 269-288.

204. Hellman C.D. 1944. *The Comet of 1577: Its Place in the History of Astronomy*, Columbia University Press, New York.

205. Heyer M.H., and Terebey S. 1998. The anatomy of the Perseus spiral arm: ^{12}CO and IRAS imaging observations of the W3-W4-W5 cloud complex. *Astrophys. J.* 502, 265-277.

206. Hills J.G. 1981. Comet showers and the steady-state infall of comets from the Oort cloud. *Astron. J.* 86, 1730-1740.

207. Hills J.G. 1982. The formation of comets by radiation pressure in the outer protosun. *Astron. J.* 87, 906-910.

208. Ho Peng Yoke 1962. Ancient and medieval observations of comets and novae in Chinese sources. *Vistas in Astronomy* 5, 127-225.

209. Hollenbach D.J., Yorke H.W., and Johnstone D. 2000. Disk dispersal around young stars. In *Protostars and Planets IV* (V. Mannings, A.P. Boss, and S.S. Russell, eds.), pp. 401-428, Univ. Arizona Press, Tucson.

210. Holmberg J., and Flynn C. 2000. The local density of matter mapped by Hipparcos. *Mon. Not. Roy. Astr. Soc.* 313, 209-216.

Holman M.J., and Wisdom J. 1993. Dynamical stability in the outer Solar System and the delivery of short period comets. *Astron. J.* 105, 1987-1999.

211. Hoyle F., and Wickramasinghe C. 1979. On the nature of interstellar grains. *Astrophys. Space Sci.* 66, 77-90.

212. Hubbard W.B. 1989. Structure and composition of Giant Planet interiors. In *Origin and Evolution of Planetary and Satellite Atmospheres* (S.K. Atreya, J.B. Pollack, and M.S. Matthews, eds.), pp. 539-563, Univ. of Arizona Press, Tucson.

213. Hubbard W.B., Podolak M., and Stevenson D.J. 1995. The interior of Neptune. In *Neptune and Triton* (D.P. Chruikshank, ed.), pp. 37-108, Univ. Arizona Press, Tucson.

214. Huebner W.F. 1970. Dust from cometary nuclei. *Astron. Astrophys.* 5, 286-297.

215. Hughes D.W. 1988. Cometary magnitude distribution and the ratio between the number of long- and short-period comets. *Icarus* 73, 149-162.

216. Hughes D.W., and McBride N. 1989. The mass of meteoroid streams. *Mon. Not. Roy. Astr. Soc.* 240, 73-79.

217. Hunten D.M., Pepin R.O., and Owen T.C. 1988. Planetary atmospheres. In *Meteorites and the Early Solar System* (J.F. Kerridge, and M.S. Matthews, eds.), pp. 565-591, Univ. Arizona Press, Tucson.

218. Hut P., and Tremaine S. 1985. Have interstellar clouds disrupted the Oort comet cloud?. *Astron. J.* 90, 1548-1557.

219. Ida S., Larwood J., and Burkert A. 2000. Evidence for early stellar encounters in the orbital distribution of Edgeworth-Kuiper belt objects. *Astrophys. J.* 528, 351-356.

Ip W.-H. 1989. Dynamical processes of macro-accretion of Uranus and Neptune: A first look. *Icarus* 80, 167-178.

220. Ip W.-H., and Fernández J.A. 1988. Exchange of condensed matter among the outer and terrestrial protoplanets and the effect on surface impact and atmospheric accretion. *Icarus* 74, 47-61.

221. Ip W.-H., and Fernández J.A. 1997. On dynamical scattering of Kuiper belt objects in 2:3 resonance with Neptune into short-period orbits. *Astron. Astrophys.* 324, 778-784.

222. Irvine W.M., Schloerb F.P., Crovisier J., Fegley Jr. B., and Mumma M.J. 2000. Comets: A link between interstellar and nebular chemistry. In *Protostars and Planets IV* (V. Mannings, A.P. Boss, and S.S. Russell, eds.), pp. 1159-1200, Univ. Arizona Press, Tucson.

223. Irwin M., Tremaine S., and Zitkow A.N. 1995. A search for slow-moving objects and the luminosity function of the Kuiper belt. *Astron. J.* 110, 3082-3092.

224. Jackson A.A., and Killen R.M. 1988. Infrared brightness of a comet belt beyond Neptune. *Earth, Moon and Planets* 42, 41-47.

225. Jeans J.H. 1919. The origin of binary systems. *Mon. Not. Roy. Astr. Soc.* 79, 409-416.

226. Jenniskens P. 2003. *IAU Circ.* No.8252.

227. Jewitt D.C. 2002. From Kuiper belt object to cometary nucleus: The missing ultrared matter. *Astron. J.* 123, 1039-1049.

228. Jewitt D. 2003. Project Pan-STARRS and the outer solar system. *Earth, Moon and Planets* 92, 465-476.

229. Jewitt D., Aussel H., and Evans A. 2001. The size and albedo of the Kuiper-belt object (20000) Varuna. *Nature* 411, 446-447.

230. Jewitt D., and Danielson G.E. 1984. Charge-coupled device photometry of comet P/Halley. *Icarus* 60, 435-444.

231. Jewitt D., and Kalas P. 1998. Thermal observations of Centaur 1997 CU26. *Astrophys. J.* 499, L103-L106.

232. Jewitt D., and Luu J. 1993. Discovery of the candidate Kuiper belt object 1992 QB1. *Nature* 362, 730-732.

Jewitt D.C., and Luu J.X. 1995. The solar system beyond Neptune. *Astron. J.* 109, 1867-1876.

233. Jewitt D.C., and Luu J.X. 2001. Colors and spectra of Kuiper belt objects. *Astron. J.* 122, 2099-2014.

234. Jewitt D.C., and Luu J. 2004. Crystalline water ice on the Kuiper belt object (50000) Quaoar. *Nature* 432, 731-733.

235. Jewitt D., Luu J., and Chen J. 1996. The Mauna Kea - Cerro Tololo (MKCT) Kuiper belt and Centaur survey. *Astron. J.*. 112, 1225-1238.

236. Jewitt D., Luu J., and Trujillo C. 1998. Large Kuiper belt objects: The Mauna Kea 8K CCD survey. *Astron. J.* 115, 2125-2135.

237. Jewitt D.C., and Meech K.J. 1987. CCD photometry of comet P/Encke. *Astron. J.* 93, 1542-1548.

238. Jewitt D.C., and Sheppard S.S. 2002. Physical properties of trans-neptunian object (20000) Varuna. *Astron. J.* 123, 2110-2120.

239. Jewitt D., and Sheppard S. 2004. The nucleus of 48P/Johnson. *Astron. J.* 127, 1784-1790.

240. Jewitt D., Sheppard S., and Fernández Y. 2003. 143P/Kowal-Mrkos and the shapes of cometary nuclei. *Astron. J.* 125, 3366-3377.

241. Jewitt D.C., Trujillo C.A., and Luu J.X. 2000. Population and size distribution of small Jovian Trojan asteroids. *Astron. J.* 120, 1140-1147.

242. Jorda L., Crovisier J., and Green D.W.E. 1992. The correlation between water production rates and visual magnitudes in comets. In *Asteroids, Comets, Meteors 1991*, pp. 285-288, Lunar and Planetary Institute, Houston.

243. Joss P.C. 1973. On the origin of short-period comets. *Astron. Astrophys.* 25, 271-273.

244. Kamoun P.G., Campbell D.B., Ostro S.J., Pettengill G.H., and Shapiro I.I. 1982a. Comet Encke: Radar detection of nucleus. *Science* 216, 293-295.

245. Kamoun P.G., Pettengill G.H., and Shapiro I.I. 1982b. Radar observations of cometary nuclei. In *Comets* (L.L. Wilkening, ed.), pp. 288-296, Univ. Arizona Press, Tucson.

246. Kawakami S., Mizutani H., Takagi Y., Kato M., and Kumazawa M. 1983. Impact experiments on ice. *J. Geophys. Res.* 88, 5806-5814.

247. Kawakita H., Watanabe J., Ando H., Aoki W., Fuse T., Honda S., Izumiura H., Kajino T., Kambe E., Kawanomoto S., Noguchi K., Okita K., Sadakane K., Sato B., Takada-Hidai M., Takeda Y., Usuda T., Watanabe E., and Yoshida M. 2001. The spin temperature of NH_3 in comet C/1999 S4 (LINEAR). *Science* 294, 1089-1091.

248. Keller H.U. 1990. The nucleus. In *Physics and Chemistry of Comets* (W.F. Huebner, ed.), pp. 13-68.

249. Keller H.U., Delamere W.A., Huebner W.F., Reitsema H.J., Schmidt H.U., Whipple F.L., Wilhelm K., Curdt W., Kramm R., Thomas N., Arpigny C., Barbieri C., Bonnet R.M., Cazes S., Coradini M., Cosmovici C.B., Hughes D.W., Jamar C., Malaise D., Schmidt K., Schmidt W.K.H., and Seige P. 1987. Comet P/Halley's nucleus and its activity. *Astron. Astrophys.* 187, 807-823.

250. Kenyon S.J., and Bromley B.C. 2004. The size distribution of Kuiper belt objects. *Astron. J.* 128, 1916-1926.

251. Kenyon S.J., and Luu J.X. 1999. Accretion in the early Kuiper belt. II. Fragmentation. *Astron. J.* 118, 1101-1119.

252. Kerr R.H. 1961. Perturbations of cometary orbits. *Proc. 4th Berkeley Symp. of Mathematical Statistics and Probability* Vol. 3, pp. 149-164, Univ. of California Press, Berkeley.

Knežević Z., Milani A., Farinella P., Froeschlé Ch., and Froeschlé Cl. 1991. Secular resonances from 2 to 50 AU. *Icarus* 93, 316-330.

253. Kowal C. 1989. A solar system survey. *Icarus* 77, 118-123.

254. Krasnopolsky V.A., Mumma M.J., Abbott M., Flynn B.C., Meech K.J., Yeomans D.K., Feldman P.D., Cosmovici C.B. 1997. Detection of soft X-

rays and a sensitive search for noble gases in comet Hale-Bopp (C/1995 O1). *Science* 277, 1488-1491.

255. Kresák L. 1975. The bias of the distribution of cometary orbits by observational selection. *Bull. Astron. Inst. Czech.* 26, 92-111.
256. Kresák L. 1978a. Passages of comets and asteroids near the Earth. *Bull. Astron. Inst. Czech.* 29, 103-114.
257. Kresák L. 1978b. The comet and asteroid population in the Earth's environment. *Bull. Astron. Inst. Czech.* 29, 114-125.
258. Kresák L. 1987. Dormant phases in the aging of periodic comets. *Astron. Astrophys.* 187, 906-908.
259. Kresák L. 1992a. Are there any comets coming from interstellar space?. *Astron. Astrophys.* 259, 682-691.
260. Kresák L. 1992b. The peculiar orbit of comet 1976 I Sato. *Astron. Astrophys.* 259, 692-695.
261. Kresák L., and Kresáková M. 1989. The absolute magnitudes of periodic comets. *Bull. Astron. Inst. Czech.* 40, 269-284.
262. Kresák L., and Kresáková M. 1990. Secular brightness decrease of periodic comets. *Icarus* 86, 82-92.
263. Kresák L., and Kresáková M. 1994. Updating the catalogue of absolute magnitudes of periodic comets. *Planet. Space Sci.* 42, 199-204.
264. Kresák L., and Pittich E.M. 1978. The intrinsic number density of active long-period comets in the inner solar system. *Bull. Astron. Inst. Czech.* 29, 299-309.
265. Kronk G.W. 1999. *Cometography. A Catalog of Comets. Volume 1: Ancient-1999*, Cambridge University Press, Cambridge.
266. Kuijken K., and Gilmore G. 1989. The mass distribution in the galactic disk - III The local volume mass density. *Mon. Not. Roy. Astr. Soc.* 239, 651-664.
267. Kührt E., and Keller H.U. 1994. The formation of cometary surface crusts *Icarus* 109, 121-132.
268. Kuiper G.B. 1951. On the origin of the solar system. In *Astrophysics* (J.A. Hynek, ed.), pp. 357-424, McGraw-Hill.
269. Lada E.A., Strom K.M., and Myers P.C. 1993. Evironments of star formation: relationship between molecular clouds, dense cores and young stars. In *Protostars and Planets III* (E.H. Levy, and J.I. Lunine, eds.), pp. 245-278, Univ. Arizona Press, Tucson.
270. Lamy P.L., Toth I., Fernández Y.R., and Weaver H.A. 2004. The sizes, shapes, albedos, and colors of cometary nuclei. In *Comets II* (M. Festou, H.U. Keller, and H. Weaver, eds.), Univ. Arizona Press, in press.
271. Larsen J.A., Gleason A.E., Danzi N.M., Descour A.S., McMillan R.S., Gehrels T., Jedicke R., Montani J.L., and Scotti J.V. 2001. The Spacewatch wide-area survey for bright Centaurs and trans-Neptunian objects. *Astron. J.* 121, 562-579.
272. Larson S.M., Edberg S.J., and Levy D.H. 1991. The modern role of visual observations of comets. In *Comets in the Post-Halley Era* (R.L. Newburn Jr., M. Neugebauer, and J. Rahe, eds.), pp. 209-223, Kluwer, Dordrecht.

273. Latham D.W., Mazeh T., Davis R.J., Stefanik R.P., and Abt H.A. 1991. A search for wide binaries in the north galactic pole using precise radial velocities. *Astron. J.* 101, 625-636.

274. Laufer D., Notesco G., Bar-Nun A., and Owen T. 1999. From the interstellar medium to Earth's oceans via comets - An isotopic study of HDO/H_2O. *Icarus* 140, 446-450.

275. Leonard F.C. 1930. The new planet Pluto. *Astron. Soc. Pacific Leaflet 30*, pp. 121-124.

276. Levison H.F. 1991. The long-term dynamical behavior of small bodies in the Kuiper belt. *Astron. J.* 102, 787-794.

277. Levison H.F., Dones L., and Duncan M.J. 2001. The origin of Halley-type comets: Probing the inner Oort cloud. *Astron. J.* 121, 2253-2267.

Levison H.F., and Duncan M.J. 1990. A search for proto-comets in the outer regions of the solar system. *Astron. J.* 100, 1669-1675.

278. Levison H.F., and Duncan M.J. 1993. The gravitational sculpting of the Kuiper belt. *Astrophys. J.* 406, L35-L38.

279. Levison H.F., and Duncan M.J. 1997. From the Kuiper belt to Jupiter-family comets: The spatial distribution of ecliptic comets. *Icarus* 127, 13-32.

280. Levison H.F., Lissauer J.J., and Duncan M.J. 1998. Modeling the diversity of outer planetary systems. *Astron. J.* 116, 1998-2014.

281. Levison H.F., Duncan M.J., Dones L., and Gladman B.J. 2005. The scattered disk as a source of Halley-type comets. *Icarus*, in press.

282. Levison H.F., and Morbidelli A. 2003. The formation of the Kuiper belt by the outward transport of bodies during Neptune's migration. *Nature* 426, 419-421.

283. Levison H.F., Morbidelli A., and Dones L. 2004. Sculpting the Kuiper belt by a stellar encounter: Constraints from the Oort cloud and scattered disk. *Astron. J.* 128, 2553-2563.

284. Levison H.F., Morbidelli A., Dones L., Jedicke R., Wiegert P.A., and Bottke Jr. W.F. 2002. The mass disruption of Oort cloud comets. *Science* 296, 2212-2215.

285. Li A., and Greenberg J.M. 1998. From interstellar dust to comets: Infrared emission from comet Hale-Bopp (C/1995 O1). *Astrophys. J.* 498, L83-L87.

286. Licandro J., Oliva E., and Di Martino M. 2001. NICS-TNG infrared spectroscopy of trans-Neptunian objects 2000 EB173 and 2000 WR106. *Astron. Astrophys.* 373, L29-L32.

287. Licandro J., and Pinilla-Alonso N. 2005. The inhomogeneous surface of Centaur (32522) Thereus (2001 PT13). Preprint.

288. Lin D.N.C., Bodenheimer P., and Richardson D.C. 1996. Orbital migration of the planetary companion of 51 Pegasi to its present location. *Nature* 380, 606-607.

289. Lindgren M., Tancredi G., Lagerkvist C.-I., and Hernius O. 1996. Searching for comets encountering Jupiter. Second campaign observations and further constraints on the size of the Jupiter family population. *Astron. Astrophys. Suppl. Ser.* 118, 293-301.

290. Lis D.C., Mehringer D.M., Benford D., Gardner M., Phillips T.G., Bockelée-Morvan D., Biver N., Colom P., Crovisier J., Despois D., and Rauer H. 1997. New molecular species in comet C/1995 O1 (Hale-Bopp) observed with the Caltech Submillimeter Observatory. *Earth, Moon and Planets* 78, 13-20.

291. Lissauer J.J., Pollack J.B., Wetherill G.W., and Stevenson D.J. 1995. Formation of the Neptune system. In *Neptune and Triton* (D.P. Chruikshank, ed.), pp. 37-108, Univ. Arizona Press, Tucson.

292. Lisse C.M., Fernández Y.R., A'Hearn M.F., Grün E., Käufl H.U., Osip D.J., Lien D.J., Kostiuk T., Peschke S.B., and Walker R.G. 2004. A tale of two very different comets: ISO and MSX measurements of dust emission from 126P/IRAS (1996) and 2P/Encke (1997). 2004. *Icarus* 171, 444-462.

293. Lodders K., and Fegley Jr. B. 1998. *The Planetary Scientists Companion*, Oxford University Press, Oxford.

294. Lowrey B.E. 1973. The effect of multiple encounters on short-period comet orbits. *Astron. J.* 78, 428-437.

295. Lowry S.C., Fitzsimmons A., and Collander-Brown S. 2003. CCD photometry of distant comets III. Ensemble properties of Jupiter-family comets. *Astron. Astrophys.* 397, 329-343.

296. Lunine J.I., Owen T.C., and Brown R.H. 2000. The outer solar system: Chemical constraints at low temperatures on planet formation. In *Protostars and Planets IV* (V. Mannings, A.P. Boss, and S.S. Russell, eds.), pp. 1055-1080, Univ. Arizona Press, Tucson.

297. Lüst R. 1984. The distribution of the aphelion directions of long-period comets. *Astron. Astrophys.* 141, 94-100.

298. Luu J.X., and Jewitt D.C. 1988. A two-part search for slow-moving objects. *Astron. J.* 95, 1256-1262.

299. Luu J.X., and Jewitt D.C. 1990. Cometary activity in 2060 Chiron. *Astron. J.* 100, 913-932.

300. Luu J.X., and Jewitt D.C. 1992. High resolution surface brightness profiles of near-Earth asteroids. *Icarus* 97, 276-287.

301. Luu J.X., and Jewitt D.C. 1998. Deep imaging of the Kuiper belt with the Keck 10-m telescope. *Astrophys. J.* 502, L91-L94.

302. Luu J.X., and Jewitt D.C. 2002. Kuiper belt objects: relics from the accretion disk of the Sun. *Ann. Rev. Astron. Astrophys.* 40, 63-101.

303. Luu J.X., Jewitt D.C., and Trujillo C.A. 2000. Water ice in 2060 Chiron and its implications for Centaurs and Kuiper belt objects. *Astrophys. J.* 531, L151-L154.

304. Luu J., Marsden B.G., Jewitt D., Trujillo C.A., Hergenrother C.W., Chen J., and Offutt W.B. 1996. A new dynamical class of object in the outer Solar System. *Nature* 387, 573-575.

305. Lyngå G. 1982. Open clusters in our Galaxy. *Astron. Astrophys.* 109, 213-222.

306. Lyttleton R.A. 1951. On the structure of comets and the formation of tails. *Mon. Not. Roy. Astr. Soc.* 111, 268-277.

Mäkinen J.T.T., Silén J., Schmidt W., Kyrölä E., Summanen T., Bertaux J.-L., Quémerais E., and Lallement R. 2001. Water production of comets

2P/Encke and 81P/Wild 2 derived from SWAN observations during the 1997 apparition. *Icarus* 152, 268-274.

307. Malhotra R. 1993. The origin of Pluto's peculiar orbit. *Nature* 365, 819-821. Malhotra R. 1995. The origin of Pluto's orbit: Implications for the solar system beyond Neptune. *Astron. J.* 110, 420-429.
308. Marcy G.W., Cochran W.D., and Mayor M. 2000. Extrasolar planets around main-sequence stars. In *Protostars and Planets IV* (V. Mannings, A.P. Boss, and S.S. Russell, eds.), pp. 1285-1311, Univ. Arizona Press, Tucson.
309. Margot J.L., Nolan M.C., Benner L.A.M., Ostro S.J., Jurgens R.F., Giorgini J.D., Slade M.A., and Campbell D.B. 2002. Binary asteroids in the near-Earth object population. *Science* 296, 1445-1448.
310. Marsden B.G., Sekanina Z., and Yeomans D.K. 1973. Comets and nongravitational forces. V. *Astron. J.* 78, 211-225.
311. Marzari F., Farinella P., and Vanzani V. 1995. Are Trojan collisional families a source for short-period comets? *Astron. Astrophys.* 299, 267-276.
312. Marzari F., Farinella P., Davis D.R., Scholl H., and Campo Bagatin A. 1997. Collisional evolution of Trojan asteroids. *Icarus* 125, 39-49.
313. Matese J.J., Whitman P.G., Innanen K.A., and Valtonen M.J. 1995. Periodic modulation of the Oort cloud comet flux by the adiabatically changing galactic tide. *Icarus* 116, 255-268.
314. Mayer L., Quinn T., Wadsley J., and Stadel J. 2002. Formation of giant planets by fragmentation of protoplanetary disks. *Science* 298, 1756-1759.
315. McCaughrean M.J., and O'Dell C.R. 1996. Direct imaging of circumstellar disks in the Orion nebula. *Astron. J.* 111, 1977-1986.
316. McFadden L.A., Cochran A.L., Barker E.S., Cruikshank D.P., and Hartmann W.K. 1993. The enigmatic object 2201 Oljato: is it an asteroid or an evolved comet? *J. Geophys. Res.* 98, 3031-3041.
317. McFarland J. 1996. Kenneth Essex Edgeworth - Victorian Polymath and founder of the Kuiper belt? *Vistas in Astronomy* 40, 343-354.
318. McKay C.P. 1991. Urey price lecture: Planetary evolution and the origin of life. *Icarus* 91, 93-100.
319. McKay C.P. 1997. Life in comets. In *Comets and the Origin and Evolution of Life* (P.J. Thomas, C.F. Chyba, and C.P. McKay, eds.), pp. 273-282, Springer, New York.
320. Meakin P., and Donn B. 1988. Aerodynamic properties of fractal grains: Implications for the primordial solar nebula. *Astrophys. J.* 329, L39-L41.
321. Meech K., and Belton M.J.S. 1990. The atmosphere of 2060 Chiron. *Astron. J.* 100, 1323-1338.
322. Meech K.J., Hainaut O.R., and Marsden B.G. 2004. Comet nucleus size distributions from HST and Keck telescopes. *Icarus* 170, 463-491.
323. Meier R., Owen T.C., Matthews H.E., Jewitt D.C., Bockelée-Morvan D., Biver N., Crovisier J., and Gautier D. 1998. A determination of the HDO/H_2O ratio in Comet C/1995 O1 (Hale-Bopp). *Science* 279, 842-844.
324. Melita M.D., Williams I.P., Collander-Brown S.J., and Fitzimmons A. 2004. The edge of the Kuiper belt: the planet X scenario. *Icarus* 171, 516-524.

325. Menichella M., Paolicchi P., and Farinella P. 1996. The main belt as a source of near-Earth asteroids. *Earth, Moon and Planets* 72, 133-149.

326. Migliorini F., Michel P., Morbidelli A., Nesvorný D., and Zappalà V. 1998. Origin of multikilometer Earth- and Mars-crossing asteroids: A quantitative simulation. *Science* 281, 2022-2024.

327. Möhlmann D. 1995. Cometary activity and nucleus models. *Planet. Space Sci.* 43, 327-332.

328. Morbidelli A. 1997. Chaotic diffusion and the origin of comets from the 2/3 resonance in the Kuiper belt. *Icarus* 127, 1-12.

329. Morbidelli A. 1999. An overview on the Kuiper belt and on the origin of Jupiter-family comets. *Cel. Mech. Dyn. Astr.* 72, 129-156.

330. Morbidelli A., Chambers J., Lunine J.I., Petit J.M., Robert F., Valsecchi G.B., and Cyr K.E. 2000. Source regions and timescales for the delivery of water to the Earth. *Meteoritics and Planetary Science* 35, 1309-1320.

331. Morbidelli A., Thomas F., and Moons M. 1995. The resonant structure of the Kuiper belt and the dynamics of the first five trans-Neptunian objects. *Icarus* 118, 322-340.

332. Morbidelli A., and Valsecchi G.B. 1997. Neptune scattered planetesimals could have sculpted the primordial Edgeworth-Kuiper belt. *Icarus* 128, 464-468.

333. Morris D.E., and Muller R.A. 1986. Tidal gravitational forces: The infall of 'new' comets and comet showers. *Icarus* 65, 1-12.

334. Mottmann J. 1977. Origin of the late heavy bombardment. *Icarus* 31, 412-413.

335. Mukai T., Blum J., Nakamura A.M., Johnson R.E., and Havnes O. 2001. Physical processes on interplanetary dust. In Interplanetary Dust (e. Grün, B.Å. Gustafson, S.F. Dermott, and H. Fechtig, eds.), pp. 445-507, Springer, Heidelberg.

336. Mumma M.J., Weissman P.R., and Stern S.A. 1993. Comets and the origin of the solar system: Reading the Rosetta stone. In *Protostars and Planets III* (E.H. Levy and J.I. Lunine, eds.), pp. 1177-1252, Univ. Arizona Press, Tucson.

337. Murray C.D., and Dermott S.F. 1999. *Solar System Dynamics*, Cambridge University Press, Cambridge.

338. Napier W.M., and Clube S.V.M. 1979. A theory of terrestrial catastrophism. *Nature* 282, 455-459.

339. Napier W.M., and Staniucha M. 1982. Interstellar planetesimals - I. Dissipation of a primordial cloud of comets by tidal encounters with massive nebulae. *Mon. Not. Roy. Astr. Soc.* 198, 723-735.

340. Newburn, Jr. R.L., and Spinrad H. 1985. Spectrophotometry of seventeen comets. II. The continuum. *Astron. J.* 90, 2591-2608.

341. Newburn, Jr. R.L., and Spinrad H. 1989. Spectrophotometry of 25 comets: Post-Halley updates for 17 comets plus new observations for eight additional comets. *Astron. J.* 97, 552-569.

342. Nölke F. 1936. Der Ursprung der Kometen, Meteore und der Zodiakalichtmaterie. *Sterne* 16, 155.

343. Noll K.S., Stephens D.C., Grundy W.M., Millis R.L., Spencer J., Buie M.W., Tegler S.C., Romanishin W., and Cruikshank D.P. 2002. Detection of two binary trans-Neptunian objects 1997 CQ_{29} and 2000 CF_{105}, with the Hubble Space telescope. *Astron. J.* 124, 3424-3429.

344. Olsson-Steel D. 1987. Collisions in the solar system - IV Cometary impacts upon the planets. *Mon. Not. Roy. Astr. Soc.* 227, 501-524.

345. Oort J.H. 1950. The structure of the cloud of comets surrounding the solar system and a hypothesis concerning its origin. *Bull. Astr. Inst. Neth.* 11, 91-110.

346. Öpik E.J. 1932. Note on stellar perturbations on nearly parabolic orbits. *Proc. Am. Acad. Arts Sci.* 67, 1659-1683.

347. Öpik E.J. 1951. Collision probabilities with the planets and the distribution of interplanetary matter. *Proc. Roy. Irish Acad.* 54, 165-199.

348. Öpik E.J. 1963. Survival of cometary nuclei and the asteroids. *Adv. Astron. Astrophys.* 2, 219-262.

349. Öpik E.J. 1965. The dynamical aspects of the origin of comets. *Mem. Soc. R. Sci. Liège* 12, 523-574.

350. Öpik E.J. 1966. Sun-grazing comets and tidal disruption. *Irish Astron. J.* 7, 141-161.

351. Oró J. 1961. Comets and the formation of biochemical compounds on the primitive Earth. *Nature* 190, 389-390.

352. Oró J., Holzer G., and Lazcano-Araujo A. 1980. The contribution of cometary volatiles to the primitive Earth. *Life Sciences and Space Research XVIII* 67-82.

353. Oró J., and Lazcano A. 1997. Comets and the origin and evolution of life. In *Comets and the Origin and Evolution of Life* (P.J. Thomas, C.F. Chyba, and C.P. McKay, eds.), pp. 3-27, Springer, New York.

354. Owen T., and Bar-Nun A. 1995. Comets, impacts and atmospheres. *Icarus* 116, 215-226.

355. Patterson C.W. 1987. Resonance capture and the evolution of the planets. *Icarus* 70, 319-333.

356. Peale S.J. 1976. Orbital resonances in the solar system. *Annu. Rev. Astron. Astrophys.* 14, 215-246.

357. Petit J.-M., Morbidelli A., and Valsecchi G.B. 1999. Large scattered planetesimals and the excitation of the small body belts. *Icarus* 141, 367-387.

358. Podolak M., and Prialnik D. 1997. ^{26}Al and liquid water environments in comets. In *Comets and the Origin and Evolution of Life* (P.J. Thomas, C.F. Chyba, and C.P. McKay, eds.), pp. 259-272, Springer, New York.

359. Pollack J.P., and Yung Y.L. 1980. Origin and evolution of planetary atmospheres. *Ann. Rev. Earth Planet. Sci.* 8, 425-487.

360. Porter J.G. 1963. The statistics of comet orbits. In *The Moon, Meteorites, and Comets* (B.M. Middlehurst and G.P. Kuiper, eds.), pp. 550-572, The university of Chicago Press, Chicago.

361. Pravec P., and Harris A.W. 2000. Fast and slow rotation of asteroids. *Icarus* 148, 12-20.

362. Prialnik D., and Bar-Nun A. 1988. The formation of a permanent dust mantle and its effect on cometary activity. *Icarus* 74, 272-283.

363. Prialnik D., Bar-Nun A., and Podolak M. 1987. Radiogenic heating of comets by ^{26}Al and implications for their time of formation. *Astrophys. J.* 319, 993-1002.

364. Rabe E. 1972. Orbital characteristics of comets passing through the 1:1 commensurability with Jupiter. In *The Motion, Evolution of Orbits, and Origin of Comets* (G.A. Chebotatev, E.-I. Kazimirchak-Polonskaya, and B.G. Marsden, eds.), IAU Symp. 45, pp. 55-60, Reidel, Dordrecht.

365. Raup D.M., and Sepkoski J.J. 1984. Periodicity of extinctions in the geologic past. *Proc. Natl. Acad. Sci. USA* 81, 801-805.

366. Reeves H. 1978. The "Big Bang" theory of the origin of the Solar System. In *Protostars and Planets* (T. Gehrels, ed.), pp. 399-426, Univ. of Arizona Press, Tucson.

367. Richter N.B. 1963. *The Nature of Comets*, Methuen, London.

368. Rickman H. 1976. Stellar perturbations of orbits of long-period comets and their significance for cometary capture. *Bull. Astr. Inst. Czech.* 27, 92-105.

369. Rickman H. 1986. Masses and densities of comets Halley and Kopff. In *The Comet Nucleus Sample Return Mission Proc. Workshop, Canterbury, U.K.*, pp. 195-205, ESA SP-249.

370. Rickman H. 1989. The nucleus of comet Halley: Surface structure, mean density, gas and dust production. *Adv. Space Res.* 9, 59-71.

371. Rickman H. 1994. Cometary nuclei. In *Asteroids, Comets, Meteors 1993* (A. Milani, M. Di Martino and A. Cellino, eds.), pp. 297-312, Kluwer, Dordrecht.

372. Rickman H., Fernández J.A., and Gustafson B.A.S. 1990. Formation of stable dust mantles on short-period comet nuclei. *Astron. Astrophys.* 237, 524-535.

373. Rickman H., and Jorda L. 1998. Comet 46P/Wirtanen, the target of the Rosetta mission. *Adv. Space Res.* 21, 1491-1504.

374. Rickman H., Fernández J.A., Tancredi G., and Licandro J. 2001. The cometary contribution to planetary impact rates. In *Collisional Processes in the Solar System* (M. Ya. Marov, and H. Rickman, eds.), pp. 131-142, Kluwer, Dordrecht.

375. Rickman H., Kamél L, Festou M.C., and Froeschlé Cl. 1987. Estimates of masses, volumes and densities of short-period comet nuclei. In *The Diversity and Similarity of Comets* (E.J. Rolfe, and B. Battrick, eds.), pp. 471-481, ESA SP-278.

376. Robert F. 2003. The D/H ratio in chondrites. *Space Science Reviews* 106, 87-101.

377. Roemer E. 1966. The dimensions of cometary nuclei. *Mem. Soc. Roy. Sci. Liège* 12, 23-28.

378. Roy A.E. 1982. *Orbital Motion*, Adam Hilger Ltd, Bristol, England, 2nd. Edition.

379. Russell H.N. 1920. On the origin of periodic comets. *Astron. J.* 33, 49-61.

380. Ryan E.V., Davis D.R., and Giblin I. 1999. A laboratory impact study of simulated Edgeworth-Kuiper belt objects. *Icarus* 142, 56-62.

381. Safronov V.S. 1969. *Evolution of the Protoplanetary Cloud and Formation of the Earth and the Planets*. Israel Program for Scientific Translations.
382. Schidlowski M. 1988. A 3,800-million-year isotopic record of life from carbon in sedimentary rocks. *Nature* 333, 313-318.
383. Schwarzschild K., and Kron E. 1911. On the distribution of brightness in the tail of Halley's comet. *Astrophys. J.* 34, 342-352.
384. Scotti J.V. and Melosh H.J. 1993. Estimate of the size of comet Shoemaker-Levy 9 from a tidal breakup model. *Nature* 365, 733-735.
385. Sekanina Z. 1997. The problem of split comets revisited. *Astr. Astrophys.* 318, L5-L8.
386. Sekanina Z. 2002. Runaway fragmentation of sungrazing comets observed with the Solar and Heliospheric Observatory. *Astrophys. J.* 576, 1085-1089.
387. Sekanina Z. 2004. *IAU Circ. No. 8440.*
388. Sekanina Z., Jehin E., Boehnhardt H., Bonfils X., Schuetz O., and Thomas D. 2002. Recurring outbursts and nuclear fragmentation of comet C/2001 A2 (LINEAR). *Astrophys. J.* 572, 679-684.
389. Sekanina Z., and Yeomans D.K. 1984. Close encounters and collisions of comets with the Earth. *Astron. J.* 89, 154-161.
390. Sheppard S.S., and Jewitt D.C. 2004. Extreme Kuiper belt object 2001 QG_{298} and the fraction of contact binaries. *Astron. J.* 127, 3023-3033.
391. Sheppard S.S., Jewitt D.C., Trujillo C.A., Brown M.J.I., and Ashley M.C.B. 2000. A wide-field CCD survey for Centaurs and Kuiper belt objects. *Astron. J.* 120, 2687-2694.
392. Shoemaker E.M., Shoemaker C.S., and Wolfe R.F. 1989. Trojan asteroids: Populations, dynamical structure and origin of the L_4 and L_5 swarms. In *Asteroids II* (R.P. Binzel, T. Gehrels, and M.S. Matthews, eds.), pp. 921-948, Univ. of Arizona Press, Tucson.
393. Shoemaker E.M., Weissman P.R., and Shoemaker C.S. 1994. The flux of periodic comets near Earth. In *Hazards Due to Comets and Asteroids* (T. Gehrels, ed.), pp. 313-335, Univ. Arizona Press, Tucson.
394. Shoemaker E.M., and Wolfe R.F. 1982. Cratering time scales for the Galilean satellites of Jupiter. In *The Satellites of Jupiter* (D. Morrison, ed.), pp. 277-339, Univ. Arizona Press, Tucson.
395. Shoemaker E.M., Wolfe R.F., and Shoemaker C.S. 1990. Asteroid and comet flux in the neighborhood of Earth. In *Global Catastrophes in Earth History* (V.L. Sharpton and P.D. Ward, eds.), pp. 155-170, Geological Society of America Special Paper 247.
396. Shteins K.A., and Riekstyn'sh É.Ya. 1961. On the diffusion of comets. I. *Soviet Astr.* 4, 998-1004.
397. Sill G.T., and Wilkening L.L. 1978. Ice clathrate as a possible source of the atmospheres of the terrestrial planets. *Icarus* 33, 13-22.
398. Singh P.D., de Almeida A.A., and Huebner W.F. 1992. Dust release rates and dust-to-mass ratios of eight comets. *Astron. J.* 104, 848-858.
399. Sleep N.H., Zahnle K.J., Kasting J.F., and Morowitz H.J. 1989. Annihilation of ecosystems by large asteroid impacts on the early Earth. *Nature* 342, 139-142.

400. Smoluchowski R., and Torbett M. 1984. The boundary of the solar system. *Nature* 311, 38-39.
401. Soderblom L.A., Becker T.L., Bennett G., Boice D.C., Britt D.T., Brown R.H., Buratti B.J., Isbell C., Giese B., Hare T., Hicks M.D., Howington-Kraus E., Kirk R.L., Lee M., Nelson R.M., Oberst J., Owen T.C., Rayman M.D., Sandel B.R., Stern S.A., Thomas N., and Yelle R.V. 2002. Observations of comet 19P/Borrelly by the miniature integrated camera and spectrometer aboard Deep Space 1. *Science* 296, 1087-1091.
402. Solem J.C. 1994. Density and size of comet Shoemaker-Levy 9 deduced from a tidal breakup model. *Nature* 370, 349-351.
403. Stagg C.P., and Bailey M.E. 1989. Stochastic capture of short-period comets. *Mon. Not. Roy. Astr. Soc.* 241, 507-541.
404. Standish E.M. Jr. 1993. Planet X - No dynamical evidence in the optical observations. *Astron. J.* 105, 2000-2006.
405. Stern S.A. 1990. External perturbations on distant planetary orbits and objects in the Kuiper belt. *Cel. Mech. Dyn. Astron.* 47, 267-273.
406. Stern S.A. 1995. Collisional time scales in the Kuiper disk and their implications. *Astron. J.* 110, 856-868.
407. Stern S.A. 1996. On the collisional environment, accretion timescales, and architecture of the massive, primordial Kuiper disk. *Astron. J.* 112, 1203-1211.
408. Stern S.A. 2003. The evolution of comets in the Oort cloud and Kuiper belt. *Nature* 424, 639-642.
409. Stern S.A., and Colwell J.E. 1997. Accretion in the Edgeworth-Kuiper belt: Forming 100-1000 km radius bodies at 30 AU and beyond. *Astron. J.* 114, 841-849.
410. Stern S.A., Slater D.C., Festou M.C., Parker J.Wm., Gladstone G.R., A'Hearn M.F., and Wilkinson E. 2000. The discovery of argon in comet C/1995 O1 (Hale-Bopp). *Astrophys. J.* 544, L169-L172.
411. Stern S.A., and Weissman P.R. 2001. Rapid collisional evolution of comets during the formation of the Oort cloud. *Nature* 401, 589-591.
412. Strazzulla G., and Johnson R.E. 1991. Irradiation effects on comets and cometary debris. In *Comets in the Post-Halley Era* (R.L. Newburn Jr., M. Neugebauer, and J. Rahe, eds.), pp. 243-275, Kluwer, Dordrecht.
413. Strom S.E. 1995. Initial frequency, lifetime and evolution of YSO disks. *Rev. Mex. Astron. Astrof.* 1, 317-328.
414. Stuart J.S. 2001. A near-Earth asteroid population estimate from the LINEAR survey. *Science* 294, 1691-1693.
415. Supulver K.D., Bridges F.G., Tiscareno S., Lievore J., and Lin D.N.C. 1997. The sticking properties of water frost produced under various ambient conditions. *Icarus* 129, 539-554.
416. Sykes M.V., and Walker R.G. 1992. Cometary dust trails. *Icarus* 95, 180-210.
417. Tancredi G. 1994. Physical and dynamical evolution of Jupiter family comets: simulations based on the observed sample. *Planet. Space Sci.* 42, 421-433.
418. Tancredi G., and Fernández J.A. 1991. The angular momentum of the Pluto-Charon system: Considerations about its origin. *Icarus* 93, 298-315.

419. Tancredi G., Fernández J.A., Rickman H., and Licandro, J. 2000. A catalog of observed nuclear magnitudes of Jupiter family comets. *Astron. Astrophys. Suppl. Ser.* 146, 73-90.
420. Tancredi G., Fernández J.A., Rickman H., and Licandro, J. 2005. Nuclear magnitudes and size distribution of Jupiter family comets. Preprint.
421. Tancredi G., Rickman H., and Greenberg J.M. 1994. Thermochemistry of cometary nuclei. I. The Jupiter family case. *Astron. Astrophys.* 286, 659-682.
422. Thi W.F., Blake G.A., van Dishoeck E.F., van Zadelhoff G.J., Horn J.M.M., Becklin E.E., Mannings V., Sargent A.I., van den Ancker M.E., and Natta A. 2001. Substantial reservoirs of molecular hydrogen in the debris disks around young stars. *Nature* 409, 60-63.
423. Tholen D.J., and Buie M.W. 1997. Bulk properties of Pluto and Charon. In *Pluto and Charon* (S.A. Stern and D.J. Tholen, eds.), pp. 193-219, Univ. Arizona Press, Tucson.
424. Throop H.B., Bally J., Esposito L.W., McCaughrean M.J. 2001. Evidence of dust grain growth in young circumstellar disks. *Science* 292, 1686-1689.
425. Tombaugh C. 1961. The trans-Neptunian planet search. In *Planets and Satellites* (G. Kuiper and B. Middlehurst, eds.), pp. 12-30, University of Chicago Press, Chicago.
426. Torbett M.V. 1986. Injection of Oort cloud comets to the inner solar system by galactic tidal fields. *Mon. Not. Roy. Astr. Soc.* 223, 885-895.
427. Torbett M.V. 1989. Chaotic motion in a comet disk beyond Neptune: The delivery of short-period comets. *Astron. J.* 98, 1477-1481.
428. Torbett M.V., and Smoluchowski R. 1990. Chaotic motion in a primordial comet disk beyond Neptune and comet influx to the solar system. *Nature* 345, 49-50.
429. Toth, I. 2000. Impact-generated activity period of the Asteroid 7968 Elst-Pizarro in 1996: Identification of the Asteroid 427 Galene as the most probable parent body of the impactors. *Astron. Astrophys.* 360, 375-380.
430. Tremaine S. 1986. Is there evidence for a solar companion star?. In *The Galaxy and the Solar System* (R. Smoluchowski, J.N. Bahcall, and M.S. Matthews, eds.), pp. 409-416, Univ. Arizona Press, Tucson.
431. Trujillo C.A., and Brown M.E. 2002. A correlation between inclination and color in the classical Kuiper belt. *Astrophys. J.* 566, L125-L128.
432. Trujillo C.A., and Jewitt D.C. 1998. A semiautomated sky survey for slow-moving objects suitable for a Pluto Express Mission encounter. *Astron. J.* 115, 1680-1687.
433. Trujillo C.A., Jewitt D.C., and Luu J.X. 2001. Properties of the trans-Neptunian belt: Statistics from the Canada-France-Hawaii telescope survey. *Astron. J.* 122, 457-473.
434. van Dishoeck E.F., Blake G.A., Draine B.T., and Lunine J.I. 1993. The chemical evolution of protostellar and protoplanetary matter. In *Protostars and Planets III* (E.H. Levy and J.I. Lunine, eds.), pp. 163-241, Univ. Arizona Press, Tucson.
435. van de Hulst H.C. 1957. *Light Scattering by Small Particles.* Wiley, New York.

436. van Woerkom A.J.J. 1948. On the origin of comets. *Bull. Astr. Inst. Neth.* 10, 445-472.

437. Vorontsov-Velyaminov B. 1946. Structure and mass of cometary nuclei. *Astrophys. J.* 104, 226-233.

438. Vsekhsvyatskii S.K. 1958. *Physical Characteristics of Comets.* Translated from Russian by the Israel Program for Scientific Translations (1964).

439. Vsekhsvyatskii S.K. 1963. Absolute magnitudes of 1954-1960 comets. *Soviet Astr.* 6, 849-854.

440. Vsekhsvyatskii S.K. 1964. Physical characteristics of 1961-1963 comets. *Soviet Astr.* 8, 429-431.

441. Vsekhsvyatskii S.K. 1967. Physical characterisitcs of comets observed during 1961-1965. *Soviet Astr.* 10, 1034-1041.

442. Vsekhsvyatskii S.K., and Il'ichishina N.I. 1971. Absolute magnitudes of comets 1965-1969. *Soviet Astr.* 15, 310-313.

443. Wallis M.K. 1980. Radiogenic heating of primordial comet interiors. *Nature* 284, 431-433.

444. Wallis M.K., and Macpherson A.K. 1981. On the outgassing and jet thrust of snowball comets. *Astron. Astrophys.* 98, 45-49.

445. Weaver H.A., Feldman P.D., A'Hearn M.F., Arpigny C., Brandt J.C., Festou M.C., Haken M., McPhate J.B., Stern S.A., and Tozzi G.P. 1997. The activity and size of the nucleus of comet Hale-Bopp (C/1995 O1). *Science* 275, 1900-1904.

446. Weaver H.A., Sekanina Z., Toth I., Delahodde C.E., Hainaut O.R., Lamy P.L., Bauer J.M., A'Hearn M.F., Arpigny C., Combi M.R., Davies J.K., Feldman P.D., Festou M.C., Hook R., Jorda L., Keesey M.S.W., Lisse C.M., Marsden B.G., Meech K.J., Tozzi G.P., and West R. 2001. HST and VLT investigations of the fragments of comet C/1999 S4 (LINEAR). *Science* 292, 1329-1333.

447. Weidenschilling S.J. 1977. The distribution of mass in the planetary system and solar nebula. *Astrophys. Space Sci.* 51, 153-158.

448. Weidenschilling S.J. 1980. Dust to planetesimals: settling and coagulation in the solar nebula. *Icarus* 44, 172-189.
Weidenschilling S.J. 1995. Can gravitational instability form planetesimals? *Icarus* 116, 433-435.

449. Weidenschilling S.J. 1997. The origin of comets in the solar nebula: A unified model. *Icarus* 127, 290-306.

450. Weidenschilling S.J. 2002. On the origin of binary transneptunian objects. *Icarus* 160, 212-215.

451. Weidenschilling S.J., and Davis D.R. 1985. Orbital resonances in the solar nebula: implications for planetary accretion. *Icarus* 62, 16-29.

452. Weissman P.R. 1980. Stellar perturbations of the cometary cloud. *Nature* 288, 242-243.

453. Weissman P.R. 1984. The VEGA particulate shell - Comets or asteroids? *Science* 224, 987-989.

454. Weissman P.R. 1985. Dynamical evolution of the Oort cloud. In *Dynamics of Comets: Their Origin and Evolution* (A. Carusi, and G.B. Valsecchi, eds.), IAU Coll. No. 83, pp.87-96, Reidel, Dordrecht.

455. Weissman P.R. 1986. Are cometary nuclei primordial rubble piles?. *Nature* 320, 242-244.

456. Weissman P.R. 1990. The cometary impactor flux at the Earth. In *Global Catastrophes in Earth History* (V.L. Sharpton and P.D. Ward, eds.), pp. 171-180, Geological Society of America Special Paper 247.

457. Weissman P.R. 1995. The Kuiper belt. *Annu. Rev. Astron. Astrophys.* 33, 327-357.

458. Weissman P.R, and Kieffer H.H. 1981. Thermal modeling of cometary nuclei. *Icarus* 47, 302-311.

459. Weissman P.R., and Levison H.F. 1997. The population of the trans-neptunian region: the Pluto-Charon environment. In *Pluto and Charon* (S.A. Stern, and D.J. Tholen, eds.), pp. 559-604, Univ. Arizona Press.

460. Weissman P.R., and Lowry S.C. 2003. The size distribution of Jupiter-family cometary nuclei. *Lunar and Planetary Science* 34, 2003.

461. Wetherill G.W. 1975. Late heavy bombardment of the moon and terrestrial planets. *Proc. Lunar Sci. Conf.* 6, 1539-1561.

462. Wetherill G.W. 1985. Occurrence of giant impacts during the growth of the terrestrial planets. *Science* 228, 877-879.

463. Wetherill G.W. 1988. Where do the Apollo objects come from? *Icarus* 76, 1-18.

464. Wetherill G.W. 1991. End products of cometary evolution: Cometary origin of Earth-crossing bodies of asteroidal appearance. In *Comets in the Post-Halley Era* (R.L. Newburn Jr., M. Neugebauer, and J. Rahe, eds.), pp. 537-556, Kluwer, Dordrecht.

465. Wetherill G.W. 1994. Possible consequences of the absence of "Jupiters" in planetary systems. *Astrophys. Space Sci.* 212, 23-32.

466. Wetherill G.W., and ReVelle D.O. 1982. Relationships between comets, large meteors, and meteorites. In *Comets* (L.L. Wilkening, ed.), pp. 297-319, Univ. Arizona Press, Tucson.

467. Whipple F.L. 1950. A comet model. I. The acceleration of comet Encke. *Astrophys. J.* 111, 375-394.

468. Whipple F.L. 1964. Evidence for a comet belt beyond Neptune. *Proc. Natl. Acad. Sci.* 51, 711-718.

469. Whipple F.L. 1976. A speculation about comets and the Earth. *Mem. Soc. Royale Sci. Liege* 9, 101-111.

470. Whitmire D.P., and Jackson A.A. 1984. Are periodic mass extinctions driven by a distant solar companion?. *Nature* 308, 713-715.

471. Whitmire D.P., and Matese J.J. 1985. Periodic comet showers and planet X. *Nature* 313, 36-38.

472. Williams D.M., Mason C.G., Gehrz R.D., Jones T.J., Woodward C.E., Harker D.E., Hanner M.S., Wooden D.H., Witteborn F.C., and Butner H.M. 1997. Measurement of submicron grains in the coma of comet Hale-Bopp C/1995 O1 during 1997 February 15-20 UT. *Astrophys. J.* 489, L91-L94.

473. Williams I.P., and Wu Z. 1993. The Geminid meteor stream and Asteroid 3200 Phaethon. *Mon. Not. Roy. Astr. Soc.* 262, 231-248.

474. Wisdom J. 1983. Chaotic behavior and the origin of the 3/1 Kirkwood gap. *Icarus* 56, 51-74.

475. Wooden D.H. 2002. Comet grains: Their IR emission and their relation to ISM grains. *Earth, Moon and Planets* 89, 247-287.

476. Wurm K. 1943. Die Natur der Kometen. *Mitt. der Hamburger Sternwarte in Bergedorf* 8 No. 51, 57-92.

477. Wyckoff S. 1982. Overview of comet observations. In *Comets* (L.L. Wilkening, ed.), pp. 3-55, Univ. Arizona Press, Tucson.

478. Yabushita S., and Tsujii T. 1991. Near-parabolic flux in the outer solar system. *Mon. Not. Roy. Astr. Soc.* 248, 34-39.

479. Yeomans D.K. 1991. *Comets: A Chronological History of Observation, Science, Myth, and Folklore*, Wiley, New York.

480. Yeomans D.K., and Chodas P.W. 1989. An asymmetric outgassing model for cometary nongravitational accelerations. *Astron. J.* 98, 1083-1093.

481. Zahnle K. 1993. Planetary noble gases. In *Protostars and Planets III* (E.H. Levy, and J.I. Lunine, eds.), pp. 1305-1338, Univ. Arizona Press, Tucson.

482. Zahnle K.J., and Sleep N.H. 1997. Impacts and the early evolution of life. In *Comets and the Origin and Evolution of Life* (P.J. Thomas, C.F. Chyba, and C.P. McKay, eds.), pp. 175-208, Springer, New York.

483. Zuckerman B., Forveille T., and Kastner J.H. 1995. Inhibition of giant planet formation by rapid gas depletion around young stars. *Nature* 373, 494-496.

Index

A'Hearn, Michael F., 67, 68, 69, 243, 245
Aggarwal, Hans R., 253
albedo
  Bond, 44, 45, 181, 183
  geometric, 27, 44, 181, 183
Allen, R.L., 219
Alvarez, Luis W., 151, 327
Alvarez, Walter, 151
Anaxagoras of Clazomenae, 5
Anderson, John D., 194, 214
André, Philippe, 275, 276
Antonov, Vadim A., 111
Apian, Peter, 9
Apollonius of Myndus, 6
Arago, François, 49
Aristotle, 6, 8
Arnold, James R., 183
Arrhenius, Svante, 71
Asphaug, Erik, 250, 270
asteroid, 42, 43, 247
Auzout, Adrien, 12

Bahcall, John N., 112
Bailey, Mark E., 1, 95, 120, 122, 141, 151, 152, 159, 194, 198, 263, 265, 299, 326
Bainbridge, John, 5
Baldet, Fernand, 40, 50
Barnard, Edward Emerson, 19, 25
Bar-Nun, Akiva, 242, 306–308, 321–323
Bauer, James M., 228, 233
Beckwith, Steven V.W., 276, 281
Belbruno, Edward, 159
Belton, Michael J.S., 209
Benz, Willy, 250, 270
Bernstein, Gary M., 215, 225, 227
Bertaux, Jean L., 59–61
Bessel, Friedrich Wilhelm, 40, 70, 82
Beyer, Max, 25
Biermann, Ludwig, 75, 120, 154, 299
Biver, Nicolas, 49
Blake, Geoffrey A., 306
Blamont, Jaques E., 61
Blitz, Leo, 119, 120, 273, 300
Bobrovnikoff, Nicholas T., 15, 26
Bockeleé-Morvan, Dominique, 52, 55, 328
Bolatto, Alberto D., 84
Bondi, Hermann, 16
Boss, Alan P., 287
Bottke, William F. Jr., 171, 172, 325, 326
Brahe, Tycho, 9–11, 39
Brandt, John C., 258, 260
Bredikhin, Fedor, 70
Briceño, César, 277
Bridges, Frank G., 282, 283
Brin, G. David, 242
Bromley, Benjamin C., 171, 172, 214
Brooke, Timothy Y., 161
Brown, Michael E., 140, 211–213, 228, 229, 233, 234, 302
Brown, Robert H., 228
Brownlee, Donald E., 160
Brownlee particles, 281
Brunini, Adrián, 81, 186, 187, 219, 222–224, 298, 301, 320
Buffon, Georges-Louis Leclerc, Comte de, 39
Buie, Mark W., 234
Byl, John, 115

Cameron, Alastair G.W., 287
Campins, Humberto, 69, 161
Cardano, Girolamo, 9
Carruthers, George R., 60
Carusi, Andrea, 87, 155, 157, 250
Cassini, Jean-Dominique, 11
Catalina survey, 19
Centaurs, 209, 210
Ceplecha, Zdenek, 270
Chamberlin, Alan B., 247
Chambers, John E., 87

Chang, S., 320
Chapman, Clark R., 323
Chapman, Robert D., 258
Charon, 193, 217, 234
Chebotarev, Gleb A., 110
Chen, H., 274
Chen, Jun, 253, 254
Chiang, Eugene I., 213
Chodas, Paul W., 84
Chokshi, Arati, 281
CHON particles, 64
Christy, James W., 193
Chyba, Christopher F., 317, 320
Clairaut, Alexis-Claude, 12
Clark, Beth E., 243
Clube, S. Victor M., 120
Colwell, Joshua E., 222
Combes, Michel, 54
Combi, Michael R., 59
comet
  discovery rate, 20, 36
  dust tail, 71–73, 258
  dynamical losses, 92
  early records, 3–5
  Halley-type, 22, 157
  hydrogen corona, 59–63
  interstellar origin, 13–17
  ion tail, 74, 75
  Jupiter family, 22, 155, 157
  Kracht family, 263
  Kreutz family, 262, 263
  lightcurve, 26, 175
  long-period, 22, 157
  mass, 39, 40, 172–174
  Marsden family, 263
  Meyer family, 263
  naming procedure, 24
  near-parabolic, 35, 78
  new, 88, 103
  old, 88
  parallax, 9
  physical decay, 96, 256, 257
  shape, 42, 175
  short-period, 155, 195
  showers, 140, 144, 145
  size, 40–42, 159, 160
  spectrum, 49–51
  spin period, 175
comets
  C/1577 V1, 9, 10, 39
  C/1680 V1, 5
  C/1819 N1, 49
  C/1864 N1 (Tempel), 49
  C/1868 L1 (Winnecke), 49, 50
  C/1881 K1 (Great comet), 50
  C/1882 F1 (Wells), 50
  C/1882 R1 (Great September comet), 40, 50
  C/1908 R1 (Morehouse), 50
  C/1921 H1 (Dubiago), 87
  C/1940 R2 (Cunningham), 52
  C/1942 EA (Väisälä), 87
  C/1956 R1 (Arend-Roland), 73
  C/1969 T1 (Tago-Sato-Kosaka), 59, 243
  C/1969 Y1 (Bennett), 59, 60, 243
  C/1973 E1 (Kohoutek), 54, 59–61
  C/1975 N1 (Kobayashi-Berger-Milon), 51, 66
  C/1975 V1 (West), 250
  C/1975 X1 (Sato), 104
  C/1983 J1 (Sugano-Saigusa-Fujikawa), 260
  C/1984 N1 (Austin), 66
  C/1986 P1 (Wilson), 254
  C/1989 K1 (Levy), 54
  C/1989 X1 (Austin), 54
  C/1995 O1 (Hale-Bopp), 48, 49, 56, 64–66, 174, 308, 309, 328
  C/1996 B2 (Hyakutake), 54, 328
  C/1999 S4 (LINEAR), 255, 309
  C/2000 B4 (LINEAR), 210
  C/2000 WM1 (LINEAR), 2
  C/2001 T4 (NEAT), 210
  C/2002 E2 (Snyder-Murakami), 25
  C/2003 A2 (Gleason), 35
  3D/Biela, 40, 82, 257
  18D/Perrine-Mrkos, 261
  D/1770 L1 (Lexell), 39, 155
  D/1892 T1 (Barnard), 25
  D/1993 F2 (Shoemaker-Levy 9), 250, 270

1P/Halley, 4, 8, 12, 25, 40, 42, 48, 50, 54, 82, 160, 161, 174–176, 181, 194, 242, 246, 256, 257, 327, 328, 334
2P/Encke, 59, 70, 81, 156, 158, 161, 167, 176, 182, 183, 248, 257–259, 334
6P/d'Arrest, 155, 180, 258
7P/Pons-Winnecke, 40
8P/Tuttle, 157
9P/Tempel 1, 159, 168, 176, 336
10P/Tempel 2, 176
12P/Pons-Brooks, 87
13P/Olbers, 87
14P/Wolf, 180
16P/Brooks 2, 82
19P/Borrelly, 42, 160, 176, 178, 179, 181, 182, 242, 243, 246, 334
21P/Giacobini-Zinner, 82, 270
22P/Kopff, 175, 176
23P/Brorsen-Metcalf, 87
28P/Neujmin 1, 161, 168, 176, 182, 245, 327
29P/Schwassmann-Wachmann 1, 155, 186, 187, 209, 210, 255, 327
31P/Schwassmann-Wachmann 2, 176
39P/Oterma, 159, 209, 210
23P/Brorsen-Metcalf, 87
42P/Neujmin 3, 250
45P/Honda-Mrkos-Pajdusakova, 260
46P/Wirtanen, 176
48P/Johnson, 176, 177
49P/Arend-Rigaux, 161, 176
51P/Harrington, 260
53P/van Biesbroeck, 250
55P/Tempel-Tuttle, 4, 40, 49
67P/Churyumov-Gerasimenko, 337, 338
73P/Schwassmann-Wachmann 3, 244, 334
74P/Smirnova-Chernykh, 159
76P/West-Kohoutek-Ikemura, 260, 261
81P/Wild 2, 42, 43, 160, 179, 181, 242, 246, 334
82P/Gehrels 3, 159
87P/Bus, 260, 261
95P/Chiron, 209, 210, 228, 327
96P/Machholz 1, 157
107P/Wilson-Harrington, 157, 246–248
109P/Swift-Tuttle, 4, 40, 87, 257
111P/Helin-Roman-Crockett, 159
126P/IRAS, 157
133P/Elst-Pizarro, 157, 246
143P/Kowal-Mrkos, 176
147P/Kushida-Muramatsu, 260, 261
153P/Ikeya-Zhang, 22, 71
P/1983 V1 (Hartley-IRAS), 264
P/1994 X1 (McNaught-Russell), 157
P/1999 RO28 (LONEOS), 260
P/2000 G1 (LINEAR), 261
P/2001 WF2 (LONEOS), 260
P/2002 T1 (LINEAR), 261
Cometographia, 11
Copernicus, Nicholas, 10
Cotera, Angela, 274
Cremonese, Gabriele, 68, 69
Cretaceous-Tertiary (K/T) event, 151, 318, 327
Crézé, Michel, 112
Crovisier, Jacques, 243, 245
Cruikshank, Dale P., 228, 229, 232
Cysat, Johann Baptist, 5

Daniels, Paul A., 304
Danielson, G. Edward, 161
Davies, John K., 234
Davis, Donald R., 270, 292, 311
Davis, M., 151
Deep Impact mission, 336
Deep Space 1 mission, 42, 178, 181, 182
Delahodde, Catherine E., 182
Delsemme, Armand H., 41, 45, 47, 59, 84, 154, 189, 194, 243, 307, 316, 320, 329, 330
Democritus of Abdera, 5
Dermott, Stanley F., 199, 203, 292, 293
Descartes, René, 11
deuterium problem, 53, 328–331
diffusion equation, 90–91
Dohnanyi, Julius S., 171, 172, 268
Dominik, Carsten, 282, 283
Donati, Giovanni Batista, 49

Donn, Bertram, 44, 303, 304, 312
Dörffel, Georg, 11
Duncan, Martin J., 123, 126, 129, 130, 141, 155, 186, 188, 189, 191, 196–198, 205–207, 209, 211, 213, 214, 221, 225, 298, 330
dust particles, 63–66

Eberhardt, Peter, 328
Edgeworth, Kenneth Essex, 193, 194, 196
Edgeworth-Kuiper belt (see trans-neptunian belt)
Eggers, Sönke, 301
Elmegreen, Debra Meloy, 110, 117
Elst, Eric, 247
Encke, Johann Franz, 81
energy balance equation, 44–46
Euler, Leonhard, 12
Everhart, Edgar, 29–32, 37, 94, 96, 155, 189, 195, 196

Farinella, Paolo, 311
Feder, W., 93
Fegley, Bruce Jr., 258
Fernández, Julio A., 23, 28, 29, 81, 94, 98, 101, 104, 110, 122, 131, 136, 141–143, 154, 158, 165, 168, 169, 183–186, 188–191, 195, 198, 208, 212, 217, 222–228, 239, 247–249, 252, 259, 268, 269, 288–290, 294, 298, 299, 301, 317, 320
Fernández, Yanga R., 182, 234, 247
Ferraz-Mello, Sylvio, 206
Fesenkov, Vasilii G., 103
Festou, Michel, 59, 61, 174
Fink, Uwe, 243, 245
Finson, M.L., 72, 73
fluorescence, 51
Flynn, Chris, 109, 112
Fogg, Martyn J., 152
Fouchard, Marc, 116, 117
Fracastoro, Girolamo, 9
Fraundorf, P., 282
Froeschlé, Claude, 199, 249
Fulle, Marco, 68, 69, 258

Gahm, Gösta F., 277
Gaidos, Eric J., 300, 301
galactic disk potential, 111–119
Galileo mission, 43
Galilei, Galileo, 10
Gallardo, Tabaré, 98, 101, 206
García-Sánchez, Joan, 109
Gehrels, Tom, 162
Gehrz, Robert D., 65, 66
giant molecular clouds, 119, 120, 273
Giblin, Ian, 305
Gilmore, G., 112
Gladman, Brett J., 183, 213, 249, 330
Goldreich, Peter, 218, 279, 281, 284, 285
Gomes, Rodney S., 211, 296
grain temperature, 64–66
Grazier, Kevin R., 330
Green, Daniel W., 25, 50
Greenberg, J. Mayo, 66
Greenberg, Richard, 297, 310
Grieve, Richard A.F., 150, 151
Groussin, O., 166, 228
Grün, Eberhard, 166, 259
Gutiérrez, Pedro J., 238

Hagecius, Thaddaeus, 9
Haisch, Karl E. Jr., 277
Halley, Edmond, 12, 25, 323
Hamid, S. El Din, 194
Hanner, Martha S., 69, 260
Harrington, Robert S., 193
Harris, Allan W., 178–180
Hartmann, William K., 209, 247, 317
Harwit, Martin, 267
Haser, Leo, 57
Haser model, 57, 58, 62, 68
Havnes, Ove, 190, 191, 195
Heisler, Julia, 111, 116, 118, 148, 149, 152
Hellman, Clarice Doris, 10
Henry of Hesse, 8
Herschell, William, 13, 315
Hevelius, Johannes, 11, 39
Heyer, Mark H., 274
Hicks, Michael D., 243, 245
Hildas, 192

Hills, Jack G., 104, 137, 140, 299–301
Hipparcos satellite, 109, 112
Hippocrates of Chios, 5
Historia Cometarum, 4, 10, 19
Ho Peng Yoke, 3
Holetschek effect, 32
Holetschek, Johann, 32
Hollenbach, David J., 277, 278
Holman, Matthew J., 205, 208
Holmberg, Johan, 109, 112
Hoyle, Fred, 16, 331
Hubbard, William B., 300
Hubble Space Telescope, 52, 53, 215, 260, 276, 281
Huebner, Walter F., 241
Huggins, William Sir, 49, 50
Hughes, David W., 159, 258, 304
Hunten, Donald M., 322
Hut, Piet, 120, 131
hydroxyl (see water)

Ida, Shigeru, 223
Il'ichishina, N.I., 27
Ip, Wing-Huen, 131, 141, 154, 208, 219, 222, 225, 288–290, 294, 317, 320
IRAS (*Infrared Astronomical Satellite*), 20, 65, 195, 258, 275
Irvine, William M., 55, 306
Irwin, Michael, 213

Jackson, Albert A., 151, 194
Jäger, Michael, 71
Jeans, James H., 127, 275
Jeans mass, 275
Jenniskens, Peter, 247
Jessberger, Elmar K., 166
Jewitt, David C., 36, 161, 163, 171, 172, 175–177, 179, 180, 197, 209, 211, 213–215, 217–220, 228–235, 253, 254, 258
Jockers, Klaus, 252
Johnson, Robert E., 210, 228
Jorda, Laurent, 174, 245
Joss, Paul C., 189, 195
Jupiter-Saturn barrier, 136, 137, 141, 142, 148

Kalas, Paul, 234
Kamoun, Paul G., 167
Kant, Immanuel, 39
Kawakami, Shin-ichi, 270
Kawakita, Hideyo, 309
Keller, Horst Uwe, 160, 241, 242
Kenyon, Scott J., 171, 172, 213, 220
Kepler, Johannes, 10, 11
Kepler's laws, 20–22
Kerr, R.H., 80
Kieffer, Hugh H., 45
Killen, Rosemary M., 194
Kirsch, Gottfried, 5
Kleinfeld, Idit, 307
Knacke, Roger F., 161
Knežević, Zoran, 199, 209
Knuckles, Claude, 250
Kowal, Charles, 197, 213
Kozai resonance, 225, 263, 264
Krasnopolsky, Vladimir A., 308
Kresák, Ľubor, 27, 29–32, 37, 104, 239, 246, 259, 260
Kresáková, M., 27, 31, 239
Kron, Erich, 51
Kronk, Gary W., 3, 4
Kührt, Ekkehard, 242
Kuijken, Konrad, 112
Kuiper, Gerard P., 194, 196

Lada, Elizabeth A., 300
Lagrange, Joseph Louis, 17
Lamy, Philippe L., 28, 161–163, 171, 180, 181, 260, 345
Laplace, Pierre-Simon Marquis de, 13, 39, 195
Larsen, Jeffrey A., 214
Larson, Stephen M., 50
Latham, David W., 131
Latyshev, I.N., 111
Laufer, Diana, 329
Lazcano, Antonio, 315, 319, 320
Leonard, Frederick C., 193, 198
Levison, Harold F., 125, 126, 155, 186, 188, 189, 191, 192, 197, 205, 209, 211, 213, 220, 223–225, 227, 273, 295, 296, 324
Li, Aigen, 66

Licandro, Javier, 228, 230, 231
Lin, Douglas N.C., 288
Lindgren, Mats, 189
LINEAR, 19, 25, 30–32, 36, 37, 156, 261
Lis, Darek, 56
Lissauer, Jack J., 298
Lisse, Carey M., 70, 259
Lodders, Katharina, 258
LONEOS, 19, 31, 32, 36, 156, 261
loss cone, 137
Lowell, Percival, 193
Lowrey, Barbara E., 185, 191
Lowry, Stephen C., 171
Lubienietz, Stanislaus de, 4, 10, 19
Luhman, Kevin, 274
Lunine, Jonathan I., 306
Lüst, Rhea, 154
Luther, Martin, 11
Luu, Jane X., 163, 197, 209, 210, 213, 215, 219, 220, 228–230, 233, 234
Lyman-α, 59, 60, 62, 63, 67
Lyngå, Gosta, 300
Lyttleton, Raymond A., 16, 17, 41

Macpherson, A.K., 172, 174, 175
Maestlin, Michael, 9
magnitudes
  nuclear, 28
  total, 26, 160
Mäkinen, Jyrki T.T., 243, 245
Malhotra, Renu, 295, 296
Marcy, Geoffrey W., 273
Margot, Jean-Luc, 179
Markov chain, 205
Marsden, Brian G., 4, 5, 19–21, 24, 32–35, 83, 84, 99, 104, 105, 132, 142, 155, 156, 158, 159, 187, 253
Marzari, Francesco, 192
Matese, John J., 112, 144, 151
Mayer, Lucio, 287
McBride, Neil, 258
McCaughrean, Mark J., 276
McCrosky, Richard E., 270
McFadden, Lucy-Ann, 247
McFarland, John, 10, 198, 199
McKay, Christopher P., 316, 331, 334
Meakin, Paul, 304
Meech, Karen J., 161, 171, 180, 209
Meier, Roland, 328
Melita, Mario D., 219
Mellosh, H.J., 250, 270
Mendis, D. Asoka, 242
Menichella, Mario, 249
meteoroid stream, 40, 257, 258
Meteorologica, 3, 6
Michel, Karl-Wolfgang, 299
Migliorini, Fabio, 249
Miller, David C., 84
molecular band, 52
Möhlmann, Diedrich, 242
Morbidelli, Alessandro, 203–206, 208, 215, 220, 222, 224, 225, 295, 296, 319, 329, 330
Morris, Charles S., 50
Morris, D.E., 111
Morrison, David, 323
Mottmann, J., 300
Mukai, Tadashi, 48
Muller, Richard A., 111, 151
Müller, Johannes (Regiomontanus), 9
Mumma, Michael J., 64, 309
Murakami, Shigeki, 25
Murray, Carl D., 199, 203
Murrell, Scott, 250

Napier, William M., 120
near-Earth asteroid (NEA), 172, 246
NEAT, 19, 31, 32, 36, 37, 156
Newburn, Ray L. Jr., 69, 243, 245
Newton, Isaac, 12, 39, 315
Ney, Edward P., 65, 66
Nölke, Friedrich, 15
Noll, Keith S., 217
nongravitational forces, 36, 81–83, 172

Oberbeck, Verne R., 253
O'Dell, Charles R., 276
Olbers, Wilhelm, 13
Olsson-Steel, Duncan, 266
Oort cloud
  core, 138, 302
  definition, 103

inner, 138
outer, 138
stability radius, 131, 132
Oort constants, 111, 116, 117
Oort, Jan Hendrik, 91, 103, 106, 195
Öpik, Ernst J., 103, 106, 183–185, 246, 249, 266
orbit
elements, 20, 21
energy, 21
original, 23, 77, 78, 104, 105
osculating, 77
Oró, Joan, 315, 319, 320
oscillator strength, 62, 74, 75
Owen, Tobias C., 306, 308, 321–323

Palitzsch, Georg, 12
Patterson, Chris W., 292
Peale, Stanton J., 199, 200
Petit, Jean-Marc, 220, 222
Petit, Pierre, 12
phase coefficient, 28, 181, 182
phase integral, 44, 183
photodissociation, 56, 57
Pickering, William, 193
Pingré, Alexander, 19
Pinilla-Alonso, N., 230, 231
Pittich, Eduard M., 37
Pizarro, Guido, 247
Pliny the Elder, 3, 7
Pluto, 193, 217, 234
Plutinos, 205
Pluvinel, A. de la Baume, 50
Podolak, Morris, 332, 333
Pollack, James B., 320
Porter, John G., 23
Posidonius, 7
Poynting-Robertson drag, 223, 259, 286
Pravec, Petr, 178–180
Prialnik, Dina, 242, 332, 333
Principia, 12
Probstein, Ronald F., 72, 73
protoplanetary disk, 278, 279
Ptolemy, 7

Quaoar (2002 LM_{60}), 211, 234

Rabe, Eugene, 192
radiation pressure, 65, 70–73
Rahe, Jürgen H., 312
random walk, 87–89
Raup, David M., 151
reflectivity gradient, 230, 231
Reeves, Hubert, 300
resonance, 85–87, 199–204
ReVelle, Douglas O., 270
Richter, Nikolaus B., 16, 71
Rickman, Hans, 45, 84, 108, 172, 174, 175, 242, 245, 255, 326
Rieke, George H., 274
Rieke, M., 274
Riekstyn'sh, É.Ya., 91
Robert, François, 328
Roche, Éduard Albert, 253
Roche limit, 253
Roemer, Elizabeth, 40, 41, 161, 169, 243
Roeslin, Helisaeus, 9
Roland, Santiago, 2
Rosetta mission, 337, 338
Roy, Archie E., 13, 79, 203
rubble pile model, 44, 250
Rud, David A., 243
runaway accretion, 287
Russell, Henry Norris, 155
Ryan, Eileen V., 270

Safronov, Victor S., 291, 297
Salvo, Raúl, 2
Sargent, Anneila I., 276
scattered disk, 210
Schiaparelli, Giovanni, 13, 257
Schidlowski, Manfred, 319
Schneider, Glenn, 274
Scholl, Hans, 199, 249
Schwarzschild, Karl, 51, 71
Scotti, James V., 161, 250, 270
Sedna (2003 VB_{12}), 140, 211, 212, 302
Sekanina, Zdenek, 29, 30, 37, 252, 254, 255, 260
Seneca, Lucius Annaeus, 3, 7
Sepkoski, J. John Jr., 151
Sheppard, Scott S., 176, 177, 213, 217, 235

Shoemaker, Eugene M., 150, 151, 169, 192, 266, 326, 327
Shteins, K.A., 91
Sill, G.T., 320
Singh, Patan Deen, 69
Sleep, Norman H., 317, 318
Slipher, Vesto M., 40
SMM (*Solar Maximum Mission*), 20, 253, 262
Smoluchowski, Roman, 111, 204
snowline, 287, 306
Snyder, Douglas, 25
Soderblom, Laurence A., 160, 181, 182
SOHO (*Solar and Heliospheric Observatory*), 20, 23, 253, 262
solar apex, 14
solar nebula, 273
solar wind, 75
Solem, Johndale C., 250
SOLWIND, 20, 253, 262
Spacewatch, 19, 156, 162
Spinrad, Hyron, 69, 243, 245
Stagg, Chris R., 95
Standish, E. Myles Jr., 194
Staniucha, M.S., 120
Stardust mission, 42, 43, 334, 335
stellar flux, 108, 109
stellar perturbations, 106–110
Stern, S. Alan, 52, 120, 220, 222, 308, 310, 311, 313, 334
Strazzulla, Giovanni, 210, 228
Strom, Stephen E., 277
Struve, Friedrich Georg Wilhelm, 40
Stuart, Joseph Scott, 325
Supulver, Kimberley D., 304
Swings effect, 67
Swings, Polidore, 52, 67, 307
Sykes, Mark V., 65, 258, 259

Tancredi, Gonzalo, 28, 164, 168–172, 217, 238, 244–246, 303, 345
Tetrabiblos, 7
Terebey, Susan, 274
Thi, Wing-Fai, 277
Tholen, David J., 234
Thompson, R., 274
Throop, Henry B., 277, 281
Tielens, Alexander G.G.M., 282, 283
Tisserand constant, 22, 157, 158, 183, 190, 247, 248
Tombaugh, Clyde, 193
Torbett, Michael V., 111, 137, 204
Toscanelli, Paolo, 4
Toth, Imre, 247
trans-neptunian belt, 140, 191, 192, 196, 199
Tremaine, Scott, 111, 116, 118, 120, 131, 152
Trojans, 150, 172, 191, 192
Trujillo, Chadwick A., 212–215, 218, 233, 234
Tsujii, T., 136, 142, 143
T Tauri phase, 277, 300

ultra-red material, 229, 230

Valsecchi, Giovanni B., 222
van de Hulst, Hendrik C., 72
van Dishoeck, Ewine F., 119
van Woerkom, Adrianus J.J., 90, 91
Varuna (2000 WR_{106}), 235
Veillet, Christian, 216
von Helmholtz, Hermann, 315
von Hepperger, Josef, 40
von Peurbach, Georg, 9
Vorontsov-Velyaminov, Boris, 40
Vsekhsvyatskii, Sergei K., 17, 25, 27, 37

Walker, Russell G., 65, 258, 259
Wallis, Max K., 172, 174, 175, 332, 333
Ward, William R., 279, 281, 284, 285
water molecule
  infrared spectrum, 54, 228–230
  hydroxyl (OH), 52–54, 61–63
  production rate, 45, 46, 238, 243
Weaver, Harold A. Jr., 53, 252
Weidenschilling, Stuart J., 218, 279, 281, 282, 284, 285, 292, 304
Weissman, Paul R., 38, 44, 45, 131, 136, 152, 171, 194, 198, 250, 310, 313, 326, 334
Wetherill, George W., 136, 246, 248, 270, 316–318

Whipple, Fred L., 41, 82, 172, 194, 320
Whitmire, Daniel P., 151
Wickramasinghe, Chandra, 331
Wilkening, Laurel L., 320
Williams, D.M., 63, 64, 66
Williams, Gareth, 4, 5, 19–21, 24, 32–35, 99, 104, 105, 132, 142, 155, 156, 158, 159, 187, 253
Williams, Iwan P., 247
Wisdom, Jack, 205, 208, 249
Wolfe, Ruth F., 169
Wooden, Diane H., 63
Wu, Zidian, 247
Wurm, Karl, 74, 75
Wyckoff, Susan, 51

Xenophanes of Colophon, 5

Yabushita, Shin, 136, 142, 143
Yeomans, Donald K., 1, 29, 30, 37, 71, 84, 260
Young, E., 274
Yung, Yuk L., 320

Zahnle, Kevin J., 318, 321
Zanstra, Hermann, 51
Zeno of Citium, 6
zodiacal dust cloud, 259
Zuckerman, Ben, 277